INFORMATION AND INFERENCE

SYNTHESE LIBRARY

MONOGRAPHS ON EPISTEMOLOGY,

LOGIC, METHODOLOGY, PHILOSOPHY OF SCIENCE,

SOCIOLOGY OF SCIENCE AND OF KNOWLEDGE,

AND ON THE MATHEMATICAL METHODS OF

SOCIAL AND BEHAVIORAL SCIENCES

INFORMATION AND INFERENCE

Edited by

JAAKKO HINTIKKA

University of Helsinki and Stanford University

and

PATRICK SUPPES

Stanford University

D. REIDEL PUBLISHING COMPANY / DORDRECHT-HOLLAND

Library of Congress Catalog Card Number 70–118132

SBN 90 277 0155 5

Printed in The Netherlands by D. Reidel, Dordrecht

PREFACE

In the last 25 years, the concept of information has played a crucial role in communication theory, so much so that the terms *information theory* and *communication theory* are sometimes used almost interchangeably. It seems to us, however, that the notion of information is also destined to render valuable services to the student of induction and probability, of learning and reinforcement, of semantic meaning and deductive inference, as well as of scientific method in general. The present volume is an attempt to illustrate some of these uses of information concepts.

In 'On Semantic Information' Hintikka summarizes some of his and his associates' recent work on information and induction, and comments briefly on its philosophical suggestions. Jamison surveys from the subjectivistic point of view some recent results in 'Bayesian Information Usage'. Rosenkrantz analyzes the information obtained by experimentation from the Bayesian and Neyman-Pearson standpoints, and also from the standpoint of entropy and related concepts.

The much-debated principle of total evidence prompts Hilpinen to examine the problem of measuring the information yield of observations in his paper 'On the Information Provided by Observations'. Pietarinen addresses himself to the more general task of evaluating the systematizing ('explanatory') power of hypotheses and theories, a task which quickly leads him to information concepts. Domotor develops a qualitative theory of information and entropy. His paper gives what is probably the first axiomatization of a general qualitative theory of information adequate to guarantee a numerical representation of the standard sort.

In 'Learning and the Structure of Information' Jamison, Lhamon and Suppes address themselves to bringing together the ideas of mathematical learning theory and the concept of information structure, carrying this enterprise to the experimental level.

A new field of application for information concepts is briefly explored by Hintikka in 'Surface Information and Depth Information'. The central concept is that of surface information, which is not invariant with respect to logical equivalence. Hintikka suggests that this concept enables us to capture the sense of information in which deductive inference can yield fresh information.

The joint paper by Hintikka and Tuomela deals with a couple of general methodological problems – the role of auxiliary concepts in first-order theories and the varieties of definability – which *prima facie* have little to do with information theory. Nevertheless, the turn their enterprise takes brings them in touch with information concepts both overtly, since the gain obtained by using auxiliary concepts and the degree of partial definability turn out to be naturally definable in information terms, and covertly through the similarity of their basic technique to that used in defining measures of information and probability for first-order languages.

An undertow of interest in concept formation is found in several papers, though it often remains implicit. This interest is expressed in the scattered remarks on concept learning in 'Learning and the Structure of Information'. Jamison notes the potential relevance of Hintikka's two-dimensional continuum of probability assignments for concept formation, and Hintikka mentions the relevance of his surface information to deductive heuristics, which is readily extended to include the introduction of new concepts for deductive purposes. Though entirely nonpsychological, Hintikka's and Tuomela's paper is to a large extent devoted to examining the process of concept introduction. Perhaps this direction is the most important one for further work.

All our contributors are or have been connected with Stanford University or the University of Helsinki – or both. We are indebted to both these institutions for the customary aid in the form of secretarial services, sabbaticals, and leaves of absence. The more specific debts (in the form of grants, etc.) are indicated in connection with the several papers.

All the papers are previously unpublished, except that Hintikka's paper 'On Semantic Information' is scheduled to appear in *Physics, Logic, and History, Proceedings of the First International Colloquium at the University of Denver* (ed. by Wolfgang Yourgrau), Plenum Press, New York. An early version of the joint paper by Hintikka and Tuomela will appear in the mimeographed proceedings of the First Scandinavian Logic Symposium in Åbo, Finland, September 1968 (Filosofiska institutionen vid Uppsala universitet, Uppsala, Sweden). An earlier version of Hilpinen's paper appeared (in Finnish) in *Ajatus* **30** (1968) 58–81. These three papers are printed here with the permission of the editors of the other volumes. We acknowledge gratefully these permissions.

THE EDITORS

Helsinki and Stanford, December 1969

CONTENTS

PART I

INFORMATION AND INDUCTION

JAAKKO HINTIKKA

ON SEMANTIC INFORMATION

In the last couple of decades, a logician or a philosopher has run a risk whenever he has put the term 'information' into the title of one of his papers. In these days, the term 'information' often creates an expectation that the paper has something to do with that impressive body of results in communication theory which was first known as *theory of transmission of information* but which now is elliptically called *information theory* (in the United States at least).[1] For the purposes of this paper, I shall speak of it as *statistical information theory*. I want to begin by making it clear that I have little to contribute to this statistical information theory as it is usually developed.

Some comments on it are nevertheless in order. One interesting question that arises here is why a logician or a philosopher of language should want to go beyond statistical information theory or to approach it from a novel point of view. The reason usually given is that this theory seems to have little to say of information in the most important sense of the word, viz. in the sense in which it is used of whatever it is that meaningful sentences and other comparable combinations of symbols convey to one who understands them. It has been pointed out repeatedly that many of the applications of statistical information theory have nothing to do with information in this basic sense. It has also been argued that the use of the term 'information' in such contexts is apt to create misunderstanding and unrealistic hopes as to what statistical information theory can offer to a logician or a philosopher of language.

This, in any case, is why a few philosophers, notably Carnap and Bar-Hillel, outlined in the early fifties a theory which sought to catch some of the distinctive features of information in the proper, narrower sense of the term. (See [1]–[2], [19] etc.) Similar ideas had been put forward informally by Karl Popper already in the thirties; cf. [24] and [25]. This kind of information was called by them *semantic information*, and the theory they started might therefore be called theory of semantic infor-

J. Hintikka and P. Suppes (eds.), Information and Inference, 3–27.

mation. It is this sense of information that I shall be dealing with in the present paper.

The relation of this theory of semantic information to statistical information theory is not very clear.[2] On the formal level, the two theories have a certain amount of ground in common. In both theories information is defined, or can be defined, in terms of a suitable concept of probability. The basic connection between probability and information (in at least one of its senses) can also be taken to be one and the same in the two cases[3]:

(1) $\inf(h) = -\log \mathrm{p}(h)$

where p is the probability-measure in question. From (1) we obtain at once the familiar entropy expression

(2) $-\sum_i p_i \log p_i$

for the expectation of information in a situation in which we have a number of mutually exclusive alternatives with the probabilities p_i ($i = 1, 2, \ldots$). This expression can thus occur in both kinds of information theory. In general, the two theories can be said to have in common a certain calculus based on (1) and on the usual probability calculus.

It is more difficult to say what the relation of the two theories is on the interpretative level after all the obvious misunderstandings and category-mistakes have been cleared away. It is sometimes said that there is a difference between the concepts of probability that are involved in them. In statistical information theory, a frequency interpretation of probability is presupposed, while in a theory of semantic information we are presupposing a purely logical interpretation of probability. The difference between the two theories will on this view be in effect the difference between Carnap's probability$_2$ and probability$_1$, introduced in [3], pp. 29–36.

It seems to me that this way of making the distinction rests on serious oversimplifications. I shall make a few comments on them at the end of this paper.

It is nevertheless clear that this explanation catches some of the difference in emphasis between the two theories. In statistical information theory, one is typically interested in what happens in the long run in certain types of uncertainty situations that can be repeated again and again.[4] In a theory of semantic interpretation, we are primarily interested in the different alternatives which we can distinguish from each other by means of the resources of expression we have at our disposal. The more of these alternatives a sentence admits of, the more probable it is in some 'purely logical' sense of the word. Conversely, the fewer of these alternatives a sentence admits of, i.e. the more narrowly it restricts the possibilities that it leaves open, the more informative it clearly is. It is also obvious that this sense of information has nothing to do with the question which of these possibilities we believe or know to be actually true. It is completely obvious that the sentence $(h \& g)$ can be said to be more informative than $(h \vee g)$ even if we know that both of them are in fact true.

Semantic information theory accordingly arises when an attempt is made to interpret the probability measure p that underlies (1) as being this kind of 'purely logical' probability. In this paper, I shall examine what happens when such an attempt is seriously made.[5]

The basic problem is to find some suitable symmetry principles which enable us to distinguish from each other the different possibilities that certain given resources of expression enable us to distinguish from each other. It seems to me that insofar as we have an unambiguous concept of (semantic) information of which serious theoretical use can be made. it must be based on such distinctions between different cases, perhaps together with a system of 'weights' (probabilities *a priori*) given to these different possibilities.

This will become clearer when we consider certain simple examples. The simplest case is undoubtedly that given to us by the resources of expression used in propositional logic. These resources of expression are the following:

(i) A finite number of unanalyzed atomic statements $A_1, A_2, \ldots, A_K$;

(ii) Propositional connectives $\sim$ (not), & (and), $\vee$ (or) plus whatever other connectives can be defined in terms of them (e.g. material implication or $\supset$).

Here the different possibilities which we can distinguish from each other as far as the world is concerned which the atomic statements are about are perfectly obvious. They are given by the statements which we shall call *constituents*.[6] These could be said to be of the form

$$(3) \qquad (\pm)A_1 \& (\pm)A_2 \& \ldots \& (\pm)A_K$$

where each of the symbols $(\pm)$ is to be replaced by $\sim$ or by nothing at all in all the different combinations. The number of different constituents will thus be 2^K.

The sense in which the constituents give us all the possible alternatives which can be described by the given atomic statements and connectives should be obvious. For instance, if $K=2$, A_1=it is raining, A_2=the wind is blowing, the constituents will be the following:

$A_1 \,\&\, A_2$ = it is raining and the wind is blowing
$A_1 \,\&\, \sim A_2$ = it is raining but the wind is not blowing
$\sim A_1 \,\&\, A_2$ = the wind is blowing but it is not raining
$\sim A_1 \,\&\, \sim A_2$ = it is not raining nor is the wind blowing.

Each statement considered in propositional logic admits some of the alternatives described by the constituents, excludes the rest of them. It is true if one of the admitted alternatives is materialized, false otherwise. It can thus be represented in the form of a disjunction of some (perhaps all) of the constituents, provided that it admits at least one of them (i.e. is consistent):

$$(4) \qquad h = C_1 \vee C_2 \vee \cdots \vee C_{w(h)}\,.$$

Here h is the statement we are considering and $w(h)$ is called its width. For an inconsistent h, we may put $w(h)=0$.

In this simple case it is fairly straightforward to define that 'logical' concept of probability which goes together with suitable measures of information. Obviously, constituents are what represent the different symmetric 'atomic events' on which we have to base our measures of information. From this point of view, a statement h is the more probable the more alternatives represented by the constituents it admits of, i.e. the greater its width $w(h)$ is. Obvious symmetry considerations suggest the following definition:

$$(5) \qquad \mathrm{p}(h) = w(h)/2^K\,.$$

It is readily seen that this in fact creates a finite probability field. The above definition of information (1) then yields the following measure for the information of h:

$$(6) \qquad \inf(h) = -\log \mathrm{p}(h) = -\log\left(w(h)/2^K\right) = K - \log w(h)$$

where the base of our logarithms is assumed to be 2. It is of some interest to see that (6) is not the only possible definition of information in the propositional case. As was already mentioned above, the basic feature of the situation we are considering is clearly that the more alternatives a statement excludes, the more informative it is. (This idea has been emphasized particularly strongly by Karl Popper. If you do not appreciate it at once, try thinking of the different alternatives as so many contingencies you have to be prepared for. The more narrowly you can restrict their range clearly, the more you can say you know about them.)

This basic idea does not imply that the exclusiveness has to be measured as in (6), although (6) clearly gives one possible way of measuring it. An even more direct way of carrying out this basic idea might seem to be to use as the measure of information the relative number of alternatives it excludes. The notion so defined will be called the *content* of a statement:

$$\text{(7)} \qquad \text{cont}(h) = [2^K - w(h)]/2^K = 1 - \text{p}(h).$$

This is in fact a perfectly reasonable measure of the information that h conveys. It is of course related in many ways to the earlier measure inf. One such connection is given by the equation

$$\text{(8)} \qquad \text{inf}(h) = \log(1/1 - \text{cont}(h)).$$

The relation of the two measures of information is a rather interesting one. It has been pointed out that our native views of (semantic) information have sometimes to be explicated in terms of one of them and at other times in terms of the other. It has been suggested that cont might be viewed as a measure of the substantial information a statement carries, while inf might be considered, as a measure of its surprise value i.e. of the unexpectedness of its truth.[7] Some feeling for this difference between the functions cont and inf is perhaps given by the following simple results:

$$\text{(9)} \qquad \text{cont}(h \mathbin{\&} g) = \text{cont}(h) + \text{cont}(g)$$

if and only if $(h \vee g)$ is logically true;

$$\text{(10)} \qquad \text{inf}(h \mathbin{\&} g) = \text{inf}(h) + \text{inf}(g)$$

if and only if h and g are independent with respect to the probability measure p defined by (5);

$$\text{(11)} \qquad \text{inf}(h) = \text{cont}(h) = 0$$

if and only if h is logically true.

Of these (11) shows the rationale of the choice of the term 'tautology' for the logical truths of propositional logic. (10) shows the basic reason why inf is for many purposes a more suitable measure of information than cont. It is natural to require that information be additive only in the case of statements that are probabilistically independent of each other; and this is precisely what (10) says. In fact, from (10) we can easily derive definition (1) if we make a few simple additional assumptions concerning the differentiability and other formal properties of inf.

Further results are obtained in terms of relative measures of information. Their definitions are obvious:

$$\mathrm{cont}(h|e) = \mathrm{cont}(h \ \& \ e) - \mathrm{cont}(e)$$
$$\mathrm{inf}(h|e) = \mathrm{inf}(h \ \& \ e) - \mathrm{inf}(e)$$

The following results are then forthcoming:

$$\text{(12)} \qquad \mathrm{cont}(h|e) = \mathrm{cont}(e \supset h)$$
$$\text{(13)} \qquad \mathrm{inf}(h|e) = -\log \mathrm{p}(h|e),$$

where $\mathrm{p}(h|e)$ is the usual conditional probability. Both (12) and (13) are entirely natural. (The naturalness of (13) is brought out by a comparison with (1).) It is interesting to see, in view of this naturalness, that they contain different measures of information and that they cannot be satisfied by one and the same measure. This fact strongly suggests that our native intuitions concerning information are somewhat ambiguous.

Simple though these results are, they already give rise to (and material for) several more general considerations. Independently of whether we are using the measure inf or the measure cont for our concept of information, the information of an arbitrary statement *h* will be the smaller the more probable it is. This inverse relation of information to probability has been one of the cornerstones of the views Karl Popper has put forward concerning the nature of scientific method and scientific knowledge. He has criticized attempts to think of scientific methods primarily as methods of verifying or of 'probabilifying' or confirming hypotheses and theories. He has emphasized that the true aim, or at least one of the true aims, of the scientific enterprise is high information. Since this was seen to go together with low probability, at least in one of its senses, it seems to be more accurate to say that science aims at highly improbable hypotheses and theories than to say that it aims at highly probable ones. Attempted falsification is for Popper a more important idea than verification or confirmation.[8]

In the extremely simple situation we studied, these facts are reflected by the observation that one could normally make a hypothesis *h* more probable by adding more disjuncts into its expansion (4). It is obvious, however, that the adjunction of such new alternatives would only result in *h*'s becoming more permissive, hence flabbier and hence worse as a scientific hypothesis.

Popper is undoubtedly right in stressing the importance of information as an aim of the scientific procedure. It is not quite obvious, however, how exactly this search for high information-content should be conceived of. It is also not clear whether it has to be conceived of in a way that excludes high probability as another *desideratum* of science. Already at the simple-minded level on which we are far moving, we can say this much: The information of a hypothesis or a theory has obviously to be defined by reference to its *prior* probability (as was already hinted at above). It is prior probability that is inversely related to information in the way explained above. Now the high probability that a scientist desires for his hypotheses and theories is clearly *posterior* probability, probability on evidence. There is no good general reason to expect that information and posterior probability are inversely related in the same way as information and prior probability are. In some simple cases, it can be shown that they are not so related.[9]

One appealing idea here is to balance one's desire for high (prior) information and one's desire for high (posterior) probability in the same way in which we typically balance our values (or 'utilities') and the hard realities of probability, viz. by considering the *expected* values of utility and by trying to maximize them. Here information would be the sole relevant utility. The appeal of this 'Bayesian' approach to scientific method is witnessed by the number of scholars that have suggested it.[10] Unfortunately, attempts to carry out this program have not got very far so far.[11] I believe that this is basically due to unfortunate choices of the underlying measure of information.[12]

In order to find more suitable measures for cases that are at least in principle more interesting than propositional logic, let us consider some more complicated cases of this kind.

The next more complicated case considered by logicians is called monadic first-order logic. The languages considered under this heading employ the following resources of expression:

(i) a number of one-place predicates (properties) $P_1(x)$, $P_2(x)$,..., $P_k(x)$;

(ii) names or other free singular terms $a_1, a_2, \ldots, b_1, b_2, \ldots$;

(iii) connectives &, $\vee$, $\sim$, $\supset$, etc.;

(iv) quantifiers (Ex) ('there is at least one individual, call it x, such that') and (x) ('each individual, call it x, is such that') together with the variables bound to them (in this instance x).

Out of these we can form atomic statements concerning individuals of the form $P_i(a_j)$, propositional combinations of these, results of generalizing upon these, propositional combinations, of these generalizations and atomic statements, etc.

What are the different basic symmetric cases out of which we can construct suitable measures of information?

This question is an easy one to answer – far too easy, in fact. We can give several entirely different answers to it which yield several entirely different measures of information. We shall begin by considering some such measures of information and probability known from literature.[13]

(1) *Carnap's* p^+. This is obtained by extending the above considerations to monadic first-order logic without any modifications. It is assumed that every individual in our domain ('universe of discourse') has a name a_j. Let us form all the atomic statements $P_i(a_j)$ and deal with them precisely in the same way as with the atomic statements of propositional logic. The constituents of propositional logic will then be come as many *state-descriptions* (as they have been called by Carnap). They are of the form

$$(14) \qquad \begin{aligned} &(\pm)\, P_1(a_1)\ \&\ (\pm)\, P_2(a_1)\ \&\cdots\&\ (\pm)\, P_k(a_1) \\ &\&\ (\pm)\, P_1(a_2)\ \&\ (\pm)\, P_2(a_2)\ \&\cdots\&\ (\pm)\, P_k(a_2) \\ &\&\ \ldots \end{aligned}$$

The probability-measure p^+ is obtained by giving all these state-descriptions an equal probability (*a priori*).

In this way we can of course treat finite universes of discourse only. Infinite domains have to be dealt with as limits of sequences of finite ones.

Relative to a given domain, every statement in the monadic first-order logic is a disjunction of a number of state-descriptions. It is true if one of these is true, and false otherwise. It may be said to admit all these state-descriptions and to exclude the rest. The problem of assigning *a priori* probabilities to statements of the monadic first-order logic therefore reduces in any case to the problem of specifying the probabilities of the state-descriptions.

But are state-descriptions really those comparable and symmetric cases between which probabilities should be distributed evenly? This does not seem to be the case. Indications of this are the strange consequences of basing a definition of information on p^+. For instance, suppose we have observed a thousand individuals and found that all of them without exception to have the property P_i. Then it is easily seen that the probability $p(P_i(a_{1001})|e)$ that the next individual a_{1001} has the same property (e being the evidence given by our observations) is still equal to 1/2, which was the probability *a priori* that the first observed individual a_1 be of this kind. Thus we have $\inf(P_i(a_{1001})|e)=1=\inf(P_i(a_i))$. In other words, we should be as surprised to find that $P_i(a_{1001})$ as we were to find that $P_i(a_1)$. There is obviously something amiss in such a measure of information (or surprise).

(2) *Carnap's p**. Suggestions for improvements are not hard to find. Perhaps the most concise diagnosis of what is wrong with p^+ and with the associated measures of information is to say that what we are interested in here are not the different alternatives concerning the world we are talking about, but rather the different *kinds of* alternatives. It is the latter, and not the former, that should be weighted equally, i.e. given an equal *a priori* probability.[14]

This idea of a different kind of world was in effect interpreted by Carnap as meaning *structurally different* kind of world. On this view, what are to be given equal probabilities are not state-descriptions but disjunctions of all structurally similar or isomorphic state-descriptions. Two state-descriptions are isomorphic in the intended sense if and only if they can be obtained from one another by permuting names of individuals. Disjunctions of all such isomorphic state-descriptions were called by Carnap, not surprisingly, *structure-descriptions.* Each structure-description can now be given an equal *a priori* probability which is then divided evenly among all the state-descriptions compatible with it (i.e., occurring in it as disjuncts). Since the number of such state-descriptions varies according to the structure-description in question, two state-descriptions thus receive the same probability if they are members of the same structure-descriptions, but rarely otherwise. The resulting probability-measure is called p*.[15]

On the basis of this probability-measure we obtain measures of information. For statements concerning individuals these measures of information are not unnatural, but rather seem to catch fairly well some of the features of our native idea of information. However, for general statements (statements containing quantifiers) the resulting measures of probability and information have highly undesirable features. It turns out, for instance, that in an infinitely large domain every statement beginning with a universal quantifier has a zero probability independently of evidence, provided it is not logically false. Moreover, these unpleasant results are approximated by what we find in large finite universe e.g. in that a contingent universal statement can have a non-negligible probability *a posteriori* only if the sample on which this probability is based is of the same order of magnitude as the whole universe of discourse.[16] They seem to be especially perverse in view of the fact that in a large universe it would seem very difficult to find individuals to exemplify existential statements, which suggests that in a large universe statements which nevertheless assert the existence of such individuals should have a pretty high degree of information rather than a very low one.

(3) *Equal distribution among constituents.* For statements containing quantifiers we therefore need better measures of logical probability and of information. They can easily be come by if we analyze the structure of the monadic first-order logic a little further. It turns out that we can find another explication for the idea of a kind of alternative, different from the one we saw Carnap offering for it.

Basically, the situation here is as follows. By means of the given basic predicates we can form a complete system of classification, i.e. a partition of all possible individuals into pairwise exclusive and collectively exhaustive classes. These classes are defined by what Carnap has called *Q-predicates.* An arbitrary Q-predicate can be said to be of the form

$$(15) \qquad (\pm)\, P_1(x) \,\&\, (\pm)\, P_2(x) \,\&\cdots\&\, (\pm)\, P_k(x),$$

where each symbol $(\pm)$ may again be replaced either by $\sim$ or by a blank in all the different combinations. The number of Q-predicates is therefore $2^k = K$. An arbitrary Q-predicate will be called $Ct_i(x)$ where $i = 1, 2, \ldots, K$.

Q-predicates may be said to specify all the different kinds of individuals that can be defined by means of the resources of expression that were given to us. By means of them, each state-description can be expressed in the following form:

$$(16) \qquad Ct_{i_1}(a_1) \,\&\, Ct_{i_2}(a_2) \,\&\cdots\&\, Ct_{i_j}(a_j) \,\&\ldots.$$

In other words, all that a state-description does is to place each of the individuals in our domain into one of the classes defined by the Q-predicates. A structure-description says less than this: it tells us *how many* individuals belong to each of the different Q-predicates.

By means of the Q-predicates we can form descriptions of all the different kinds of worlds that we can specify by means of our resources of expression without speaking of any particular individuals. In order to give such a description, all we have to do is to run through the list of all the different Q-predicates and to indicate for each of them whether it is exemplified in our domain or not. In other words, these descriptions are given by statements of the following form:

$$(17) \qquad (\pm)(Ex)\,Ct_1(x) \,\&\, (\pm)(Ex)\,Ct_2(x) \,\&\cdots\&\, (\pm)(Ex)\,Ct_K(x).$$

The number of these is obviously $K=2^k$. It is also obvious how (17) could be rewritten in a somewhat more perspicuous form. Instead of listing both the Q-predicates that are instantiated and those that are not, it suffices to indicate which Q-predicates are exemplified, and then to add that each individual has to have one of the Q-predicates of the first kind. In other words, each statement of form (17) can be rewritten so as to be of form

$$(18) \qquad (Ex)\,Ct_{i_1}(x) \,\&\, (Ex)\,Ct_{i_2}(x) \,\&\cdots\&\, (Ex)\,Ct_{i_w}(x) \\ \&\, (x)\left[Ct_{i_1}(x) \vee Ct_{i_2}(x) \vee\cdots\vee Ct_{i_w}(x)\right]$$

Here $\{Ct_{i_1}, Ct_{i_2}, \ldots, Ct_{i_w}\}$ is a subset of the set of all Q-predicates.

The new form (18) is handier than (17) for many purposes. For one thing, (18) is often considerably shorter than (17).

The statements (18) will be called the constituents of monadic first-order logic. Among those statements of this part of logic which contain no names (or other free singular terms) they play exactly the same role as the constituents of propositional logic played there. These statements are said to be closed. Each consistent closed statement can now be represented as a disjunction of some of the constituents. It admits all these constituents, excludes the rest. Constituents (18) are thus the strongest closed statements of monadic first-order logic that are not logically false. For this reason, they might be called *strong generalizations*. The difference between constituents and structure-descriptions should be obvious by this time: A structure-description says of each Q-predicate *how many* individuals have it, whereas a constituent only tells you whether it is *empty* or not.

The analogy between the constituents of propositional logic and the constituents of monadic first-order logic suggests using the latter in the same way the former were used. We may give each of them an equal *a priori* probability and study the resulting measures of information. As far as the probabilities of state-descriptions are concerned, we have more than one course open for us. We may distribute the *a priori* probability that a constituent receives evenly among all the state-descriptions that make it true.[17] Alternatively we may first distribute the probability-mass of each constituent evenly among all the structure-descriptions compatible with it, and only secondly distribute the probability of each structure-description evenly among all the state-descriptions whose disjunction it is. Since the latter method combines some of the features of the first one with those of Carnap's p*, we shall call it the 'combined method'.[18]

Both methods yield measures of probability and information which are in certain respects more natural than those proposed by Carnap and Bar-Hillel, especially when they are applied to closed statements. It is of a special interest here to see how probabilities are affected by increasing evidence. Let us assume that we have observed a number of individuals $a_1, a_2, \ldots, a_n$ which have been found to exemplify a number of Q-predicates. Let these Q-predicates be

(19) $$Ct_{i_1}(x), Ct_{i_2}(x), \ldots, Ct_{i_c}(x).$$

This is clearly the general form of a body of evidence consisting of completely observed individuals.

Some constituents are ruled out (falsified) by this evidence while others are compatible with it. Clearly, (18) is compatible with it if and only if $w \geqslant c$. The important question is what happens when we obtain more and more evidence compatible with these constituents, i.e. what happens if c remains constant while n grows. Without entering any details (they are discussed briefly in [12] and [13]), I shall say simply that a closer analysis shows that in this case the posterior probability of precisely one constituent grows more and more and approaches one as its limit, while the posterior probability of all the other constituents converges to zero. This unique constituent which gets more and more highly confirmed is the unique simplest constituent compatible with our evidence. It is the unique constituent which says that only those Q-predicates are instantiated in the whole universe which are already instantiated in our sample. If our sample is rather large (large absolutely speaking, though it might nevertheless be small as compared with the whole universe), this constituent clearly represents on intuitive grounds too the most rational guess one can make concerning the whole universe. In other words, in monadic first-order logic the posterior probabilities of our constituents lead us to prefer that unique constituent among them in the long run which is the only rationally preferable constituent in the case of a large sample on obvious intuitive grounds. Asymptotically at least, our results in this case therefore accord with our inductive good sense.

This consideration is in a sense completely general, for when we run through all the individuals of our universe the number c of observed Q-predicates will have to stop at some finite value $1 \leqslant c \leqslant K$.

As far as comparisons between constituents (strong generalizations) are concerned, we thus obtain an asymptotically realistic model of the scientific procedure according to which we are guided in our search of hypotheses by high posterior probability, and by high posterior probability only.

This cannot be the whole story concerning our choice of hypotheses and theories, however, not even in the artificially simple test case of monadic first-order logic. For one thing, it was already noted that if we consider generalizations other than strong ones, we can always raise their probability by adjoining to their expansions (into disjunctions of constituents) 'irrelevant' new disjuncts. If our choice is not restricted to constituents, high posterior probability cannot therefore be the guide of scientific life, conceived of as a choice between different generalizations (closed statements).

A better guide is given to us by the Bayesian decision principle according to which one should strive to maximize expected utility. What results does this principle lead to here?[19]

Let us ask whether we should accept or reject a hypothesis h, which is assumed to be of the nature of a generalization (closed statements). The expected utility of accepting it will be

$$\text{(20)} \qquad \mathrm{p}(h|e)\cdot\mathrm{u}(h) + \mathrm{p}(\sim h|e)\cdot\mathrm{u}(\sim h),$$

where e is the evidence on which we are basing our decision and where $\mathrm{u}(h)$ and $\mathrm{u}(\sim h)$, respectively, are the utilities resulting from our decision when h is true and when it is false. Since we want to consider information as the sole utility here, we must obviously put $\mathrm{u}(h) = \mathrm{inf}(h)$ or $\mathrm{u}(h) = \mathrm{cont}(h)$. In other words, what we have gained by accepting h is the information it gives us. Conversely, if h is false, what we have lost is the information we would have had if we had accepted $\sim h$ instead of h. In other words, we have to put $\mathrm{u}(\sim h) = -\mathrm{inf}(\sim h)$ or $\mathrm{u}(\sim h) = -\mathrm{cont}(\sim h)$.[20]

It turns out that the choice between our two information-measures inf and cont does not make any essential difference here. For simplicity, we may consider what happens if cont is used. Its intuitive meaning as 'substantial information' seems to make it preferable here in any case.

In this case we have $\mathrm{u}(h) = 1 - \mathrm{p}(h)$ and $\mathrm{u}(\sim h) = \mathrm{p}(h)$. Substituting these into (20) and simplifying we obtain for the expected utility of our decision the very simple expression

$$\text{(21)} \qquad \mathrm{p}(h|e) - \mathrm{p}(h).$$

What this represents is simply the net gain in the probability of h that is brought about by our evidence e.

If h is a generalization (closed statement), it can be represented as a disjunction of constituents as in (4). From (21) we shall then obtain

$$\text{(22)} \qquad \sum_{i=1}^{w(h)} \mathrm{p}(C_i|e) - \sum_{i=1}^{w(h)} \mathrm{p}(C_i) = \sum_{i=1}^{w(h)} [\mathrm{p}(C_i|e) - \mathrm{p}(C_i)].$$

In the right-hand sum the different terms are the net gains that our evidence induces in the probabilities of the different constituents compatible with h. We already know how they behave asymptorically. Since the probability of one and only one constituent approaches one while the probabilities of all the others approach zero, asymptotically there is one and only one constituent whose posterior probability is greater than its probability *a priori*. This result is independent of our decision to give equal *a priori* probabilities to constituents; it holds as soon as all of them have any non-zero probabilities *a priori*.

In other words, in the last sum of (22) at most one term is positive when the number of observed individuals is large enough. Let the constituent which gives rise to this member be C_c. Then what we have to do in order to maximize (22) is to choose h to be identical with this one constituent. According to the Bayesian approach that I have sketched, this will be our preferred hypothesis asymptotically. We have already seen that this hypothesis seems to be the most reasonable thing we can in these circumstances surmise about the universe in any case.

Simple though this result is, it is very suggestive because it is one of the first clear-cut positive results which has been obtained by considering induction (choice of scientific hypotheses) as a process of maximizing expected utility. It is, in other words, one of the first clear-cut applications of decision theory to the philosophical theory of induction.

Other considerations are also possible here. Instead of looking for a generalization which maximizes expected information we can ask: Which kind of individual would contribute most to our knowledge if we could observe it? Asking this is tantamount to asking: Which Q-predicate yields the highest information-value $\inf(Ct_i(b)|e)$ to the statement $Ct_i(b)$ when b is an individual which we have not yet observed? Again easy calculations give us a simple answer.[21] The information of $Ct_i(b)$ in relation to the kind of evidence e which we have been considering is maximal when $Ct_i(x)$ is a Q-predicate which has not yet been exemplified in the evidence e, provided that this evidence is large enough. An individual b having such a Q-predicate would falsify the previously preferred strong generalization, for this was found to assert that only those Q-predicates are instantiated in the whole universe that are already exemplified in a sufficiently large e. In other words, the most informative new observations will be those which falsify the previously preferred generalization. This simple observation may go some way toward explaining and justifying Popper's emphasis on the role of falsification in scientific procedure. On the view we have reached here, however, the emphasis on falsification is not important solely because it guards us against false hypotheses, as Popper sometimes seems to suggest. It is also a direct consequence of our search for the most informative kind of evidence.

All these observations seem very important philosophically. Their theoretical foundation is somewhat shaky, however, as long as we do not understand better the reasons for the choice of the underlying probability measures on which our measures of information are based. Furthermore, it must be pointed out that although the results we have obtained are often very natural asymptotically, in almost all the cases I have so far discussed they are not very natural quantitatively. For small values of n (= the number of observed individuals) they give too high probabilities to generalizations and thus correspond to wildly overoptimistic inductive behavior with respect to inductive generalization from small samples. Furthermore, in some cases (especially when the *a priori* probabilities are divided evenly among state-descriptions) we do not obtain reasonable results for singular inductive inferences.[22]

(4) *Probability-measures dependent on parameters.* All these difficulties can be eliminated by taking a somewhat more general view of the situation. It seems to me that we have so far overlooked an important factor that affects our ideas of information. For it seems to me obvious that the information that our observation-statements are supposed to have depends heavily on the amount of order or regularity that we know (or, for a subjectivist, believe) that there is in the world. For instance, if we know or believe that the universe is perfectly regular in the sense that all its members have the same Q-predicate (though we do not know which), one single observation-statement of the form $Ct_i(a)$ gives us all the information we might want to have (in so far it can be expressed by means of the resources we have at our disposal in monadic first-order logic). Taking a less extreme case, if we know that almost all the individuals of our domain have one and the same Q-predicate, the observation of two individuals having the same Q-predicate tells us more about the probable distribution of individuals into the different Q-predicates than what it would do if we know that individuals are likely to be distributed more or less evenly among several Q-predicates.[23]

Our probability-judgements are apt to exhibit the same dependence on our knowledge of the amount of order or regularity in the universe. If we knew (or believe) that the universe is very regular, we are apt to follow the indications of observations much more boldly than we would do if we knew that it were fairly irregular. In the latter case our estimates of the distribution of unobserved individuals between the Q-predicates will be much more conservative than in the former and also much more dependent on *a priori* probabilities. Observed regularities would not be followed up as quickly as in the former case because they would be expected to be due to chance rather than the actual regularities that obtain in the universe at large. Thus the amount of regularity we think there is in the universe is reflected by the rate at which observations affect our a priori probabilities.

This effect has to be taken into account in constructing measures of information. It seems to me that it has to be taken into account on two different levels. We have to consider the amount of regularity or irregularity that there is in the world (or is thought of as being there) as far as the distribution of individuals between those Q-predicates is concerned that are instantiated in the universe. Secondly, we have to consider how regularly the individuals are apt to be concentrated into few Q-predicates rather than to be spread out among all of them.[24]

Let us consider these problems in order. The former effect is shown by the probability that the next individual will exemplify certain given Q-predicate $Ct_i(x)$, when we know that a certain definite constituent C_w which allows w Q-predicates to be instantiated (among them $Ct_i(x)$) is true. Let us assume that we have observed n individuals of which q exemplify $Ct_i(x)$. It is natural to assume that the desired probability should depend on the numbers w, n, and q only. From this assumption it follows, together with certain other very plausible assumptions, that the desired probability is expressed by

$$(23) \qquad \frac{q + \lambda/w}{n + \lambda}$$

where λ is a free parameter, $0<\lambda<\infty$.

Expression (23) may be taken to be a kind of weighted average of the *a priori* factor $1/w$ and the purely empirical (*a posteriori*) factor q/n which is nothing but the observed relative frequency. In this weighted average, λ is the weight of the *a priori* factor. (Formally, it may be thought of as the size of a fictional sample which is evenly divided between the w admissible Q-predicates and which has to be added to the actually observed sample of n individuals to bring out the effects of *a priori* factors.) For instance, if $\lambda=0$, we have the most purely empirical rule of induction ('straight rule') which tells us to follow observed relative frequencies directly.

According to what was said earlier, λ is thus also a measure of the *dis*order or *ir*regularity that we know or believe to obtain in our universe of discourse, as far as the distribution of individuals between the different Q-predicates that C_w allows to be instantiated are concerned.

In any finite universe, the choice of λ determines together with (23) the probabilities (relative to the different C_i) of all state-descriptions. This is seen most easily by pointing out that because each state-description can be written in the form (16), its probability can be expressed in the following form (omitting all reference to the C_i):

$$(24) \qquad p(Ct_{i_1}(a_1)) \cdot p(Ct_{i_2}(a_2) | Ct_{i_1}(a_1)) \times p(Ct_{i_3}(a_3) | (Ct_{i_1}(a_1) \,\&\, Ct_{i_2}(a_2))) \ldots$$

In order to find a completely defined distribution of *a priori* probabilities to state-descriptions (and therefore to everything else too) we thus have but to specify the *a priori* probabilities of the different constituents, for everything we have said in the last couple of paragraphs has been relative to the truth of one of them.

These *a priori* probabilities of constituents are precisely what is at stake when we consider the other kind of regularity that there may or may not be in the universe, for this kind of regularity was what determines how likely it is that some of the Q-predicates, or even several of them, are empty. A convenient way of telling how likely it is for a constituent C_w to be true, i.e. for certain general laws to hold in our universe, is to compare the *a priori* probability $p(C_w)$ of C_w with the probability of obtaining a sample compatible with it in a universe whose individuals have been distributed 'randomly' between the different Q-predicates in accordance with (23) except that w is replaced by K. Such a universe might be called an atomistic one. My suggestion for specifying the probabilities *a priori* that the different general laws hold in our universe is to say that they are proportionally as great (or small) as the probabilities of obtaining samples of a fixed size (say containing α individuals) compatible with these laws in an atomistic universe. The larger α is, the smaller these probabilities obviously are. In fact, for a constituent C_w admitting w Q-predicates to be instantiated the probability of obtaining an atomistic sample of size α compatible with it is by (23)

$$(25) \qquad \frac{\lambda w/K}{\lambda} \cdot \frac{1 + \lambda w/K}{1 + \lambda} \cdot \frac{2 + \lambda w/K}{2 + \lambda} \cdot \ldots \cdot \frac{\alpha - 1 + \lambda w/K}{\alpha - 1 + \lambda},$$

which generalizes as

$$(26)\qquad \frac{\Gamma(\alpha + \lambda w/K)\,\Gamma(\lambda)}{\Gamma(\alpha + \lambda)\,\Gamma(\lambda w/K)},$$

where Γ is the familiar gamma-function of analysis and α an arbitrary non-negative real number. My suggestion is to make the *a priori* probability of C_w proportional to (25) or (26).

Together with (23), these *a priori* probabilities suffice to specify our probability distribution. Strictly speaking, this works without further explanations only in the case of an infinite universe, while in a finite universe certain further explanations are needed. They are not pertinent to our present purposes, however.

The parameter α which occurs in (25) and (26) is a kind of measure of the amount of disorder (irregularity) we expect, or are entitled to expect, to obtain in the universe as far as general laws are concerned. According to what was said earlier, it therefore plays the role of an *index of caution*: it is a kind of weight of the *a priori* factor in inductive generalization, somewhat in the same way λ was a kind of weight attached to the *a priori* factor in singular inductive inference. However, caution leads us to different-looking inferences in the case of α and λ. In the case of λ (i.e. of singular inference) extreme caution leads us to reduce (23) to the simple form $1/w$.

In contrast, in the case of generalizations *a priori* considerations tend to discourage neat inferences. If we are extremely conservative, we must obviously say that all Q-predicates are pretty much on a par even after a number of non-symmetrical observations are made, and hence that in a sufficiently large universe all of them are likely to be exemplified, which means that no non-trivial general laws hold.

All the different ways of defining measures of probability and information which were discussed above (and a number of others) may be taken to be special cases of our two-dimensional continuum of probability-measures depending on α and λ. Carnap's p^+ results as a limiting case by letting $\alpha \to \infty$, $\lambda \to \infty$. His p^* results when $\alpha \to \infty$, $\lambda = K$. More generally, the λ-continuum of inductive methods considered by Carnap in some of his publications results when $\alpha \to \infty$. (No wonder it has nothing to say of inductive generalization: it embodies infinite caution *vis-à-vis* generalization.) The two-stage procedure of first distributing *a priori* probabilities evenly among constituents and then as evenly among state-descriptions as is compatible with the first distribution amounts to letting $\alpha=0$, $\lambda\to\infty$. The 'combined method' is tantamount to putting $\alpha=0$, $\lambda=w$.

All the undesirable features of these particular probability-measures and of the associated measures of information can now be seen to be consequences of the particular choices of α and λ that they are implicitly based on. If $\alpha=\infty$, inductive generalizations are judged on *a priori* grounds only. Unsurprisingly, it then turns out that all the different Q-predicates are in a large universe highly likely to be instantiated, and that a statement to the effect that some of them are instantiated is therefore highly uninformative. It is not any more surprising that a system with $\lambda=\infty$ should fail to yield a reasonable treatment of singular inductive inference, for this choice of λ amounts to treating singular inductive inferences completely on *a priori* grounds. Likewise, it is now seen to be inevitable that the systems in which probabilities are distributed evenly among constituents should give wildly overoptimistic results as far as inductive generalizations are concerned, for they are based on the choice $\alpha=0$ which means that we are expecting a great deal of uniformity in the world as far as generalizations are concerned. All these 'mistakes' can be corrected by choosing α (or λ) differently.

Independently of these choices, the asymptotic results mentioned above hold as long as α and λ are both non-negative finite numbers and $\lambda\neq0$. Furthermore, they remain valid when $\lambda\to\infty$. Hence in all these cases the connection between reasonable inductive generalization and the maximization of expected information will remain the same. Here we cannot study in detail the consequences of the different choices of λ and α in other respects.

A couple of matters of principle should be taken up here, however, First of all, it may be pointed out that there is another, possibly still more persuasive way of justifying the use of the parameter α. Several philosophers of science have discussed what they call the *lawlikeness* of some generalizations.[25] There are many problems connected with this notion, but one of the few things most parties agree on is that the degree of lawlikeness of a general hypothesis (say h) is shown by the rate at which positive evidence for h increases our confidence that the generalization applies also to so far unexamined cases. This is a conception which we already used above in introducing probability-measures dependent on parameters by reference to "the rate at which observations affect our probabilities". Now the line of thought used there can be viewed in a larger perspective. In so far as the degree of our confidence in new applications of a tentative generalization can be expressed as an exchangeable probability in de Finetti's sense [26], it follows from de Finetti's representation theorem that assumptions concerning degrees of lawlikeness can be expressed in terms of an *a priori* probability distribution. If generalizations themselves are ever to receive non-zero probabilities *a posteriori*, the *a priori* distribution has to agree with ours in giving a non-zero probability *a priori* to all constituents. The only further element in our treatment is a particular way of calibrating one's assumptions of lawlikeness by comparing the *a priori* probability of a constituent with the probability of obtaining from a completely atomistic Carnapian universe a sample of a fixed size α compatible with this constituent. These simple ideas are all we need to arrive straight at (25) or (26). They also show the additional theoretical interest of our parameter α, due to its very close connection with the (probabilistically interpreted) notion of lawlikeness.

Another general point seems even more important. I have suggested that it is natural and even inevitable that the factors which our parameters α and λ bring out should infiuence one's measures of semantic information. At the same time, it seems impossible to justify any particular choice of these parameters by means of purely logical considerations. For what they express is, from an objectivistic point of view, the amount of disorder that there probably is in the universe, or, from the subjectivistic point of view, one's expectations concerning the amount of this disorder. An objectivist (though not necessarily a frequentist) can thus understand what goes into these two parameters, but it does not seem to be possible even to associate a clear sense to them on the basis of a strictly logical interpretation of probability. In so far as the considerations that are presented in this paper concerning the relation of information and probability and concerning the role of the parameters α and λ are acceptable, they therefore suggest rather strongly that a strictly logical interpretation of probability cannot be fully carried out, at least not in the simple languages we have been considering. Rather, the force of circumstance (more accurately speaking, the role of the idea of order or regularity in our judgements of information and probability) pushes one inevitably toward a much more 'Bayesian' position in the sense that we are forced to consider the dependence of our *a priori* probabilities on non-logical factors and also to make them much more flexible than a hardcore defender of a logical interpretation of probability is likely to accept. It also seems to me that a closer investigation of the factors affecting our judgements of semantic information and probability will strengthen this view and perhaps necessitate an even greater flexibility in our distribution of *a priori* probabilities.

Among other things, this implies that the relation of a theory of semantic information to statistical information theory is more problematic than earlier discussions of this relation are likely to suggest. This relation cannot be identified with any clear-cut difference between senses of probability.

The distinctive feature of a theory of semantic information and of the associated theory of probability will perhaps turn out not to be any special sense of probability but rather the need of developing methods for discussing the probabilities and information-contents of generalizations (closed statements). For I do not see how one could avoid speaking of the information that different generalizations convey to us in any serious theory of semantic information. And if this talk is to be understood along the lines of any concept of information that is related to a suitable notion of probability in a natural way, we need a concept of probability which can assign finite non-zero probabilities to generalizations in all domains. Above, some such senses of probability were examined in the context of certain especially simple languages. The task of studying these senses of probability and similar ones seems especially important from a subjectivistic point of view, for it seems to me obvious that people do in fact associate degrees of belief with general statements. They even seem to be able to compare these degrees of belief with the degrees of their belief in various singular statements.

Further evidence for our general point of view is forthcoming from an examination of what happens when relations are introduced in addition (or instead of) monadic predicates.[27] It was suggested above that already in the monadic case purely logical considerations (symmetry principles) do not yield any unique answer as to how the different alternatives one can distinguish in one's language are to be weighted, i.e. what *a priori* probabilities they ought to be given. Although few discussions of the relational cases have reached the print, it is nevertheless already clear that this indeterminacy becomes more blatant when we go beyond the monadic case. Again, the weighting principles can only be understood as embodying assumptions concerning the degree of regularity in one's universe of discourse. What is new in the relational case is that we now have different kinds of regularities competing against each other (e.g. the regularity one's individuals exhibit in their classification into the generalized Q-predicates *versus* the regularity they exhibit in their relations to this or that particular individual). Here it is especially patent that no mere symmetry considerations can be expected to suffice to adjudicate between these competing regularity principles so as to yield a unique measure of 'logical probability', and the same seems to go for all the other general *a priori* requirements that our 'inductive intuition' might suggest. Thus a consideration of the relational case will reinforce our case against the self-sufficiency of a purely logical interpretation of probability and against its mirror-image concept of (uniquely defined) semantic information based solely on symmetry principles and on other *a priori* considerations.

BIBLIOGRAPHY

[1] Y. Bar-Hillel, *Language and Information*, Addison-Wesley and The Jerusalem Academic Press, Reading, Mass., and Jerusalem 1964.

[2] Y. Bar-Hillel and R. Carnap, 'Semantic Information', *British Journal for the Philosophy of Science* **4** (1953) 147–157.

[3] R. Carnap, *Logical Foundations of Probability*, University of Chicago Press, Chicago 1950, 2nd ed., 1963.

[4] R. Carnap, *Continuum of Inductive Methods*, University of Chicago Press, Chicago 1952.

[5] R. Carnap, 'Probability and Content Measure', in *Mind, Matter and Method* (ed. by P. K. Feyerabend and G. Maxwell), University of Minnesota Press, Minneapolis 1966, pp. 248–260.

[6] R. Carnap and Y. Bar-Hillel, 'An Outline of a Theory of Semantic Information', *Technical Report No.* 247, M.I.T. Research Laboratory in Electronics 1952; reprinted in [1], Ch. 15.

[7] C. Cherry, *On Human Communication*, M.I.T. Press, Cambridge, Mass., 1957.

[8] N. Goodman, *Fact, Fiction, and Forecast*, University of London Press, London 1955.

[9] C. G. Hempel, 'Inductive Inconsistencies', *Synthese* **12** (1960) 439–469; reprinted in C. G. Hempel, *Aspects of Scientific Explanation and Other Essays in the Philosophy of Science*, The Free Press, New York 1965, pp. 53–79, especially pp. 73–78.

[10] C. G. Hempel, 'Deductive-nomological vs. Statistical Explanation', in *Minnesota Studies in the Philosophy of Science*, vol. III (ed. by H. Feigl and G. Maxwell), University of Minnesota Press, Minneapolis 1962, pp. 98–169, especially pp. 153–156.

[11] R. Hilpinen, 'On Inductive Generalization in Binary Functional Calculus', *Third International Congress of Logic, Methodology and Philosophy of Science, Abstracts of Papers* (mimeographed), Amsterdam 1967.

[12] K. J. Hintikka, 'Towards a Theory of Inductive Generalization', in *Proceedings of the 1964 International Congress for Logic, Methodology and Philosophy of Science* (ed. by Y. Bar-Hillel), North-Holland Publishing Company, Amsterdam 1965, pp. 274–288.

[13] K. J. Hintikka, 'On a Combined System of Inductive Logic', in *Studia Logico-Mathematica et Philosophica in honorem Rolf Nevanlinna, Acta Philosophica Fennica* **18** (1965) 21–30.

[14] K. J. Hintikka, 'A Two-Dimensional Continuum of Inductive Methods', in *Aspects of Inductive Logic* (ed. by K. J. Hintikka and P. Suppes), North-Holland Publishing Company, Amsterdam 1966, pp. 98–117.

[15] K. J. Hintikka, 'Induction by Enumeration and Induction by Elimination', in *The Problem of Inductive Logic, Proceedings of the International Colloquium in Logic, Methodology, and Philosophy of Science, London, 1965* (ed. by I. Lakatos), vol. II, North-Holland Publishing Company, Amsterdam 1967.

[16] K. J. Hintikka, 'The Varieties of Information and Scientific Explanation', in *Logic, Methodology and Philosophy of Science III, Proceedings of the 1967 International Congress* (ed. by B. van Rootselaar and J. F. Staal), North-Holland Publishing Company, Amsterdam 1968, pp. 151–171.

[17] K. J. Hintikka and R. Hilpinen, 'Knowledge, Acceptance, and Inductive Logic', in *Aspects of Inductive Logic* (ed. by K. J. Hintikka and P. Suppes), North-Holland Publishing Company, Amsterdam 1966, pp. 1–20.

[18] K. J. Hintikka and J. Pietarinen, 'Semantic Information and Inductive Logic', in *Aspects of Inductive Logic* (ed. by K. J. Hintikka and P. Suppes), North-Holland Publishing Company, Amsterdam 1966, pp. 81–97.

[19] J. G. Kemeny, 'A Logical Measure Function', *Journal of Symbolic Logic* **18** (1953) 289–308.

[20] H. E. Kyburg and H. E. Smokler (eds.), *Studies in Subjective Probability*, John Wiley and Sons, New York 1964.

[21] I. Levi, 'Decision Theory and Confirmation', *Journal of Philosophy* **58** (1961) 614–624.

[22] R. D. Luce and H. Raiffa, *Games and Decisions*, John Wiley and Sons, New York 1957, pp. 318–324, especially p. 324.

[23] E. Nagel, *The Structure of Science*, Harcourt, Brace and World, Inc., New York 1961.

[24] K. R. Popper, *The Logic of Scientific Discovery*, Hutchinson, London 1959. The bulk of this work consists of a translation of *Logik der Forschung*, Springer, Vienna 1934.

[25] K. R. Popper, 'Degree of Confirmation', *British Journal for the Philosophy of Science* **5** (1954) 143–149. (Correction, *ibid.* 334.)

[26] P. A. Schilpp (ed.), *The Philosophy of Rudolf Carnap*, The Library of Living Philosophers, La Salle, Illinois, 1963.

[27] C. E. Shannon and W. Weaver, *The Mathematical Theory of Communication*, The University of Illinois Press, Urbana, Illinois, 1949.

REFERENCES

[1] For this theory, see e.g. [27]. A popular survey is given in [7].

[2] On this subject, cf. [1], Chapters 15–18; also [7], pp. 229–255.

[3] E.g., [1], p. 242.

[4] It is therefore natural that such expected values or estimates of information as (2) should play an important role in this theory. Its statistical character is not the only reason, however, for the prevalence of expressions similar to (2); they have a place in the semantic theory, too.

[5] I shall draw on the works I have published elsewhere. See [12–18].

[6] The term (and the idea) goes back all the way to Boole.

[7] Cf. [1], p. 307.

[8] [24] *passim.*

[9] An example to this effect is given by Carnap in [5]. Further examples are found in my survey [16], especially the expected content $\mathrm{p}(h|e) - \mathrm{p}(h)$, which is as likely to vary with $\mathrm{p}(h|e)$ as with $1 - \mathrm{p}(h)$.

[10] See e.g. [9–10].

[11] See e.g. [21–22].

[12] This is in effect surmised by Hempel in [9], p. 77.

[13] The pioneering work in this area has been done by Rudolf Carnap. See [3–4]. Cf. also [26], especially Carnap's own contribution, pp. 966–998.

[14] The line of thought presented here is not Carnap's.

[15] This measure was for a while preferred by Carnap to other probability measures, and it appears to be an especially simple one in certain respects.

[16] Cf. my criticism of Carnap in the early pages of [12].

[17] This was the course followed in [12].

[18] It was proposed in [13].

[19] This question was asked by Hintikka and Pietarinen [18], whose line of thought we shall here follow.

[20] The choice of an appropriate utility here is a very tricky matter. Further comments on the subject are made in [16].

[21] Notice, however, that other kinds of results might be forthcoming if we considered other kinds of relative information instead of the added surprise value $\mathrm{inf}(h|e)$.

[22] See the last few pages of [12] and [13].

[23] These informal arguments can easily be reconstructed on the basis of the distinctions between different kinds of information discussed in [16].

[24] For the following, cf. [4] and [14].

[25] See e.g. Nelson Goodman [8] or E. Nagel [23], Ch. 4.

[26] See de Finetti's classical paper reprinted in [20].

[27] A preliminary report on some of the simplest cases is given in [11].

DEAN JAMISON*

BAYESIAN INFORMATION USAGE

I. INTRODUCTION

We might distinguish between inductive and deductive inferences in the following way: Deductive inferences refer to the implications of coherence for a given set of beliefs, whereas inductive inferences follow from conditions for 'rational' *change* in belief. Change in belief, I shall argue in the following section, is perhaps the most philosophically relevant notion of semantic information. Thus rules governing inductive inferences may be regarded as rules for the acquisition of semantic information.

I have four purposes in this paper. First I shall attempt to provide a definition of semantic information that is adequate from a subjectivist point of view and that is based on the concept of information as change in belief. From this I shall turn to a subjectivistic theory of induction; the second purpose of the paper is to suggest a solution to the inductive problem that Suppes [37, pp. 514–515] points out to lie at the foundations of a subjectivistic theory of decision. (By this I do not mean to suggest a solution to the inductive problem of Hume; I would agree with Savage [32] that the subjective theory of probability simply cannot do this.) The third thing I wish to do is to show how Carnap's continuum of inductive methods may be easily interpreted as a special case of the subjectivistic theory of induction to be presented. Finally, I provide a subjectivistic interpretation of Hintikka's two dimensional inductive continuum, and show how this is related to the problem of concept formation.

II. SEMANTIC INFORMATION AND INDUCTION

A. *Two Notions of Semantic Information*

Two alternative notions of semantic information are *reduction in uncertainty* and *change in belief.* Reduction in uncertainty is, clearly, a special

J. Hintikka and P. Suppes (eds.), Information and Inference, 28–57.

case of change in belief. Information is defined in terms of probabilities; hence, one's view of the nature of probability is inevitably an input to his theory of information. As there are three prominent views concerning the nature of probability – the relative frequency, logical, and subjectivist views – and there are the two concepts of information just mentioned, we can distinguish six alternative theories of information. Table I arrays these theories.

TABLE I
Theories of information

Concept of Information	Concept of Probability		
	Relative Frequency	*Logical*	*Subjective*
Change in Belief	*CR*	*CL*	*CS*
Reduction of Uncertainty	*RR*	*RL*	*RS*

RS, for example, would be a theory of information based on a subjectivist view of probability and a reduction of uncertainty approach to information. The development of the *RR* theory by Shannon and Weaver [33] has provided the formal basis for most later work. Carnap and Bar-Hillel [1] developed *RL* and Bar-Hillel [3, 4] hints at the potential value of developing what I would call *RS* or *CS*, though his precise meaning is unclear. Sneed's [36] discussion of 'pragmatic informativeness' is related. Smokler [34] as well as Hintikka and Pietarinen [13] have further developed *RL*.

An undesirable feature of *RL* is that in it logical truths carry no information. For example, solving (or being told the solution of) a difficult differential equation gives you no new information. This is a result of accepting the 'equivalence condition', ramifications of which are discussed by Smokler [35]. Wells [41] has made an important contribution to the development of *RS* by beginning a theory of the information content of *a priori* truths. To continue the example above, Wells allows that the solution to the differential equation may, indeed, give information. Howard's [16] paper on 'information value theory' develops *RS* in a decision-theoretic context, deriving the value of *clairvoyance* and using that value as the upper bound to the value of any information. McCarthy

[26] has also developed a class of measures of the value of *RS* information.

Two further works concerning semantic information and change in belief should be noted. MacKay [25] has developed techniques of information theory to analyze scientific measurement and observation. His view may be considered a change in belief view. In a more recent work Ernest Adams [1] has developed a theory of measurement in which information theoretic considerations play an important role. It seems to me that one interpretation of his approach would be that the purpose of measurement is simply the attainment of semantic information, though Adams would not agree with this. Throughout Adams uses a frequency interpretation of probability.

B. *Initiating a CS Theory of Information*

What seems to me to be the most natural notion of semantic information is change in belief as reflected in change in subjective probabilities. That is, I would regard *CS* as the most fundamental entry in the table shown above, at least from a psychologist's or philosopher's point of view. There are two primary reasons for this. The first is that change in belief is a more general notion than reduction of uncertainty, subsuming reduction in uncertainty as a special case. The second is that reality is far too rich and varied to be adequately reflected in a logical or relative frequency theory of probability.[1] Let me now turn to definitions of belief and information.

Consider a situation in which there are m mutually exclusive and collectively exhaustive possible states of nature. Define an $m-1$ dimensioned simplex, Ξ, in m dimensioned space in the following manner: $\Xi = \{\boldsymbol{\xi} \mid \sum_{i=1}^{m} \xi_i = 1 \text{ and } \xi_i \geqslant 0 \text{ for } 1 \leqslant i \leqslant m\}$. The vector $\boldsymbol{\xi} = (\xi_1, \xi_2, \ldots, \xi_m)$ intuitively corresponds to a probability distribution over the states of nature with ξ_i the probability of the ith state of nature. Ξ is the set of all possible probability distributions over the m states of nature. For these purposes a belief may be simple defined as a subjectively held vector $\boldsymbol{\xi}$. Measurement of belief is an example of 'fundamental' measurement and the conditions under which such measurement is possible are simply the conditions that must obtain in order that a qualitative probability relation on a set may be represented by a numerical measure. Information is an example of 'derived' measurement.

Roby [30] has an interesting discussion of belief states defined in this way. Let ξ be a person's beliefs before he receives some information (or message) M, and ξ' his beliefs afterwards. The notion of message here is to be interpreted very broadly – it may be the result of reading, conversation, observation, experimentation, or simply reflection. The primary requirement of a definition of the amount of information in the message M, $\inf(M)$, is that it be a (strictly) increasing function of the 'distance' between ξ and ξ'. Perhaps the simplest definition that satisfies this requirement is:

$$\text{(1)} \qquad \inf(M) = |\xi - \xi'| = \sum_{i=1}^{m} (\xi_i - \xi_i')^2 .$$

A drawback to this definition is that the amount of information is relatively insensitive to m. Consider two cases where in the first $m=4$ and in the second $m=40$. In each $\xi_i = 1/m$ for $1 \leqslant i \leqslant m$ and $\xi_1' = 1$ and $\xi_i' = 0$ for $i > 1$. It would seem that in some sense in the case where m equaled 40 a person would have received much more information than if m had equaled 4, and the Shannon measure of information, for example, reflects this intuition. However, for $m=4$, the information received as defined in (1) is .876, and for $m=40$ it is .989, a rather small difference. An alternative definition, that takes care of this defect, is:

$$\text{(2)} \qquad \inf(M) = \frac{m\sqrt{m(m-1)}}{2(m-1)} \, |\xi - \xi'| .$$

The apparent complexity makes some numbers come out nicely; from the preceeding example, when $m=4$ the information conveyed as measured by (2) is 2. For $m=40$, it is 20.

The definitions in Equations (1) and (2) are meant merely to show that a *CS* theory of information can be discussed in a clear and formal way. Implications of these definitions – or alternatives to them – must await another time, as the rest of this paper will be concerned primarily with induction.

C. *Semantic Information and Induction*

For purposes of discussing induction we might consider three levels of inductive inference. The first and simplest level is simply conditionalization or the updating of subjective probabilities by means of Bayes' theorem. That this is the normatively proper way to proceed in some instances seems undeniable. A more complicated level of inductive inference concerns inferences made on the basis of the formation of a concept. The highest level of inductive inferences are inductions made from scientific laws, by which I simply mean mathematical models of natural phenomena. The distinction between the second and third levels of inference is that models have parameters to be evaluated whereas concepts do not.

A question of some interest concerning philosophical theories of induction is whether some form of Bayesian updating will suffice for a normative account of inductive behavior at the second and third levels. Suppes [38] answers the question just asked with a clear 'no'. He summarizes his position in the following way:

> The core of the problem is developing an adequate psychological theory to describe, analyze, and predict the structure imposed by organisms on the bewildering complexities of possible alternatives facing them. [I hope I have made it clear that] the simple concept of an *a priori* distribution over these alternatives is by no means sufficient and does little toward offering a solution to any complex problem (p. 47).

Suppes even suggests that in cases where Bayes' theorem would fairly obviously be applicable a person might not be irrational to act in some other way. While I cannot see the rationale for this, the points he makes about concept formation and, implicitly, about the construction of scientific laws seem well taken. To put this in the context of our discussion of semantic information I would suggest that a concept had been formed when a person acquires much semantic information (i.e., radically rearranges his beliefs) on the basis of small evidence.

In the following two sections of this paper I deal with inductive inference of the simplest sort. In the final section of the paper I attempt to show that Suppes' pessimism concerning a Bayesian theory of concept formation is partially unjustified.

III. A SUBJECTIVISTIC THEORY OF INDUCTION

My discussion of induction will be formulated in a decision theoretic framework, and I will digress to problems of decision theory here and there. The discussion of decisions under total ignorance forms the basis for the later discussion of inductive inference, and the intuitive concepts of that subsection should be understood, though the mathematical details are not of major importance.

All essentials of a subjectivistic theory of induction are contained in Bruno de Finetti's [9] classic paper. The probability of probabilities approach developed here can be translated (though not always simply) into the de Finetti framework; the only real justification for using probabilities of probabilities is their conceptual simplicity. The importance of this simplicity will, I think, be illustrated in Sections IV and V.

A triple $P = \langle D, \Omega, U \rangle$ may be considered a finite decision problem when: (i) D is a finite set of alternative courses of action available to a decision-maker, (ii) Ω is a finite set of mutually exclusive and exhaustive possible states of nature, and (iii) U is a function on $D \times \Omega$ such that $u(d_i, \omega_j)$ is the utility to the decision-maker if he chooses d_i and the true state of nature turns out to be ω_j. A decision procedure (solution) for the problem P consists either of an ordering of the d_i's according to their desirability or of the specification of a subset of D that contains all d_i that are in some sense optimal and only those d_i that are optimal.

If there are m states of nature, a vector $\boldsymbol{\xi} = (\xi_1, \ldots, \xi_m)$ is a *possible* probability distribution over Ω (with $\text{prob}(\omega_j) = \xi_j$) if and only if $\sum_{j=1}^{m} \xi_j = 1$ and $\xi_j \geqslant 0$ for $1 \leqslant j \leqslant m$. The set of all possible probability distributions over Ω, that is, the set of all vectors whose components satisfy the above equation and set of inequalities will, as in the preceeding section, be denoted by Ξ. Atkinson *et al.* [2] assume our knowledge of $\boldsymbol{\xi}$ to be completely specified by asserting that $\boldsymbol{\xi} \in \Xi_0$, where $\Xi_0 \subseteq \Xi$. If $\Xi_0 = \Xi$, they say we are in complete ignorance of ξ. In the manner of Chernoff [8] and Milnor [28], Atkinson *et al.* [2] give axioms stating desirable properties for decision procedures under complete ignorance. A class of decision procedures that isolates an optimal subset of D is shown to exist and satisfy the axioms. These procedures are non-Bayesian in the sense that the criterion for optimality is not maximization of expected utility. Other non-Bayesian procedures for complete ignorance (that fail to satisfy some axioms that most people would consider reasonable) include the following: minimax regret, minimax risk (or maximin utility), and Hurwicz's α procedure for extending the minimax risk approach to non-pessimists.

The Bayesian alternative to the above procedures attempts to order the d_i according to their expected utility; the optimal act is, then, simply the one with the highest expected utility. Computation of the expected utility of d_i, $Eu(d_i)$, is straightforward if the decision-maker knows that Ξ_0 is a set with but one element – ξ^*: $Eu(d_i)=\sum_{j=1}^{m} u(d_i, \omega_j)\, \xi_j^*$. Only in the rare instances when considerable relative frequency data exist will the decision-maker be able to assert that Ξ_0 has only one element. In the more general case the decision-maker will be in 'partial' or 'total' ignorance concerning the probability vector ξ. It is the purpose of the next two subsections to characterize total and partial ignorance from a Bayesian point of view and to show that decision procedures based on maximization of expected utility extend readily to these cases.

A. *Decisions Under Total Ignorance*

Rather than saying that our knowledge of the probability vector ξ is specified by asserting that $\xi \in \Xi_0$ for some Ξ_0, I suggest that it is natural to say that our knowledge of ξ is specified by a density, $f(\xi_1, \ldots, \xi_m)$, defined on Ξ. If the probability distribution over Ω is known to be ξ^*, then f is a δ function at ξ^* and computation of $Eu(d_i)$ proceeds as in the introduction. At the other extreme from precisely knowing the probability distribution over Ω is the case of total ignorance. In this subsection a meaning for total ignorance of ξ will be discussed. In the following subsection decisions under partial ignorance – anywhere between knowledge of ξ and total ignorance – will be discussed.

If $H(\xi)$ is the Shannon and Weaver [33] measure of uncertainty concerning which ω in Ω occurs, then $H(\xi)=\sum \xi_i \log_2(1/\xi_i)$, where $H(\xi)$ is measured in bits. When this uncertainty is a maximum, we may be considered in total ignorance of ω and, as one would expect, this occurs when we have no reason to expect any one ω more than another, i.e., when for all i, $\xi_i = 1/m$. By analogy, we can be considered in total ignorance of ξ when

$$H(f) = \int\int \underset{\Xi}{\cdots} \int f(\xi) \log_2 (1/f(\xi))\, d\Xi$$

is a maximum. This occurs when f is a constant, that is, when we have no reason to expect any particular value of ξ to be more probable than any other (see Chapter 3 of Shannon and Weaver [33]). If there is total ignorance concerning ξ, then it is reasonable to expect that there is total ignorance concerning ω – and this is indeed true (if we substitute the expectation of ξ_i, $E(\xi_i)$, for ξ_i).[3] Let me now prove this last assertion, which is the major result of this subsection. While this could be proved using the rather general theorems to be utilized in my discussion of Carnap, I think it is intuitively useful to go into a little more detail here.

Proving that under total ignorance $E(\xi_i)=1/m$ involves, first, determination of the appropriate constant value of f, then determination of the marginal density functions for the ξ_i's and, finally, integration to find $E(\xi_i)$.

Let the constant value of f equal K; since f is a density the integral of K over Ξ must be unity:

$$\int\int_{\Xi}\ldots\int K d\Xi = 1\,, \tag{3}$$

where $d\Xi = d\xi_1 \ldots d\xi_m$. Our first task is to solve this equation for K. Since f is defined only on a section of a hyperplane in m dimensioned space, the above integral is a many dimensioned 'surface' integral.

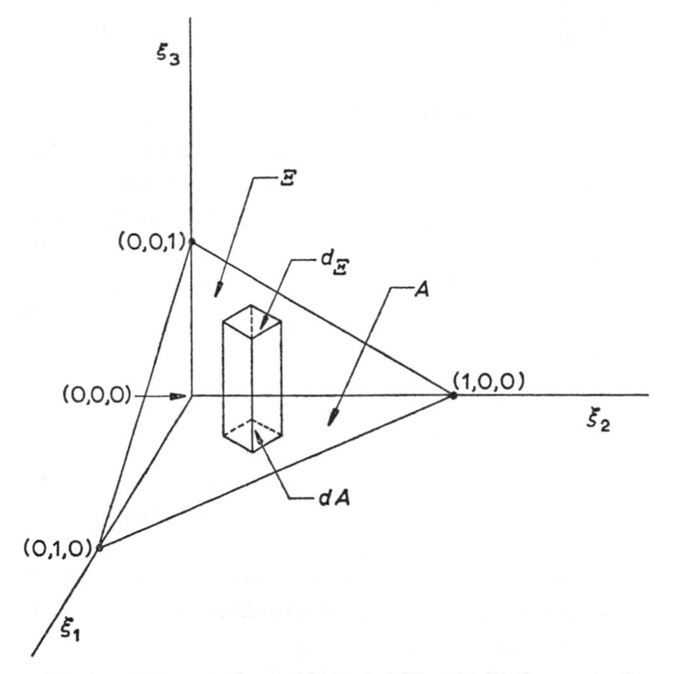

Fig. 1. Ξ, the set of possible probability distributions over Ω.

Figure 1 depicts the three dimensional case. As $\sum_{i=1}^{m} \xi_i = 1$, ξ_m is determined given the previous $m-1$ ξ_i's and the integration need only be over a region of $m-1$ dimensioned space, the region A in Figure 1. It is shown in advanced calculus that $d\Xi$ and dA are related in the following way:

$$d\Xi = \sqrt{\left(\frac{\partial (x_2, \ldots, x_m)}{\partial (\xi_1, \ldots, \xi_{m-1})}\right)^2 + \cdots + \left(\frac{\partial (x_1, \ldots, x_{m-1})}{\partial (\xi_1, \ldots, \xi_{m-1})}\right)^2} \, dA$$

where x_i is the function of $\xi_1, \ldots, \xi_{m-1}$ that gives the ith component of $\boldsymbol{\xi}$, that is $x_i(.) = \xi_i$ for i less than or equal to $m-1$ and $x_i(.) = 1 - \xi_1 - \cdots - \xi_{m-1}$ if $i = m$. It can be shown that each of the m quantities that are squared under the radical above is equal to either plus or minus one; thus $d\Xi = \sqrt{m}\, dA$. Therefore (3) may be rewritten as follows:

$$(4) \qquad \int\!\!\int_A \cdots \int K \sqrt{m} dA = 1, \quad \text{or}$$

$$\int_0^1 \int_0^{1-\xi_1} \cdots \int_0^{1-\xi_1-\cdots-\xi_{m-2}} d\xi_{m-1} d\xi_{m-2} \ldots d\xi_1 = 1/K\sqrt{m}.$$

The multiple integral in (4) could conceivably be evaluated by iterated integration; it is much simpler, however, to utilize a technique devised by Dirichlet. Recall that the gamma function is defined in the following way: $\Gamma(n) = \int_0^\infty x^{n-1} e^{-x} dx$ for $n \geqslant 0$. If n is a positive integer, $\Gamma(n) = (n-1)!$ and $0! = 1$. Dirichlet showed the following (see Jeffreys and Jeffreys [23], pp. 468–470): If A is the closed region in the first ortant bounded by the coordinate hyperplanes and by the surface $(x_1/c_1)^{p_1} + (x_2/c_2)^{p_2} + \cdots + (x_n/c_n)^{p_n} = 1$, then

$$(5) \qquad \int\!\!\int_A \cdots \int x_1^{\alpha_1 - 1} x_2^{\alpha_2 - 1} \ldots x_n^{\alpha_n - 1} dA$$
$$= \frac{c_1^{\alpha_1} c_2^{\alpha_2} \ldots c_n^{\alpha_n}}{p_1 p_2 p_3 \ldots p_n} \cdot \frac{\Gamma(\alpha_1/p_1)\, \Gamma(\alpha_2/p_2) \ldots \Gamma(\alpha_n/p_n)}{\Gamma(1 + \alpha_1/p_1 + \alpha_2/p_2 + \cdots + \alpha_n/p_n)}.$$

For our purposes, $c_i = p_i = \alpha_i = 1$, for $1 \leqslant i \leqslant m$ and the $m-1$ ξ_i's replace the n x's. The result is that the integral in (4) becomes $1/\Gamma(m) = 1/(m-1)!$. Therefore $K = (m-1)!\, \sqrt{m}/m$.

Having determined the constant value, K, of f we must next determine the densities $f_i(\xi_i)$ for the individual probabilities. By symmetry, the densities must be the same for each ξ_i. The densities are the derivatives of the distribution functions which will be denoted $F_i(\xi_i)$. $F_1(c)$ gives the probability that ξ_1 is less than c; denote by $F_1^*(c)$ the probability that $\xi_1 \geqslant c$, that is, $F_1(c) = 1 - F_1^*(c)$ is simply the integral of f over Ξ_c, where

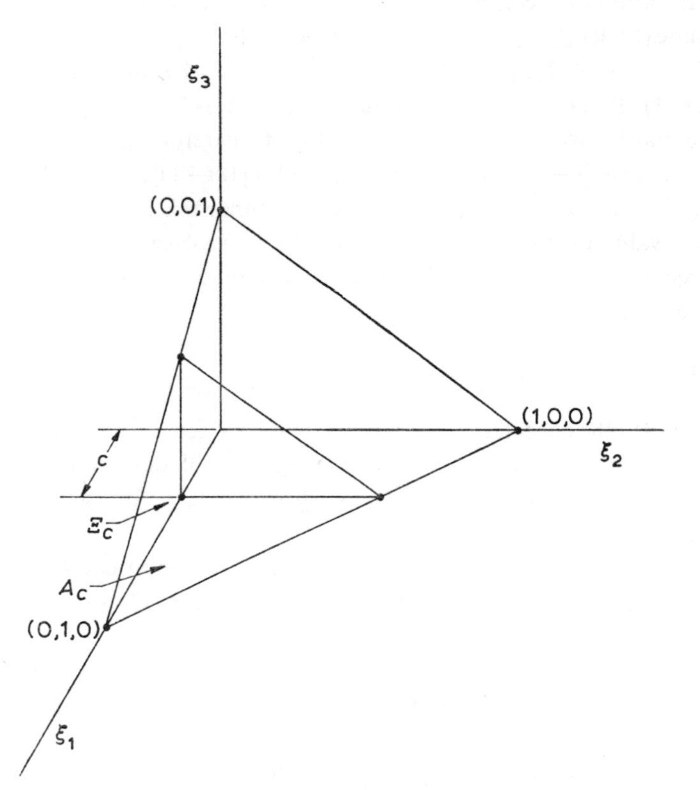

Fig. 2. Ξ_c, the subset of Ξ such that $\xi_1 \geqslant c$.

Ξ_c is the subset of Ξ including all points such that $\xi_1 \geqslant c$. See Figure 2. $F_1^*(c)$ is given by:

$$F_1^*(c) = \int\int_{\Xi_c} \cdots \int f(\xi)\, d\Xi = \int\int_{A_c} \cdots \int KmdA_c. \tag{6}$$

Since $K = (m-1)!\, \sqrt{m}/m$, (6) becomes (after inserting the limits of integration):

$$(7) \qquad F_1^*(c) = (m-1)! \int_c^1 \int_0^{1-\xi_1} \cdots \int_0^{1-\xi_1-\cdots-\xi_{m-2}} d\xi_{m-1} d\xi_{m-2} \ldots d\xi_1 .$$

A translation of the ξ_1 axis will enable us to use Dirichlet integration to evaluate (5); let $\xi_1' = \xi_1 - c$. Then $\xi_1' + \xi_2 + \cdots + \xi_{m-1} = 1 - c$, or $\xi_1'/(1-c) + \xi_2/(1-c) + \cdots + \xi_{m-1}/(1-c) = 1$ (since $\sum_{i=1}^{m-1} \xi_1 = 1$ is the boundary of the region A). Referring back to Equation (5) is can be seen that the c_i's in that equation are all equal to $1-c$ and that, therefore, the integral on the r.h.s. of (7) is $(1-c)^{m-1}/\Gamma(m)$. Thus $F_1^*(c) = [(m-1)!(1-c)^{m-1}]/\Gamma(m) = (1-c)^{m-1}$. Therefore $F_1(c) = 1-(1-c)^{m-1}$. Since this holds if c is set equal to any value of ξ_1 between 0 and 1, ξ_1 can replace c in the equation; differentiation gives the probability density function of ξ_1 and hence of all the ξ_i's:

$$(8) \qquad f_i(\xi_i) = (m-1)(1-\xi_i)^{m-2} .$$

From (8) the expectation of ξ_i is easily computed – $E(\xi_i) = \int_0^1 \xi_i (m-1)(1-\xi_i)^{m-2}$. Recourse to a table of integrals will quickly convince the reader that $E(\xi_i) = 1/m$. Figure 3 shows $f_i(\xi_i)$ for several values of m.

Jamison and Kozielecki [21] have determined empirical values of the function $f_i(\xi_i)$ for m equal to two, four, and eight. The experiment was run under conditions that simulated total uncertainty. The results were that subjects underestimated density in regions of relatively high density and overestimated it in regions of low density – an interesting extension of previous results.

Let $u(d_i, \xi) = \sum_{j=1}^m \xi_j u(d_i, \omega_j)$. Then the expected utility of d_i is given by:

$$(9) \qquad Eu(d_i) = \int\int_{\Xi} \cdots \int Ku(d_i, \xi) \, d\Xi .$$

This is equal to $\sum_{j=1}^m E(\xi_j) u(d_i, \omega_j) = (1/m) \sum_{j=1}^m u(d_i, \omega_j)$, since $u(d_i, \xi)$ is a linear function of the random variables ξ_i. Thus, taking the view of total ignorance adopted herein, we arrive by a different route at the decision rule advocated by Bernoulli and Laplace and axiomatized in Chernoff [8].

B. *Decisions Under Partial Ignorance*

Partial ignorance exists in a given formulation of a decision if we neither know the probability distribution over Ω nor are in total ignorance of it. If we are given $f(\xi_1, \ldots, \xi_m)$, the density over Ξ, computation of $Eu(d_i)$

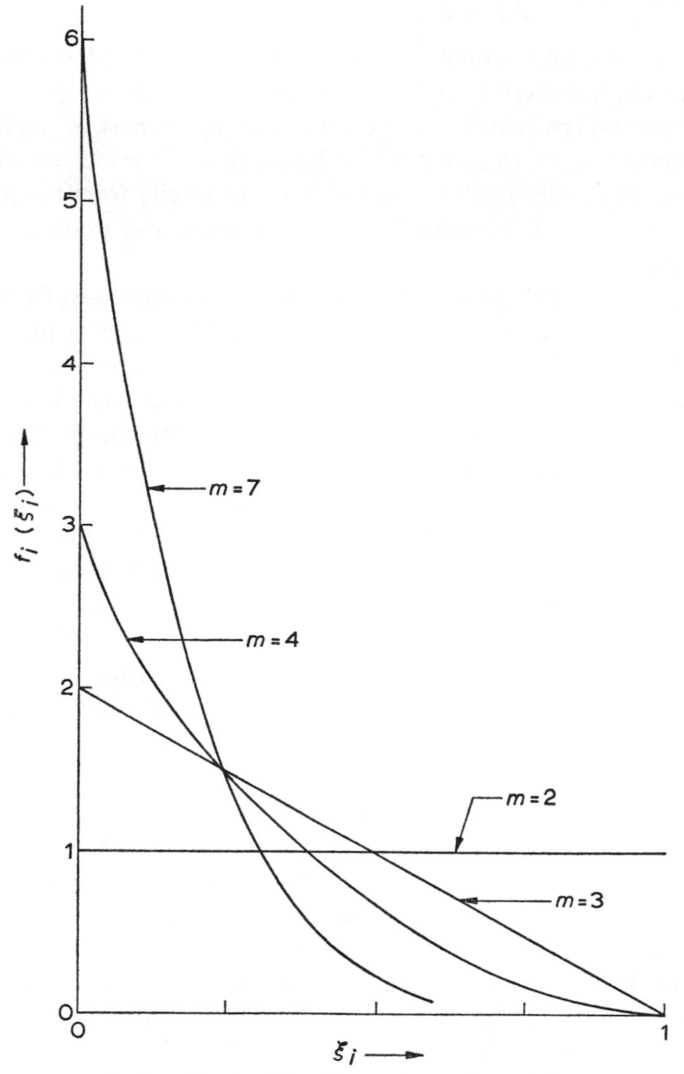

Fig. 3. Marginal densities under total uncertainty.

under partial ignorance is in principle straightforward and proceeds along lines similar to those developed in the previous section. Equation (9) is modified in the obvious way to:

$$(10) \qquad Eu(d_i) = \int\int_{\Xi} \cdots \int f(\xi)\, u(d_i, \xi)\, d\Xi .$$

If f is any of the large variety of appropriate forms indicated just prior to Equation (5), the integral in (10) may be easily evaluated using Dirichlet integration; otherwise more cumbersome techniques must be used.

In practice it seems clear that unless the decision-maker has remarkable intuition, the density f will be most difficult to specify from the partial information at hand. Fortunately there is an alternative to determining f directly.

Jeffrey [22, pp. 183–190], in discussing degree of confidence of a probability estimate, describes the following method for obtaining the distribution function, $F_i(\xi_i)$, for a probability.[4] Have the decision-maker indicate for numerous values of ξ_i what his subjective estimate is that the 'true' value of ξ_i is less than the value named. To apply this to a decision problem the distribution function – and hence $f_i(\xi_i)$ – for each of the ξ_i's must be obtained. Next, the expectations of the ξ_i's must be computed and, from them, the expected utilities of the d_i's can be determined. In this way partial information is processed to lead to a Bayesian decision under partial ignorance.

It should be clear that the decision-maker is *not* free to choose the f_i's subject only to the condition that for each f_i, $\int_0^1 f_i(\xi_i)\, d\xi_i = 1$. Consider the example of the misguided decision-maker who believed himself to be in total ignorance of the probability distribution over 3 states of nature. Since he was in total ignorance, he reasoned, he must have a uniform p.d.f. for each ξ_i. That is, $f_1(\xi_1) = f_2(\xi_2) = f_3(\xi_3) = 1$ for $0 \leqslant \xi_i \leqslant 1$. If he believes these to be the p.d.f.s., he should be willing to simultaneously take even odds on bets that $\xi_1 > \frac{1}{2}$, $\xi_2 > \frac{1}{2}$, and $\xi_3 > \frac{1}{2}$. I would gladly take these three bets, for under no conditions could I fail to have a net gain. This example illustrates the obvious – certain conditions must be placed on the f_i's in order that they be *coherent*. A necessary condition for coherence is indicated below; I have not yet derived sufficient conditions.

Consider a decision, d_k, that will result in a utility of 1 for each ω_j. Clearly, then $Eu(d_k) = 1$. However, $Eu(d_k)$ also equals $E(\xi_1)u(d_k, \omega_1) +$

$\cdots + E(\xi_m)u(d_k, \omega_m)$. Since for $1 \leqslant i \leqslant m$, $u(d_k, \omega_i)=1$, a necessary condition for coherence of the f_i's is that $\sum_{i=1}^{m} (\xi_i)=1$, a reasonable thing to expect. That this condition is not sufficient is easily illustrated with two states of nature. Suppose that $f_1(\xi_1)$ is given. Since $\xi_2=1-\xi_1, f_2$ is uniquely determined given f_1. However, it is obvious that infinitely many f_2's will satisfy the condition that $E(\xi_2)=1-E(\xi_1)$, and if a person were to have two distinct f_2's it would be easy to make a book against him; his beliefs would be incoherent.

If m is not very large, it would be possible to obtain conditional densities of the form $f_2(\xi_2 \mid \xi_1), f_3(\xi_3 \mid \xi_1, \xi_2)$, etc., in a manner analogous to that discussed by Jeffrey. If the conditional densities were obtained, then $f(\boldsymbol{\xi})$ would be given by the following expression:

$$(11) \qquad f(\xi) = f_1(\xi_1)\, f_2(\xi_2 \mid \xi_1) \ldots f_m(\xi_m \mid \xi_1, \xi_2, \ldots, \xi_{m-1}).$$

A sufficient condition that the f_i's be coherent is that the integral of f over Ξ be unity; if it differs from unity, one way to bring about coherence would be to multiply f by the appropriate constant and then find the new f_i's. If m is larger than 4 or 5, this method of insuring coherence will be hopelessly unwieldy. Something better is needed.

At this point I would like to discuss alternatives and objections to the theory of decisions under partial information that is developed here. The notion of probability distributions over probability distributions has been around for a long time; Knight, Lindall, and Tintner explicitly used the notion in economics some time ago – see Tintner [40].[5] This work has not, however, been formulated in terms of decision theory. Hodges and Lehmann [14] have proposed a decision rule for partial ignorance that combines the Bayesian and minimax approaches. Their rule chooses the d_i that maximizes $Eu(d_i)$ for some best estimate (or expectation) of $\boldsymbol{\xi}$, subject to the condition that the minimum utility possible for d_i is greater than a preselected value. This preselected value is somewhat less than the minimax utility; the amount less increases with our confidence that $\boldsymbol{\xi}$ is the correct distribution over Ω. Ellsberg [10], in the lead article of a spirited series in the *Quarterly Journal of Economics*, provides an elaborate justification of the Hodges and Lehmann procedure, and I will criticize his point of view presently.

Hurwicz [18] and Good (discussed in Luce and Raiffa [24], p. 305) have suggested characterizing partial ignorance in the same fashion that was later used by Atkinson *et al.*, [2]. That is, our knowledge of ξ is of the form $\xi \in \Xi_0$ where Ξ_0 is a subset of Ξ. Hurwicz then proposes that we proceed as if in total ignorance of where ξ is in Ξ_0. In the spirit of the second section of this paper, the decision rule could be Bayesian with $f(\xi) = K$ for $\xi \in \Xi_0$ and $f(\xi) = 0$ elsewhere. Hurwicz suggests instead utilization of non-Bayesian decision procedures; difficulties with non-Bayesian procedures were alluded to in the introduction to Section III.

Let me now try to counter some objections that have been raised against characterizing partial ignorance as probability distributions over probabilities. Ellsberg [10, p. 659] takes the view that since representing partial ignorance (ambiguity) as a probability distribution over a distribution leads to an expected distribution, ambiguity must be something different from a probability distribution. I fail to understand this argument; ambiguity is high, it seems to me, if f is relatively flat over Ξ, otherwise not. The 'reliability, credibility, or accuracy' of one's information simply determines how sharply peaked f is. Even granted that probability is somehow qualitatively different from ambiguity or uncertainty, the solution devised by Hodges and Lehmann [14] and advocated by Ellsberg relies on the decision-maker's completely arbitrary judgment of the amount of ambiguity present in the decision situation. Ellsberg would have us hedge against our uncertainty in ξ by rejecting a decision that maximized utility against the expected distribution but that has a possible outcome with a utility below an arbitrary minimum. By the same reasoning one could 'rationally' choose d_1 over d_2 in the non-ambiguous problem below if, because of our uncertainty in the outcome, we said (arbitrarily) that we would reject any decision with a minimum gain of less than 3.

	ω_1	ω_2
d_1	5	5
d_2	1	25

$$E(\xi_1) = E(\xi_2) = .5 .$$

I would reject Ellsberg's approach for the simple reason that its pessimistic bias, like any minimax approach, leads to decisions that fail to fully utilize one's partial information.

Savage [31, pp. 56–60] raises two objections to second-order probabilities. The first, similar to Ellsberg's, is that even with second-order probabilities expectations for the primary probabilities remain. Thus we may as well have simply arrived at our best subjective estimate of the primary probability, since it is all that is needed for decision-making. This is correct as far as it goes but, without the equivalent of second-order probabilities, it is impossible to specify how the primary probability should change in the light of evidence.

Savage's second objection is that "... once second-order probabilities are introduced, the introduction of an endless hierarchy seems inescapable. Such a hierarchy seems very difficult to interpret, and it seems at best to make the theory less realistic, not more." Luce and Raiffa [24, p. 305] express much the same objection. An endless hierarchy does not seem inescapable to me; we simply push the hierarchy back as far as is required to be 'realistic'. In making a physical measurement we could attempt to specify the value of the measurement, the probable error in the measurement, the probable error in the probable error, and on out the endless hierarchy. But it is not done that way; probable errors usually seem to be about the right order of realism. Similarly, I suspect that second-order probabilities will suffice for most circumstances.[6] However, in discussing concept formation in Section V, I shall have occasion to use what are essentially third-order probabilities.

C. *Induction*

The preceding discussion has been limited to situations in which the decision-maker has no option to experiment or otherwise acquire information. When the possibility of experimentation is introduced, the number of alternatives open to the decision-maker is greatly increased, as is the complexity of his decision problem, for the decision-maker must now decide which experiments to perform and in what order, when to stop experimenting, and which course of action to take when experimentation is complete. The problem of using the information acquired is the problem of induction.

If we are quite certain that ξ is very nearly the true probability distribution over Ω, additional evidence will little change our beliefs. If, on the other hand, we are not at all confident about ξ – if f is fairly flat – new evidence can change our beliefs considerably. (New evidence may leave the expectations for the ξ_i's unaltered even though it changes beliefs by making f more sharp. In general, of course, new evidence will both change the sharpness of f *and* change the expectations of the ξ_i's.) Without the equivalent of second-order probabilities there appears to be no answer to the question of exactly how new evidence can alter probabilities. Suppes [37] considers an important defect of both his and Savage's [31] axiomatizations of subjective probability and utility to be their failure to specify how prior information is to be used. Let us consider an example used by both Suppes and Savage.

A man must decide whether to buy some grapes which he knows to be either green (ω_1), ripe (ω_2), or rotten (ω_3). Suppes poses the following question: If the man has purchased grapes at this store 15 times previously, and has never received rotten grapes, and has no information aside from these purchases, what probability should he assign to the outcome of receiving rotten grapes the 16th time?

Prior to his first purchase, the man was in total ignorance of the probability distribution over Ω. Thus from Equation (8) we see that the density for ξ_3, the prior probability of receiving rotten grapes, should be $f_3(\xi_3)=2-2\xi_3$. Let X be the event of receiving green or ripe grapes on the first 15 purchases; the probability that X occurs, given ξ_3, is $p(X \mid \xi_3)=(1-\xi_3)^{15}$. What we desire is $f_3(\xi_3 \mid X)$, the density for ξ_3 given X, and this is obtained by Bayes' theorem in the following way:

$$(12) \qquad f_3(\xi_3 \mid X) = p(X \mid \xi_3)\, f_3(\xi_3) \Big/ \int_0^1 p(X \mid \xi_3)\, f_3(\xi_3)\, d\xi_3 .$$

After inserting the expressions for $f_3(\xi_3)$ and $p(X \mid \xi_3)$, Equation (12) becomes:

$$f_3(\xi_3 \mid X) = (1-\xi_3)^{15}(2-2\xi_3) \Big/ \int_0^1 (1-\xi_3)^{15}(2-2\xi_3)\, d\xi_3 .$$

Performing the integration and simplifying gives $f_3(\xi_3 \mid X) = 17(1-\xi_3)^{16}$; from this the expectation of ξ_3 given X can be computed – $E(\xi_3 \mid X) = 17 \int_0^1 \xi_3(1-\xi_3)^{16} = 1/18$. (Notice that this result differs from the 1/17 that Laplace's law of succession would give. The difference is due to the fact that the Laplacian law is derived from consideration of only two states of nature – rotten and not rotten.[7])

Let us consider another example discussed by Savage [31, p. 65]: "... it is known of an urn that it contains either two white balls, two black balls, or a white ball and a black ball. The principle of insufficient reason has been invoked to conclude that the three possibilities are equally probable, so that in particular the probability of one white and one black ball is concluded to be $\frac{1}{3}$. But the principle has also been applied to conclude that there are four equally probable possibilities, namely, that the first is white and the second black, etc. On that basis, the probability of one white and one black is, of course, $\frac{1}{2}$." Let us consider the second case more carefully. Analogously to the problem of purchasing grapes, we may consider this a sequential problem with two states of nature – ω_1, the event of drawing black and ω_2, the event of drawing white. Before the first draw we are in total ignorance of ξ_1 and ξ_2, i.e., $f_1(\xi_1) = f_2(\xi_2) = 1$. Denote by $\omega_1\omega_2$ the event of black followed by white; then the probability of $\omega_1\omega_2$ given ξ_1 is $\xi_1(1-\xi_1)$ and, therefore, $p(\omega_1\omega_2) = \int_0^1 \xi_1(1-\xi_1) f_1(\xi_1) d\xi_1 = \frac{1}{6}$. Likewise $p(\omega_2\omega_1) = \frac{1}{6}$ so that the probability of one black and one white is $\frac{1}{3}$. Thus the apparent paradox is resolved. However, there were two critical assumptions in deriving this result: (i) There was total ignorance of the prior distribution over Ω, and (ii) The decision-maker chooses to utilize evidence in accord with Bayes' theorem. Drop either assumption and you can easily obtain a value of $\frac{1}{2}$ (or just about anything else) for the probability of one black and one white.

In this section I have tried to show why second-order probability distributions are useful in order to have a subjectivistic theory of induction, and I have outlined the nature of such a theory.

IV. SUBJECTIVISTIC INTERPRETATION OF CARNAP'S INDUCTIVE SYSTEM

Rudolf Carnap [6] has devised a system of inductive logic that fits within the framework of the logical theory of probability. The purpose of this section is to show that Carnap's system can be interpreted in a straightforward way as a special case of the subjectivistic theory of induction presented in the preceeding section. That it *can* be so interpreted does not imply, of course, that it *must* be so interpreted. Let me begin by informally sketching Carnap's λ continuum of inductive methods.

A. *Carnap's λ System*

Carnap's system is built around a 'language' that contains names of n

individuals – $x_1, x_2, \ldots, x_n$ – and π one place primitive predicates – $P_1, P_2, \ldots, P_\pi$. Of each individual it may be said that it either does or does not instantiate each characteristic, i.e., for all $i\,(1 \leqslant i \leqslant n)$ and all $j\,(1 \leqslant j \leqslant \pi)$, either $P_j(x_i)$ or $\sim P_j(x_i)$. If, for example, P_j is 'is red', then x_i is either red or it is not.

A 'Q-predicate' is defined as a conjunction of π primitive characteristics such that each primitive predicate or its negation appears in the conjunction. Let κ be the number of Q-predicates; clearly, $\kappa = 2^\pi$. The following are the Q-predicates if $\pi = 2$:

$$\begin{aligned} P_1 \,\&\, P_2 &= Q_1 \\ P_1 \,\&\, \sim P_2 &= Q_2 \\ \sim P_1 \,\&\, P_2 &= Q_3 \\ \sim P_1 \,\&\, \sim P_2 &= Q_4 . \end{aligned} \tag{13}$$

If P_1 is 'is red' and P_2 is 'is square' then, for example, $Q_4(x_i)$ means that x_i is neither red nor square, etc.

The Q-properties represent the strongest statements that can be made about the individuals in the system; once an individual has been asserted to instantiate a Q-predicate, nothing further can be said about it within the language. Weaker statements about individuals may be formed by taking disjunctions of Q-predicates. To continue the preceeding example, if we let $M = Q_1 \vee Q_2 \vee Q_3$, then $M(x_i)$ is true if x_1 is either red or square or both. Any non-selfcontradictory characteristic of an individual that can be described in the language can be expressed as a disjunction of Q-predicates.

The logical width, w, of a predicate, say M, is the number of Q-predicates in the disjunction of Q-predicates equivalent to M. Its relative width is defined to be w/κ. If M is as defined in the preceeding paragraph, its logical width would be 3 and its relative width $\frac{3}{4}$. A predicate equivalent to the conjunction of all the Q-predicates in the system is tautologically true and its relative width is 1. The logical width of a predicate that cannot be instantiated (like $P_1 \,\&\, \sim P_1$) is zero. In some sense, then, the greater the relative width of a predicate the more likely it is to be true of any given individual. Notice that the relative width of any primitive predicate, P_j, is $\frac{1}{2}$, whatever the value of π.

Let us turn now to the inductive aspects of the system. Suppose that we are interested in some property M and have seen a sample of size s

of individuals, s_i of whom had the property M. What are we to think of the (logical) probability that the next individual that we observe will have the property M? Carnap suggests that two factors enter into assessing this probability. The first is an empirical factor, s_i/s, which is the observed fraction of individuals having property M. The second is a logical factor, independent of observation, and equal to the relative width of M – w/κ. A weighted average of these two factors gives the probability that the $s+1$st individual, x_{s+1}, will have M. One of the factor weightings may be arbitrarily chosen and, for convenience, Carnap chooses the weight of the empirical factor to be s. The weight of the logical factor is given by a parameter λ (λ may be some function $\lambda(\kappa)$, but we need not go into that). Thus we have:

$$\text{(14)} \qquad \text{prob}\left(M(x_{s+1}) \text{ is true}\right) = (s_i + \lambda w/\kappa)/(s + \lambda).$$

The limiting value of the expression in (14) as λ gets very large is w/κ, i.e., only the logical factor counts. If, on the other hand, $\lambda=0$, then the logical factor has no weight at all and only empirical considerations count. Thus the parameter λ indexes a continuum of inductive methods – from those giving all weight to the logical factor to those giving it none.

B. *A Subjectivistic Interpretation of the λ System*

There are $\kappa=2^\pi$ Q-predicates in the Carnap system. The Q-predicates may be numbered $Q_1, \ldots, Q_\kappa$. Let ξ_i be the (subjective) probability that any individual will instantiate Q_i. The probabilities ξ_i may be unknown and, following the precedent of the preceeding section, we may represent our knowledge of these probabilities by a density f defined on Ξ. Since $\xi_\kappa = 1 - \xi_1 - \cdots - \xi_{\kappa-1}$, the density need only be defined on a $\kappa-1$ dimensioned region analogous to the region A in Figure 1. The densities we shall consider will be Dirichlet densities, so let us now define these densities and examine some of their properties.

The $\kappa-1$ variate Dirichlet density is defined for all points $(\xi_1, \ldots, \xi_{\kappa-1})$ such that $\xi_i \geqslant 0$ and $\sum_{i=1}^{\kappa-1} \xi_i \leqslant 1$. The density has κ parameters – $v_1, \ldots, v_\kappa$ – and is defined as follows:

$$\text{(15)} \qquad f(\xi_1, \ldots, \xi_{\kappa-1}) = \frac{\Gamma\left(\sum v_i\right)}{\Pi\Gamma(v_i)} \xi_1^{v_1-1} \ldots \xi_{\kappa-1}^{v_{\kappa-1}-1} \times (1 - \xi_1 - \cdots - \xi_{\kappa-2})^{v_\kappa - 1},$$

where the sums ($\sum$) and products ($\prod$) are over all the v_i, and the Γ denotes the gamma function. Let us let $v_i = \lambda/\kappa$ for $1 \leqslant i \leqslant \kappa$ and see what happens. First we need two theorems proved in Wilks [42, pp. 177–182]:

THEOREM 1. If ξ_i is a random variable in the density given in (15) then $E(\xi_i) = v_i / \sum_{i=1}^{\kappa} v_i$.

THEOREM 2. If $(\xi_1, \ldots, \xi_{\kappa-1})$ is a vector random variable having a $\kappa - 1$ variate Dirichlet density with parameters $v_1, \ldots, v_\kappa$, then the random variable $(z_1, \ldots, z_s)$ where $z_1 = \xi_1 + \cdots + \xi_{j_1}$, $z_2 = \xi_{j_1+1}, + \cdots + \xi_{j_1+j_2}, \ldots,$ $z_s = \xi_{j_1+\cdots+j_{s-1}+1}, + \cdots + \xi_{j_1+\cdots+j_s}$, and $j_1 + \cdots + j_s \leqslant \kappa - 1$, has an s variate Dirichlet distribution with parameters $\theta_1, \ldots, \theta_{s+1}$, where $\theta_1 = v_1 + \cdots + v_{j_1}, \ldots, \theta_s = v_{j_1+\cdots+j_{s-1}+1} + \cdots + v_{j_1+\cdots+j_s}$, and $\theta_{s+1} = v_{j_1+\cdots+j_s+1}, + \cdots + v_\kappa$.

Finally we need one more standard theorem about Dirichlet distributions that concerns modification of the density by Bayes' theorem in the light of new evidence. This theorem too I will state without proof.

THEOREM 3. If $\xi_1, \ldots, \xi_{\kappa-1}$ are the probabilities of the Q-predicates $Q_1, \ldots, Q_{\kappa-1}$, and if $1 - \xi_1 - \cdots - \xi_{\kappa-1}$ is the probability of Q_κ, if the prior density for the ξ_i's is a Dirichlet with parameters $v_1, \ldots, v_\kappa$, and if an observation of s individuals is made in which s_i have property Q_i ($\sum s_i = s$), then the posterior density for the ξ_i's is a Dirichlet with parameters $v'_1, \ldots, v'_\kappa$ where $v'_i = v_i + s_i$ for $1 \leqslant i \leqslant \kappa$.

With this mathematical apparatus at hand I will now show that Carnap's λ continuum is formally identical to a subjectivist inductive system when the prior on the ξ_i's is a Dirichlet density with all its parameters equal to λ/κ, i.e., $v_i = \lambda/\kappa$ for all i.[8]

Consider first induction involving only Q-predicates rather than more general predicates. When $s=0$ – before we make any observations – by Theorem 1 $E(\xi_i) = 1/\kappa$ for all i. If we observe a sample, X, of size s, in which Q_i appears s_i times then, by Theorem 3, $v'_i = (\lambda/\kappa) + s_i$ and $\sum v'_i = \kappa(\lambda/\kappa) + s$. By Theorem 1 again:

$$(16) \qquad E(\xi_i \mid X) = \frac{v'_i}{\sum_{i=1}^{\kappa} v'_i} = \frac{s_i + \lambda/\kappa}{s + \lambda}.$$

Since the logical width, w, of a Q-predicate is 1, (16) is clearly the same as (14) when the predicate M referred to there is a Q-predicate.

To deal with predicates more complicated than Q-predicates we need Theorem 2. Consider a predicate M with logical width w; $\sim M$, then, has logical width $\kappa - w$. By Theorem 2 the prior density function for ξ_M (the probability of M) will be the one variate Dirichlet (or beta) density with parameters $v_1 = w\lambda/\kappa$ and $v_2 = (\kappa - w)\,\lambda/\kappa$. By Theorem 1 the prior expectation of ξ_M is what it should be:

$$E(\xi_M) = (w\lambda/\kappa)/(w\lambda/\kappa + (\kappa - w)\,\lambda/\kappa) = w/\kappa.$$

If we observe a sample X, of size s, that has a total of s_M instances of M (and, therefore, $s - s_M$ instances of $\sim M$) then $v_1' = w\lambda/\kappa + s_M$ and $v_2' = (\kappa - w)\,\lambda/\kappa + s - s_M$. Using Theorem 1 again we obtain:

$$(17) \qquad E(\xi_M \mid X) = v_1'/(v_1' + v_2') = (s_M + \lambda w/\kappa)/(s + \lambda),$$

which is essentially the same as (14).

Leaving aside debate concerning the relative philosophical merits of the logical vs. subjective views, the subjectivist approach has two important advantages over the λ system. These are:

(1) In the Carnapian system $v_i = v_j$ for all i and j; this clearly much reduces the range of possible prior distributions. Or, to put this another way, Carnap's 1-dimensional continuum of inductive methods is a special case of a κ-dimensional continuum.

(2) It may be desirable to have predicates in the language that are not dichotomous. For example, instead of saying of x_i that it is red or not red, we may wish to say that it is red, puce, or ultramarine. If we denote by $v(P_j)$ the number of alternatives P_j may take on, then the number of Q-predicates we have, κ, is given by:

$$(18) \qquad \kappa = \prod_{j=1}^{\pi} v(P_j),$$

where, as before, π is the number of predicates. Clearly, the subjective approach can handle any finite value of $v(.)$.

V. CONCEPT FORMATION AND INDUCTION

My purpose in this section is to provide an essentially Bayesian mechanism for certain types of concept formation. It turns out that this task is closely related to providing a subjectivistic generalization of Hintikka's [12] two-dimensional continuum of inductive methods, and I shall begin by briefly describing his work. Next I shall provide a subjectivistic interpretation of it, then show how all this relates to concept formation.

A. *Hintikka's Two-Dimensional Continuum of Inductive Methods*

Consider a predicate M (a disjunction of several Q-predicates) and suppose that we have observed several thousand individuals, that all of them have instantiated M, and that there exists no M' with logical width less than M such that all the observed individuals also instantiated M'. Having seen several thousand instances of M, and none of $\sim M$, we may very well wish to assign a non-zero probability to the assertion that *all* of the (infinite number of) individuals in this series exemplify M. This cannot be done in the Carnapian system (unless M is tautologous) or in the subjectivistic generalization of it that I outlined; that is, what is known as *inductive generalization* is impossible in these systems. Hintikka's [12] purpose is to generalize the Carnapian system in such a way that inductive generalization is possible.

Hintikka defines a 'constituent' in the following way: the constituent $C(i, j, k)$ is true if and only if

$$(\exists x)\, Q_i(x)\ \&\ (\exists x)\, Q_j(x)\ \&\ (\exists x)\, Q_k(x)\ \&\ (x) \times [Q_i(x) \vee Q_j(x) \vee Q_k(x)]$$

is true. Referring back to Equation (13), $C(1, 3)$ would mean that all individuals have the property P_2, some have P_1, and some do not have P_1. $C(.)$ may have any number of arguments from 1 to κ; let us denote by C_w any constituent that asserts that exactly w Q-predicates are instantiated. $\binom{\kappa}{w}$ is the number of different constituents there are with exactly w Q-predicates instantiated. The total number of constituents, N, is, therefore, given by:

$$N = \sum_{w=1}^{\kappa} \binom{\kappa}{w} = 2^{\kappa} - 1\,. \tag{19}$$

Assume that a total of c different Q-predicates have been observed in a sample, e, of size n. Consider a constituent C^*. Following Hintikka, we obtain by Bayes' theorem the posterior probability for C^* given e, under the assumption that the prior probability of a constituent depends only on the number of Q-predicates in it:

$$(20) \qquad p(C^* \mid e) = \frac{p(C^*)\, p(e \mid C^*)}{\sum_{w=1}^{\kappa-c+1} \binom{\kappa - c + 1}{w - 1} p(C_w)\, p(e \mid C_w)},$$

where $p(C_w)$ is the prior probability of a constituent containing w Q-predicates. (Equation (20) corrects some typographical mistakes in Hintikka's Equation (2).)

Hintikka makes two assumptions to obtain the prior probabilities $p(C_w)$ and the likelihood $p(e \mid C_w)$. As noted, unless $w = \kappa$, $p(C_w) = 0$ in the Carnapian system with an infinite number of individuals. Hintikka uses as $p(C_w)$ the (non-zero) number that $p(C_w)$ would be in a Carnapian universe with α individuals. Thus he obtains a family of priors indexed by α running from 0 to ∞. To obtain $p(e \mid C_w)$ he makes the same assumptions as in the Carnapian system except that he allows only w instead of κ Q-predicates. In this way Hintikka allows for the possibility of inductive generalization. A low α corresponds to a prior expectation of a highly ordered universe in which but few Q-predicates are instantiated; a high α corresponds to a prior expectation that almost all the Q-predicates will be instantiated. Carnap's system is the special case of Hintikka's obtained by letting $\alpha \to \infty$.

B. *Subjectivistic Interpretation of Hintikka's System*

From (19) we see that there are $N = 2^\kappa - 1$ different constituents; let us label them $C_1, \dots, C_N$ letting C_N be the constituent containing all κ Q-predicates. To each C_i let us assign a w-variate Dirichlet density where, as before, w is the number of Q-predicates C_i asserts to exist. (A 1-variate Dirichlet density is assumed to be an impulse or δ function.) The Dirichlet density corresponding to C_i, which I shall call D_i, is assumed to hold given that C_i is true. D_i is a p.d.f. for the probabilities ξ_j of the Q-predicates contained in C_i. Let $\zeta = (\zeta_1, \dots, \zeta_N)$ be a vector that gives the prior probabilities of the C_i's, i.e., $p(C_i) = \zeta_i$. We thus have third order probabilities – ζ_i corresponds to the probability that D_i is the correct p.d.f. for the probabilities ξ_j. If $\zeta_N = 1$ and, hence, all the other ζ_i's equal zero, we have the subjective system outlined previously in this paper. If all the D_i's are equal for constituents containing the same number of Q-predicates, if each D_i has all its parameters equal to one another, if all the predicates are dichotomous, and if ζ is contained in a certain subset of Ξ_N, then the system outlined here reduces to Hintikka's two-dimensional continuum.

C. *Concept Formation and Induction*

In lectures at Stanford University, Professor Patrick Suppes developed what he calls the 'template' representation of a concept. This has been further developed in a recent paper by Roberts and Suppes [29]. His lectures centered around the psychological problem of describing how people actually do acquire concepts. A typical experimental paradigm would be something like the following: A subject is shown geometrical figures that differ in size, form, and color. After he is shown a figure he must say whether the figure belongs to class 'A' or whether it does not. After making his response, the subject is told the correct answer, then shown a new figure.

Let us assume there are three sizes, three colors, and three forms. Each figure can then be described by a Q-predicate; by Equation (18) the total number of Q-predicates is 27. To the three natural predicates – size, form, and color – we can add the predicate 'is a member of class "A"'. Thus we have a new system with 54 Q-predicates. Suppose the concept to be learned is 'is aquamarine or triangular'; exactly one of the $2^{54}-1$ constituents exemplifies this concept. More specifically, that constituent is $(\exists x)[R(x) \,\&\, A(x)] \,\&\, (\exists x)[\sim R(x) \,\&\, \sim A(x)]$ and $(x)\{[R(x) \,\&\, A(x)] \vee [\sim R(x) \,\&\, \sim A(x)]\}$, where $R(x)$ is 'x is aquamarine or triangular' and $A(x)$ is 'x is in class "A"'. An important question then is whether or not the subjectivistic generalization of Hintikka's system can provide an adequate empirical account of human concept formation. The possibility of a low value for α (or its subjectivistic equivalent) makes it conceivable that this approach could be adequate to account for the extremely rapid concept learning that humans exhibit.

Let me now suggest a fairly specific two-parameter model for human concept formation. The assumptions of the model are:

ASSUMPTION 1. On trial n the subject's state may be represented by a vector $S_n = (s_1, \ldots, s_N)$ where N is the number of constituents in the system and s_i may be considered the subject's estimate of the probability that constituent C_i holds.

ASSUMPTION 2. With probability θ_1, S_{n+1} is computed from S_n and the most recently observed figure by means of (20); with probability $1-\theta_1$, $S_{n+1}=S_n$.

ASSUMPTION 3. When on trial n, the subject is given a new figure to respond to he computes from S_n the probability that the figure is in class 'A'. If this probability exceeds .5 he responds 'A'; otherwise, he responds '$\sim A$'.

ASSUMPTION 4. All constituents containing an equal number of Q-predicates have equal prior probabilities. The prior probability that the true constituent will have $j\,(1 \leqslant j \leqslant \kappa)$ Q-predicates is given by $\binom{\kappa-1}{j-1}\theta_2^{j-1}(1-\theta_2)^{\kappa-j}$. (Large θ_2 implies rapid inductive generalization or, in Hintikka's system, it corresponds to small α.) This assumption determines S_1.

Given these four assumptions and estimated values of the parameters θ_1 and θ_2, the subject's responses can be predicted from the figures he has been shown and their classifications. It should be clear, of course, that the model just outlined is but one of many possible similar models.

I will close this section by posing two questions: (i) To what extent can existing empirical models of concept formation be shown to be special cases (or generalizations) of the model I have described? (ii) What, if anything, would estimated values of θ_2 tell us about the true regularity of the universe we live in?

VI. CONCLUDING COMMENTS

I have attempted in this paper to extend a subjectivistic theory of induction in a way that allows the logical systems of Carnap and Hintikka to appear as special cases. In the course of this effort I have attempted to provide a definition of information that is adequate from a subjective point of view and have extended the subjectivist approach to account for certain types of concept formation. Yet there is *nothing* in what I have said that would provide any fundamental justification for utilizing information from the past to make inferences concerning the future.

I will conclude by suggesting that theories of induction may be lexicographically ordered according to how satisfactory they are. Along the first dimension the criterion is "How well does the theory deal with the problem posed by Hume?" All inductive systems are equally (and totally) unsatisfactory from this point of view. Along the secondary dimension the subjective theory is, though problems remain, probably the best. But unsatisfactory is unsatisfactory: Hume's intellectual successors are Sartre and Dylan.

Harvard University

BIBLIOGRAPHY

[1] E. Adams, 'On the Nature and Purpose of Measurement', *Synthese* **16** (1966) 125–169.
[2] F. Atkinson, J. Church, and B. Harris, 'Decision Procedures for Finite Decision Problems Under Complete Ignorance', *Annals of Mathematical Statistics* **35** (1964) 1644–1655.
[3] Y. Bar-Hillel, 'Semantic Information and Its Measures' (1955), reprinted in Bar-Hillel [5].
[4] Y. Bar-Hillel, 'An Examination of Information Theory', *Philosophy of Science* **22** (1955) 86–105.
[5] Y. Bar-Hillel, *Language and Information*, Reading, Mass., 1964.
[6] R. Carnap, *The Continuum of Inductive Methods*, Chicago 1952.
[7] R. Carnap and Y. Bar-Hillel, 'An Outline of a Theory of Semantic Information', Technical Report 247, Research Laboratory of Electronics, M.I.T., 1952. Reprinted in Bar-Hillel [5].
[8] H. Chernoff, 'Rational Selection of Decision Functions', *Econometrica* **22** (1954) 422–443.
[9] B. de Finetti, 'La prévision: Ses lois logiques, ses sources subjectives', *Annales de l'Institute Henri Poincaré* **7** (1937). Translated as 'Foresight: Its Logical Laws, Its Subjective Sources', in *Studies in Subjective Probability* (ed. by H. Kyburg and H. Smokler), pp. 97–158, New York and London 1964.
[10] D. Ellsberg, 'Risk, Ambiguity, and the Savage Axioms', *Quarterly Journal of Economics* **75** (1961) 643–669.
[11] I. J. Good, *The Estimation of Probabilities: An Essay on Modern Bayesian Methods*, Research Monograph No. 30, Cambridge, Mass. 1965.
[12] K. J. Hintikka, 'A Two-Dimensional Continuum of Inductive Methods', in *Aspects of Inductive Logic* (ed. by K. J. Hintikka and P. Suppes), Amsterdam 1966, pp. 113–132.
[13] K. J. Hintikka and J. Pietarinen, 'Semantic Information and Inductive Logic', in *Aspects of Inductive Logic* (ed. by K. J. Hintikka and P. Suppes), Amsterdam 1966, pp. 96–112.
[14] J. Hodges and E. Lehmann, 'The Use of Previous Experience in Reaching Statistical Decisions', *Annals of Mathematical Statistics* **23** (1952) 396–407.
[15] R. Howard, 'Prediction of Replacement Demand', in *Proceedings of the Third International Conference on Operational Research* (ed. by G. Kreweras and G. Morlat), London 1963, pp. 905–918.

[16] R. Howard, 'Information Value Theory', *IEEE Transactions on Systems Science and Cybernetics* **2** (1966) 22–26.
[17] D. Hume, *A Treatise of Human Nature*, London 1962. Originally published in 1739.
[18] L. Hurwicz, 'Some Specification Problems and Applications to Econometric Models' (abstract), *Econometrica* **19** (1951) 343–344.
[19] D. Jamison, 'Information and Subjective Probability', unpublished senior thesis. Department of Philosophy, Stanford University, 1966.
[20] D. Jamison, 'Bayesian Decisions Under Total and Partial Ignorance', in Technical Report 121, Institute for Mathematical Studies in the Social Sciences, Stanford University, 1967.
[21] D. Jamison and J. Kozielecki, 'Subjective Probabilities Under Total Uncertainty', *American Journal of Psychology* **81** (1968) 217–225.
[22] R. Jeffrey, *The Logic of Decision*, New York 1965.
[23] H. Jeffreys and B. S. Jeffreys, *Mathematical Physics*, Cambridge 1956.
[24] R. Luce and H. Raiffa, *Games and Decisions: Introduction and Critical Survey*, New York 1956.
[25] D. MacKay, 'Quantal Aspects of Scientific Information', *Philosophical Magazine* **41** (1949) 289–311.
[26] J. McCarthy, 'Measures of the Value of Information', *Proceedings of the National Academy of Sciences of the USA* **42** (1956) 654–655.
[27] G. Miller, 'The Magical Number Seven, Plus or Minus Two, Some Limitations on Man's Capability To Process Information', reprinted in *Readings in Mathematical Psychology*, vol. I (ed. by R. Luce, R. Bush, and E. Galanter), New York 1963.
[28] J. Milnor, 'Games Against Nature', in *Decision Processes* (ed. by C. Coombs, R. Davis, and R. Thrall), New York 1954.
[29] F. S. Roberts and P. Suppes 'Some Problems in the Geometry of Visual Perception', *Synthese* **17** (1967) 173–201.
[30] R. Roby, 'Belief States and the Uses of Evidence', *Behavioral Science* **10** (1965) 255–270.
[31] L. J. Savage, *The Foundations of Statistics*, New York 1954.
[32] L. J. Savage, 'Implications of the Theory of Personal Probability for Induction', *Journal of Philosophy* **64** (1967) 593–607.
[33] C. E. Shannon, and W. Weaver, *The Mathematical Theory of Communication*, Urbana, Ill., 1947.
[34] H. Smokler, 'Informational Content: A Problem of Definition', *Journal of Philosophy* **63** (1966) 201–210.
[35] H. Smokler, 'The Equivalence Condition', *American Philosophical Quarterly* **4** (1967) 300–307.
[36] J. Sneed, 'Entropy, Information, and Decision', *Synthese* **17** (1967) 392–407.
[37] P. Suppes, 'The Role of Subjective Probability and Utility in Decision-Making', in *Proceedings of the Third Berkeley Symposium on Mathematical Statistics and Probability, 1955*. Reprinted in *Readings in Mathematical Psychology*, vol. II (ed. by R. Luce, R. Bush, and E. Galanter), New York 1965.
[38] P. Suppes, 'Concept Formation and Bayesian Decisions', in *Aspects of Inductive Logic* (ed. by K. J. Hintikka and P. Suppes), Amsterdam 1966.
[39] P. Suppes, *Set Theoretical Structures in Science*, mimeographed notes from a forthcoming book, 1967.

[40] G. Tintner, 'The Theory of Choice Under Subjective Risk and Uncertainty', *Econometrica* 9 (1941) 298–304.
[41] R. Wells, 'A Measure of Subjective Information', in *Proceedings of Symposia in Applied Mathematics*, vol. XII, Providence, Rhode Island, 1961.
[42] S. S. Wilks, *Mathematical Statistics*, New York 1962.

REFERENCES

* Professors Ronald Howard and Howard Smokler have made valuable comments concerning aspects of this work and the author had a long and very helpful conversation with Professor L. J. Savage concerning related topics. The author is particularly indebted to Professor Patrick Suppes, the influence of whose writings will be evident throughout, and whose advice and comments have been of great value. This paper is based in part on the author's previous work – Jamison [19, 20] – and was completed by the author in his capacity as consultant to the RAND Corporation.

[1] I realize that this is treating rather briefly a still ongoing debate concerning the nature of probability. But entering into that discussion here would take me too far afield.

[2] Two applications to psychology of the notion of information discussed here should be mentioned; both relate to problems posed by David Hume [17]. The first relates to Hume's distinction between simple and complex impressions. Work reviewed by Miller [27] suggests a way of making this distinction precise. Miller describes work that indicates that the amount of information a human can process is strictly limited and about the same for different dimensions; combining dimensions provides means for increasing the information input. Simple impressions might be defined, then, as impressions involving only one perceptual dimension, and complex ones defined as involving more than one. The problem here is to construct an algebra for combining perceptual dimensions and one approach to this (that resolves an apparent contradiction in the experimental literature) is suggested in Jamison [19]. The second application of the notions of semantic information to psychological problems posed by Hume is to the problem of distinguishing between memory and imagination. Here we might say that something is imagined if the amount of information concerning that something that a person can supply is virtually unlimited. Otherwise, it is a memory. This definition suffers from the defect, as Professor Suppes has pointed out to me, that the more vivid a memory is, the more difficult will it be to separate it from imagination.

[3] Usually we can characterize the uncertainty in a decision situation as the sum of $H(E(\xi))$ and $H(f)$. If, however, f itself is not precisely known, the uncertainty associated with alternative possible f's must be added in, and so on.

[4] An important practical problem for the theory of subjective probability is the problem of measuring subjective probabilities. Suppes [39] suggests that a problem with the method of using wagers is that persons will change the odds at which they will bet as the size of their bet increases. A solution to this problem is to fix the size of the person's bet, let him choose the odds, and have the experimenter choose the side of the bet the subject must take (the 'you divide, I choose' principle). If the situation is such that the subject believes that the experimenter knows more about the odds than he does, the subject will be strongly motivated to give an accurate probability assessment regardless of the amount he has at stake.

[5] Ronald Howard [15] utilizes what are essentially probability distributions over probability distributions by considering a probability density function for the param-

eters of another probability density function. The notion of probabilities of probabilities is regularly used in applied Bayesian work.

[6] Professor Suppes points out to me that, though there is a rich body of results in meta-mathematics, mathematicians apparently feel no need to derive formal results concerning meta-mathematics in a meta-meta-mathematics. I might add, however, concerning the probable error example, that several years ago when I was helping design an experiment to measure the astronomical unit, I found the notion of probable error in probable error rather useful.

[7] Laplace's law of succession is derived from Bayes' theorem and the assumption of a uniform prior for ξ_i. If the uniform prior is changed to any of the possibilities given in Equation (8), the following generalization of the law of succession can be derived: $P_{r+1}(\omega_i) = (n+1)/(r+m)$, where $p_{r+1}(\omega_i)$ is the (expectation of) the probability that on the $r+1$st trial ω_i will occur, n is the number of times it has occurred in the previous r trials, and m is the number of states of nature. Since completing a draft of this paper, Raimo Tuomela has pointed out to me that Good [11] has discussed notions that are formally analogous to $f(\xi)$. Good mentions that this generalized version of the law of succession was known to Lidstone in 1925.

[8] This assertion must be slightly qualified; the Dirichlet density is undefined for $\nu_i = 0$. Hence, though the inductive method characterized by $\lambda = 0$ may be approached with arbitrary closeness, it cannot be attained in the subjective system. This point is of some importance, since $\lambda = 0$ is the inductive system implicit in the 'maximum likelihood' estimation principle that is rather widely used, at least in psychology.

ROGER ROSENKRANTZ

EXPERIMENTATION AS COMMUNICATION WITH NATURE

I. INTRODUCTION

There is an obvious analogy between experimentation and communication over a noisy channel. The parameter space of an assumed model plays the role of source or message ensemble from which the input, the true state of nature, is selected. The latter is transmitted by the experiment to the target, the experimenter, who then decodes the message. Noise enters in the form of sampling error, the masking effects of hidden variables, and uncontrolled variation in the experimental materials.

The information transmitted from source to target measures the average amount by which the experiment reduces the experimenter's uncertainty regarding the parameters. Thus, for a fixed model, each of the available experiments has an associated expected yield of information. *Ceteris paribus*, the experiment with the highest expected yield should be performed; the experimenter should select the least noisy of the attainable channels. In a straightforward sense, the associated experiment can be regarded as most sensitive, or as providing the weightiest evidence.

Transmitted information does not depend on the direction of flow, that is, on whether the parameter space of the model or the outcome space of the experiment plays the role of source. The symmetry of the situation suggests that, for a fixed experiment, that model should be preferred which transmits, on the average, a maximal quantity of information regarding the outcome space. Such models can be said to possess maximal explanatory power with respect to the contemplated experiment.

Nor does the analogy end here. A model may be considered a means of recoding data. Such coding can be more or less efficient, depending on the degree to which the model exploits redundancies present in the data. An efficient code reserves the shortest sequences for the most frequently occurring messages. Analogously, an economical or highly systematized

J. Hintikka and P. Suppes (eds.), Information and Inference, 58–93.

theory reserves its free parameters for the salient or non-redundant features of the data. In this way we are led to a new treatment of theoretical simplicity.

When experimentation is undertaken with a practical decision in view, as in acceptance sampling or aptitude testing, an additional valuational component governs the choice of experiment. The increase in utility which results when the decision maker observes an outcome z, modifying his initial choice of optimal act accordingly, measures the value of the information z. The weighted average of these utility increments measures the expected value of sample information per unit cost. Let us turn to a formal development of these ideas.

II. THE EXPECTED VALUE OF SAMPLE INFORMATION

We call the two-stage problem of selecting a pair (e, d), where e is an experiment and d is a decision rule mapping experimental outcomes onto terminal actions, the *general decision problem.* In acceptance sampling, e.g., the general decision problem involves choice of an optimal sample (usually a random sample of fixed size) and an acceptance scheme for the sample selected.

Mathematically, such a problem may be defined as a quadruple (E, Z, A, Ω) comprising spaces of experiments, outcomes, terminal acts, and states of nature, respectively. In addition we assume the existence of a utility characteristic $u\,(e, z, a, \omega)$ on the Cartesian product of these spaces, and a data distribution $P_{z/d,\,\omega}$ defined for each e in E. The utility $u\,(e, z, a, \omega)$ represents the utility to the decision maker of performing e, observing z, and selecting a, when $\tilde{\omega}=\omega$.[1] The existence of the measure $P_{z/d,\,\omega}$ amounts to a sufficiently detailed specification of e to permit determination of the data distribution $P\,(z/\omega)$ over its outcome space Z_e. (When the experiment intended is clear from the context, we suppress the subscript 'e' in 'Z_e' and '$P_{z/e,\omega}$'.)

If, in addition to the utility characteristic and the data distribution, a prior measure $\pi(\omega)$ over the parameter space is assumed, we speak of the *general Bayesian decision problem.* A more compact specification of such problems may be given by replacing the conditional measure $P_{z/e,\omega}$ and the marginal measure $\pi(\omega)$ by a single joint measure $P_{\omega,z/e}$ over the possibility space $\Omega \times Z$. Our discussion is largely confined to Bayesian decision problems, their non-Bayesian couterparts being introduced only incidentally and for purposes of comparison.

The definition of the general decision problem in terms of E, Z, A, Ω may be replaced by an equivalent definition in terms of E, Z, D, Ω, where D is the space of all decision rules $d:Z\rightarrow A$ mapping Z into A. The former definition is associated with the *extensive form analysis*, which involves working backwards from the terminal nodes $u(e, z, a, \omega)$ of the decision tree, successively expecting out $\tilde{\omega}$ and $\tilde{z}$ to obtain the (expected) utility of e. Expecting first with respect to $\pi(\omega/z)$,

$$u^*(e, z, a) = E''_{\omega/z}\, u(e, z, a, \tilde{\omega})$$

we obtain the expected utility of each triple (e, z, a). Then defining a_z to be a *Bayes act* against $\pi(\omega/z)$ (i.e., an act which maximizes posterior expected utility $E''_{\omega/z}u(e, z, a, \tilde{\omega})$ when $\tilde{z}=z$) we have

$$u^*(e, z) = E''_{\omega/z}\, u(e, z, a_z, \tilde{\omega}),$$

whence, by expecting over $\tilde{z}$,

$$u^*(e) = E_{z/e}\, E''_{\omega/z}\, u(e, \tilde{z}, \tilde{a}_z, \tilde{\omega}),$$

or, by combining all three steps,

$$u^*(e) = E_{z/e} \max_a E''_{\omega/z}\, u(e, \tilde{z}, a, \tilde{\omega}). \qquad (1)$$

The use of D, on the other hand, is associated with the *normal form analysis* of the decision tree, wherein one considers the class of all possible decision rules for each e, and then computes the optimal d for the given e. From this point of view, the decision maker's objective is to choose an (e, d) pair which maximizes his expected utility

$$u_*(e, d) = E_{\omega,\, z/e}\, u(e, \tilde{z}, d(\tilde{z}), \tilde{\omega}),$$

whence

$$u_*(e) = \max_d u_*(e, d). \qquad (2)$$

Here the expectation $E_{\omega, z/d}$ is over the entire possibility space $\Omega+Z$, and can be accomplished iteratively in two equivalent ways given by

$$E_{\omega,\, z/e} = E_{z/e}\, E''_{\omega/z} \quad \text{or} \quad E_{\omega,\, z/e} = E'_{\omega}\, E_{z/e,\, \omega}\,.$$

By the latter path we successively obtain

$$u_*(e, d, \omega) = E_{z/e, \omega}\, u(e, \tilde{z}, d(\tilde{z}), \omega)$$
$$u_*(e, d) = E'_\omega\, u_*(e, d, \tilde{\omega})$$

and

$$u_*(e) = \max_d E'_\omega\, E_{z/e, \omega}\, u(e, \tilde{z}, d(\tilde{z}), \tilde{\omega}). \tag{3}$$

However, the alternative path yields

$$u_*(e) = \max_d E_{z/e}\, E''_{\omega/z}\, u(e, \tilde{z}, d(\tilde{z}), \tilde{\omega}),$$

from which it is clear that the best d maximizes $E''_{\omega/z}u(e, z, d(z), \tilde{\omega})$ for every z. But a decision rule d accomplishes this iff it maps each z onto an a_z (a posterior Bayes act). Consequently,

$$\begin{aligned} u_*(e) &= \max_d E_{z/e}\, E''_{\omega/z}\, u(e, \tilde{z}, d(\tilde{z}), \tilde{\omega}) \\ &= E_{z/e}\, E''_{\omega/z}\, u(e, \tilde{z}, \tilde{a}_z, \tilde{\omega}) \\ &= u^*(e), \end{aligned}$$

so that the extensive and normal form analyses are equivalent.

The decision whether or not to perform a given e will depend not on $u^*(e)$ alone, but on its value relative to the utility of immediate terminal action without benefit of experimentation. If we introduce the convenient fiction of a 'null experiment' with dummy outcome z_0, the latter quantity can be written as

$$u^*(e_0) = E'_\omega\, u(e_0, z_0, a', \tilde{\omega}),$$

where $a' = \max_a E'_\omega u(e_0, z_0, a, \tilde{\omega})$ is the *prior Bayes act*. Then e will be worth performing iff $u^*(e) > u^*(e_0)$. If we select $u^*(e_0)$ as the zero point of the utility scale, then e will be worth performing iff $u^*(e) > 0$.

Of course, cost-free experimentation is always of value. The possibility that $u^*(e) - u^*(e_0)$ be negative is attributable to the fact that sampling costs are included in the appraisals $u(e, z, a, \omega)$. The analysis is greatly simplified when sampling costs associated with an (e, z) pair can be separated from the utility associated with an (a, ω) pair. When utility appraisals can be decomposed in this way, so that

$$u(e, z, a, \omega) = u_t(a, \omega) - c_s(e, z),$$

where $u_t(a, \omega)$ is called the *terminal utility* of (a, ω) and $c_s(e, z)$ the *sampling cost* of (e, z), we shall speak of *additive utilities*. This condition implies not only that sampling costs and terminal utilities are decomposable, but that the two can be measured on the same utility scale. Thus, if sampling costs are measured in terms of man-hours expended, and terminal utilities are roughly identifiable with monetary gains, the condition demands that the decision maker be able to equate a monetary value with any given expenditure of man-hours within the ranges considered.

Apart from the familiar difficulty of comparing incomparables, utilities may fail to be additive for at least three reasons.

(i) The utility of an (a, ω) pair may depend on z. Thus, if sampling is destructive, the utility of placing the remainder of a lot of electric fuses on the market at ten cents a piece with a double-your-money-back guarantee will depend on the proportion of defectives among them. But this quantity is a function of both $\tilde{\omega}$, the lot proportion defective, and $\tilde{z}$, the number of defectives in the sample. In fact, if the lot contains 10000 fuses, and a is the decision to put the remaining fuses on the market, then, identifying utility with dollars, $u(z, a, \omega) = (1000 - 0.1\ N)\ (1 - 2\ \gamma)$, where N is the sample size and $\gamma = (10000\ \omega - z)/(10000 - N)$ is the proportion of defectives in the residue of $10000 - N$ fuses.

(ii) The utility of an (e, z) pair may depend on $\tilde{\omega}$, as when an experimenter innoculates himself with a drug of unknown potency $\tilde{\omega}$ to ascertain its affect on reaction time $\tilde{z}$. Unforeseen side effects whose intensity depends on $\tilde{\omega}$ may drastically alter the utility of submitting to the innoculation and observing a given z.

(iii) The utility of an (e, z) pair may also depend on a, the terminal action. Suppose that a pollster is required to publish the results of a sample of voter preference. Before conducting the poll he is offered a bet on the outcome of the election at a substantial stake (action a consists in accepting the bet). Then, clearly, if he selects a and places his bet on a specified candidate, the utility for him of obtaining a sample in which that candidate captures a large share of the votes will be very much greater for him had he not selected a – the more so if he believes that the poll will enhance the chances of his candidate. If polls influence elections, then, in addition, the utility of (a, ω), where $\tilde{\omega}$ is the proportion of voters favoring his candidate prior to publication of the poll, will depend on $\tilde{z}$, the share of the sampled votes captured by each candidate. So we have another case of type (ii).

Nevertheless, it is possible to regard (i) and (ii) as satisfying the condition of additive utilities to good approximation by averaging out the additional variable, $\tilde{z}$ or $\tilde{\omega}$, upon which the utility of an (a, ω) or (e, z) pair depends, so that

$$u^*(a, \omega) = E_{z/e, \omega}\, u(\tilde{z}, a, \omega)$$

in the former case, and

$$u^*(e, z) = E'_{\omega}\, u(e, z, \tilde{\omega})$$

in the latter. By and large, the assumption of additive utilities is quite realistic in a wide variety of contexts, and we shall continue to maintain it in what follows.

When utilities are additive, we can express the expected utility of e additively as

$$u^*(e) = u_t^*(e) - c_s^*(e), \tag{4}$$

where $u_t^*(e) = E_{z/e} \max_a E''_{\omega/z} u_t(a, \tilde{\omega})$ and $c_s^*(e) = -E_{z/e} u_s(e, \tilde{z})$. For the null experiment e_0 we have

$$u^*(e_0) = u_t^*(e_0) - c_s^*(e_0)$$

where $u_t^*(e_0) = E'_{\omega} u_t(a', \tilde{\omega}) = \max_a E'_{\omega} u_t(a, \tilde{\omega})$, and where we set $c_s^*(e_0) = 0$. Then e will be worth performing iff

$$u_t^*(e) - u_t^*(e_0) > c_s^*(e). \tag{5}$$

The quantity $u_t^*(e) - u_t^*(e_0)$ can be further analysed as follows. Writing a' and a_z for acts which are Bayes against $\pi(\omega)$ and $\pi(\omega/z)$, respectively, the quantity

$$v_t(e, z) = E''_{\omega/z} \left[u_t(a_z, \tilde{\omega}) - u_t(a', \tilde{\omega})\right] \tag{6}$$

measures the value of the outcome z. It is called the *conditional value of sample information* (the *CVSI*) when the outcome is z. It is obviously non-negative in view of the definition of a_z as an act which maximizes utility against $\pi(\omega/z)$. The expected value of (6),

$$v_t^*(e) = E_{z/e}\, v_t(e, \tilde{z}) \tag{7}$$

measures the value of e, called the *expected value of sample information* (the *EVSI*). Evidently,

$$\begin{aligned} v_t^*(e) &= E_{z/e}\, E''_{\omega/z}\,[u_t(\tilde{a}_z, \tilde{\omega}) - u_t(a', \tilde{\omega})] \\ &= E_{z/e}\, E''_{\omega/z}\, u_t(\tilde{a}_z, \tilde{\omega}) - E'_\omega\, u_t(a', \tilde{\omega}), \end{aligned}$$

since $E_{z/e}E''_{\omega/z}u_t(a', \tilde{\omega}) = E'_\omega u_t(a', \tilde{\omega})$. Thus, $v_t^*(e)$ can be thought of as the utility increment obtained by subtracting the expected utility of any prior Bayes act a' from a weighted average of the expected utilities of the posterior Bayes acts a_z:

$$v_t^*(e) = u_t^*(e) - u_t^*(e_0). \tag{8}$$

In general, $u_t^*(e)$ will lie between $u_t^*(e_0)$ and $u_t^*(e_\infty)$, where e_∞ is an experiment that yields perfect information about $\tilde{\omega}$ at a cost of $c_s^*(e_\infty)$. If one defines a_ω as an act which is optimal when $\tilde{\omega} = \omega$, so that $u_t(a_\omega, \omega) \geqslant u_t(a, \omega)$ for all a in A, then

$$v_t(e_\infty, \omega) = u_t(a_\omega, \omega) - u_t(a', \omega) \tag{9}$$

measures the value of perfect information when $\tilde{\omega} = \omega$. (9) is therefore called the *conditional value of perfect information* (the *CVPI*), and its expected value

$$v_t^*(e_\infty) = E'_\omega\, v_t(e_\infty, \tilde{\omega}) \tag{10}$$

measures the *expected value of perfect information* (*EVPI*). This quantity can be equivalently expressed as

$$v_t^*(e_\infty) = u_t^*(e_\infty) + u_t^*(e_0),$$

where $u_t^*(e_\infty) = E'_\omega u_t(\tilde{a}_\omega, \tilde{\omega})$.

A decisive experiment might consist of sampling an entire population, in which case the expected cost could be realistically appraised. It is clear that the definition of e_∞ in terms of the a_ω justifies the remark that for any e in E we have the inequalities:

$$u_t(e_0) \leqslant u_t(e) \leqslant u(e_\infty). \tag{11}$$

To avoid trivial cases of equality in (11) we need to exclude *null states*, viz. states ω for which $u_t(a, \omega)$ is a constant. Clearly, if all states are null, equality holds throughout (11), and no experiment will be worth performing at however small a cost. In what follows we assume that all null states (as well as states of zero probability) together with all inadmissible acts have been eliminated, leaving what may be called the *kernel* of the original decision problem. For such reduced decision problems, equality cannot hold on the right in (11) unless $e=e_\infty$.

On the other hand, $u_t^*(e_0)=u_t^*(e)$ is possible even when $e \neq e_0$. Call an outcome z of e *effective* if $a_z \neq a'$, so that the *CVSI* of z is positive; otherwise call z *non-effective*. (We reserve the term 'irrelevant' for z such that $\pi(\omega/z)=\pi(\omega)$, in harmony with the use of the term in connection with sufficient statistics and the likelihood principle.) Our assertion is now clear from the fact that a non-null experiment can lack effective outcomes.

We remark, finally, that the definitions given here can all be framed in terms of regrets instead of utilities, where the *regret* $r(e, z, a, \omega)$ is defined by

$$r(e, z, a, \omega) = u(e, z, a_\omega, \omega) - u(e, z, a, \omega).$$

The theory framed in terms of regret parallels that presented here in terms of utility.

Having defined the *EVSI*, the optimal experiment to perform is evidently that which maximizes the increment

$$v^*(e) = v_t^*(e) - c_s^*(e) \tag{12}$$

called the *expected net gain of sampling* (*ENGS*). Equation (12) lends precision to the intuitive characterization of an optimum e as that which yields the highest expected value of sample information per unit cost. Since $u^*(e)=u_t^*(e)-c_s^*(e)$, $u_t^*(e)=u_t^*(e_0)+v^*(e)+c_s^*(e)$, whence

$$u^*(e) = u_t^*(e_0) + v^*(e). \tag{13}$$

But $u_t^*(e_0)$ is a constant, and therefore (13) shows that maximization of the expected utility of sampling is equivalent to maximization of *ENGS*. We illustrate these ideas with a simple example from which it will be clear how to routinize the computation of the *EVSI*.

Example: In this problem both $\tilde{\omega}$ and $\tilde{z}$ are discrete. Let the joint matrix with typical entry $P(\omega_i, z_j)=p_{ij}$ be

$$P = \begin{bmatrix} .08 & .02 & .15 \\ .12 & .21 & .01 \\ .10 & .19 & .12 \end{bmatrix}$$

where

$$p_{i.} = \Sigma_j p_{ij} = \pi(\omega_i),\ p_{.j} = \Sigma_i p_{ij} = P(z_j),$$

are the associated marginal measures over the parameter and outcome spaces. Let $P' = [.25, .34, .41]$ and $P_z = [.30, .42, .28]$ be the column matrices associated with the prior measure $\pi(\omega) = [p_{1.}, p_{2.}, p_{3.}]$ and $P(z) = [p_{.1}, p_{.2}, p_{.3}]$. Then if

$$P'' = \begin{bmatrix} \frac{8}{30} & \frac{2}{41} & \frac{15}{28} \\ \frac{12}{30} & \frac{21}{41} & \frac{1}{28} \\ \frac{10}{30} & \frac{19}{41} & \frac{12}{28} \end{bmatrix}$$

is the posterior matrix with typical entry $\pi(\omega_i/z_j) = p_{ij}/p_{.j}$, direct computation shows that $P''P_z = P'$ (the equality holds in general). Let the utility matrix be

$$U = \begin{bmatrix} 1 & -1 & 0 \\ 3 & -2 & 1 \\ 1 & 1 & 0 \\ 2 & 0 & 2 \end{bmatrix}$$

with typical entry $u_{ij} = u_t(a_i, \omega_j)$. Then

$$UP' = [-.09, .48, .59, 1.32],$$

so that $a_4 = a'$ is the prior Bayes act with expected utility 1.32. On the other hand,

$$UP'' = \begin{bmatrix} -\frac{4}{30} & -\frac{19}{42} & \frac{14}{28} \\ \frac{10}{30} & -\frac{17}{42} & \frac{55}{28} \\ \frac{20}{30} & \frac{23}{42} & \frac{16}{28} \\ \frac{36}{30} & \frac{42}{42} & \frac{54}{28} \end{bmatrix}$$

so that a_4, a_4, a_2 are the posterior Bayes acts against $\pi(\omega/z_1)$, $\pi(\omega/z_2)$, and $\pi(\omega/z_3)$, respectively, which gives

$$u_t^*(e) = (36/30)P(z_1) + (42/42)P(z_2) + (55/28)P(z_3) = 1.33.$$

Thus $u_t^*(e)$ is the product of the row matrix (36/30, 42/42, 55/28) comprising the maximum column entries of UP'' by the column matrix P_z. Consequently, the *EVSI* is $1.33 - 1.32 = .01$ utiles.

To routinize the computation, first normalize the joint matrix by multiplying through by the l.c.m. of the denominators of its entries. Next, sum the maximal column entries of $m(UP)$, where m is the l.c.m. in question, and substract the maximal entry of $m(UP')$. The result expresses the *EVSI* in a linear transform of the original utility scale, which is restored after a division by m.

The routine is justified by the fact that the maximal column entries of UP'' are just those of UP divided by $P(z_j)$, while $u_t^*(e)$ is the sum of the former, each weighted by $P(z_j)$; i.e., the sum of the maximal column entries of UP.

Rather than consider other concrete applications of the theory, which are fully treated in Raiffa and Schlaifer (1961), we apply it to the analysis of more orthodox approaches to the general decision problem.

III. NEYMAN-PEARSON THEORY

The orthodox approach conceived by Neyman and Pearson selects e so as to reduce the risks of type I and type II errors (the rejection of a true hypothesis and the acceptance of a false hypothesis) to a tolerable level. The decision maker's tolerance for error of the two kinds presumably depends upon the losses to which they severally lead, although these are considered only qualitatively (see below). Consequently, the theory has a rather informal utility-free character. We propose to show here that when utilities or losses are explicitly taken into account, the optimum (e, d) pair from a Neyman-Pearson standpoint is the one yielded by the Bayesian analysis of the last section – irrespective of whether or not the prior measure $\pi(\omega)$ is utilized. (Prior measures are often introduced as formal weighting factors for the purpose of selecting one among several 'best' Neyman-Pearson tests.)

Consider once again the problem of acceptance sampling, with $\tilde{\omega}$ the unknown proportion of defectives in the lot. Often there will be a cutoff ω_0 at which the losses incurred by acceptance of the lot (a_1) and rejection of the lot (a_2) are equal. This means that the parameter space Ω can be sub-divided into three zones Ω_i, $i=1, 2, 3$, for which $a_1 \succ a_2$, $a_2 \succ a_1$, and $a_1 \sim a_2$. (Notation: '$\succ$' is strict preference, and '$\sim$' is indifference.) These subregions are defined in terms of the cutoff by

$$\begin{aligned}\Omega_1 &= \{\omega : \omega < \omega_0\}\\ \Omega_2 &= \{\omega : \omega > \omega_0\}\\ \Omega_3 &= \{\omega : \omega = \omega_0\}.\end{aligned}$$

Suppose, for simplicity, that the utility of a_1 is linear in $\tilde{\omega}$:

$$u_t(a_1, \omega) = -k(\omega - \omega_0).$$

Then if e is fixed, as in Neyman-Pearson theory, we may compute the expected utility of acceptance in the light of the outcome z at

$$u_t^*(a_1, z) = \int_0^1 u_t(a_1, \omega)\, \pi(\omega/z)\, d\omega = -(\bar{\tilde{\omega}}'' - \omega_0), \qquad (14)$$

where $\bar{\tilde{\omega}}'' = E''_{\omega/z}\tilde{\omega} = \int_0^1 \omega\pi(\omega/z)\, d\omega$ is the posterior mean of $\tilde{\omega}$ (i.e., the mean of $\pi(\omega/z)$). Consequently, if $u_t(a_2, \omega)$, the utility of rejecting the lot, is set equal to zero for all ω, then $u_t^*(a_2, z) = 0$ for all z, and acceptance will be preferable just in case $\bar{\tilde{\omega}}'' < \omega_0$. Thus, the partition of Ω generates a parallel partition of the outcome space Z into zones

$$\begin{aligned} Z_1 &= \{z : \bar{\tilde{\omega}}'' < \omega_0\} \\ Z_2 &= \{z : \bar{\tilde{\omega}}'' > \omega_0\} \\ Z_3 &= \{z : \bar{\tilde{\omega}}'' = \omega_0\}, \end{aligned} \qquad (15)$$

for which $a_1 \succ a_2$, $a_2 \succ a_1$, and $a_1 \sim a_2$, respectively.

If the decision maker is chary of prior probabilities, he is free to dispense with them without thereby altering the optimality of the test (15). For to dispense with the prior measure is to utilize a rectangular prior measure, in which case $\pi(\omega/z) = P(z/\omega)/P(z)$. Since $P(z)$ can be absorbed into the constant of proportionality, integration with $\pi(\omega/z)$ and $P(z/\omega)$ yield identical partitions (15) of Z. The only difference in the resulting tests is that, in the latter, $\bar{\tilde{\omega}}''$ is computed against a uniform prior (which simplifies the calculations). Note also that the Bayesian acceptance scheme (15) depends on the data distribution only through $\bar{\tilde{\omega}}''$.

Neyman-Pearson theory is restricted to dichotomies (simple or composite) in which Ω partitions into subregions of acceptance and rejection (ignoring a possibly non-empty indifference zone). The analogous n-chotomy involves n terminal actions $a_1, \ldots, a_n$ which partition Ω into n zones such that a_i is the preferred action when ω lies in Ω_i (again ignoring possible indifference zones). The analysis for this case is facilitated by computing the *error characteristics* $P_d(a_i/\Omega_j)$ which yield, for each decision rule d, the probability that d will select a_i when $\tilde{\omega}$ lies in Ω_j:

$$P_d(a_i/\Omega_j) = \int_{\Omega_j} P_d(a_i/\omega)\, \mathrm{d}\omega\,, \tag{16}$$

where $P_d(a_i/\omega)$, the probability that d selects a_i when $\tilde{\omega}=\omega$, is computed from the data distribution. In harmony with Neyman-Pearson theory let us assume that for all i, j, the regret incurred by choice of a_i is constant for all ω in Ω_j. We write $r_{ij}=r_t(a_i,\omega)$ for ω in Ω_j. Then the expected regret associated with the pair (e, d) is given by

$$r_t^*(e, d) = \Sigma_{ij}\, P_d(a_i/\Omega_j)\, r_{ij}\,. \tag{17}$$

Consequently, for fixed e, we see that the best d minimizes (17).

The Bayes solution therefore extends to non-linear losses and to the general n-chotomy in a way that preserves the operational flavor of Neyman-Pearson theory, but avoids what I take to be several shortcomings of their original formulation. First, in testing a simple null hypothesis against a composite alternative, the severity of the loss incurred by rejecting a true null hypothesis will in general depend on which of the alternatives is adopted: our analysis makes this dependence explicit. Secondly, (17) embodies the symmetrical treatment of terminal actions at which the requirement of invariance is directed. In the original formulation of the theory, an optimal d need not remain optimal under all relabellings of the terminal actions.

Given the motivation of Neyman-Pearson theory that a good test maximize expected income (leading with high probability to profitable decisions and with low probability to unprofitable decisions) we see that, for fixed e, the Bayes solution fits the bill more directly than does Neyman-Pearson theory itself. For one thing, the mathematical problems associated with the general non-existence of uniformly most powerful tests in non-simple dichotomies do not even arise on a Bayesian approach. The incorporation of a quantitative utility characteristic into the analysis imposes sufficiently strong constraints to generate a solution to the optimization problem that is fully adequate in the sense that two optimum partitions of Z provide equally good insurance against loss.

The approach of Neyman and Pearson, on the other hand, is to compare qualitatively the severity of the losses incurred by the two kinds of error – in our example, accepting the lot when $\tilde{\omega} > \omega_0$, and rejecting the lot when $\tilde{\omega} < \omega_0$. Let α be a tolerable level of risk for that one of the two errors whose consequences are adjudged more severe. The theory labels α the *size* of any test whose probability of committing the associated error does not exceed α. Subject to this constraint, one then seeks a test which has a low probability of committing the error whose consequences are considered less severe. If the regrets associated with either kind of error vary greatly with $\tilde{\omega}$, the decision maker will want to take these variations explicitly into account in selecting his test. Consequently, Neyman-Pearson theory, which disregards variations in the regret function, is at best a rule-of-thumb. It is strictly appropriate only in dichotomous contexts for which the regrets are constant over the acceptance and rejection subregions of parameter space. If in our sampling inspection problem we set the regrets equal to zero and one, i.e., $r_t(a_1, \omega) = 0$ or 1 according as $\tilde{\omega} < \omega_0$ or $\tilde{\omega} > \omega_0$, and $r_t(a_2, \omega) = 0$ or 1 according as $\tilde{\omega} > \omega_0$ or $\tilde{\omega} < \omega_0$, then

$$\begin{aligned} r_t^*(a_1, z) &= \int_0^1 r_t(a_1, \omega)\, \pi(\omega/z)\, \mathrm{d}\omega \\ &= \int_{\omega_0}^1 \pi(\omega/z)\, \mathrm{d}\omega \\ &= \pi(\tilde{\omega} > \omega_0/z), \end{aligned}$$

and similarly $r_t^*(\mathrm{a}_2, z) = \pi(\tilde{\omega} < \omega_0/z)$. Hence the appropriate acceptance, rejection, and indifference subregions of Z are given by

$$\begin{aligned} Z_1 &= \{z : \pi''(z) > 1\} \\ Z_2 &= \{z : \pi''(z) < 1\} \\ Z_3 &= \{z : \pi''(z) = 1\}, \end{aligned} \tag{18}$$

where $\pi''(z) = \pi(\tilde{\omega} < \omega_0/z) : \pi(\tilde{\omega} > \omega_0/z)$ are the posterior odds on $\tilde{\omega} < \omega_0$.

Thus we are led to a Bayes test. Using the fact that the posterior odds are the prior odds $\pi(\tilde{\omega} < \omega_0) : \pi(\tilde{\omega} > \omega_0)$ multiplied by the likelihood ratio

$$L(z) = P(z/\tilde{\omega} < \omega_0)/P(z/\tilde{\omega} > \omega_0)$$

we see that our Bayes test is also a likelihood ratio test with critical ratio

$\pi(\tilde{\omega}>\omega_0):\pi(\tilde{\omega}<\omega_0)$, the prior odds on $\tilde{\omega}>\omega_0$. In case the regrets incurred by the two kinds of error are constant but unequal, a different critical ratio will be obtained. We conclude that where Neyman-Pearson testing is at all appropriate, it is likelihood ratio testing, stressing, once again, that our conclusion does not depend on the existence of the prior measure $\pi(\omega)$.

The critical ratio $\gamma=\pi(\tilde{\omega}>\omega_0):\pi(\tilde{\omega}<\omega_0)$ determines the decision rule d_0 given by

$$d_0(z)=\begin{cases} a_1 & \text{when } L(z)>\gamma \\ a_2, & \text{when } L(z)<\gamma \\ \text{either,} & \text{when } L(z)=\gamma\,. \end{cases}$$

The optimal (e, d) pair is thereby determined up to the choice of e. Neyman and Pearson originally considered only samples of pre-assigned size. Let e_n be the experiment which consists in taking a random sample of n items. Let α_n, β_n be the error probabilities associated with (e_n, d_0):

$$\alpha_n = P_{d_0}(a_1/\tilde{\omega} > \omega_0) = \Sigma P(z/\tilde{\omega} > \omega_0)$$
$$\beta_n = P_{d_0}(a_2/\tilde{\omega} < \omega_0) = \Sigma P(z/\tilde{\omega} < \omega_0),$$

where the first summation is over $Z_1=\{z:d_0(z)=a_1\}$ and the second over $Z_2=\{z:d_0(z)=a_2\}$. Let $r=r_t(a_1,\ \tilde{\omega}>\omega_0)$ and $s=r_t(a_2,\tilde{\omega}<\omega_0)$ be the constant regrets associated with the two kinds of error, and set

$$\delta_n = \alpha_n r + \beta_n s\,.$$

Then δ_n measures the security afforded by $(e_n,\ d_0)$ and is called the *risk*.

In an extended sense of the term, an optimum *insurance policy* can be defined by the following conditions: (i) it provides enough insurance, and (ii) it maximizes the security/premium ratio among all the policies that meet condition (i). If we set $c_s(e_0,d_0)=u_s(e_0,d_0)=0$, then $r_s(e_n,d_0)= u_s(e_0,d_0)-u_s(e_n,d_0)=-u_s(e_n,d_0)=c_s(r_n,d_0)$, whence $c_s^*(e_n)=r_s^*(e_n)=c_n$. Suppose that Δ is a tolerable level of risk (a minimal security level). Then the optimal sample size n^0 is the largest n, among those whose risk lies below Δ, for which the security increment exceeds the cost increment when one additional item is sampled; i.e.,

$$n^0 = \max_n\{n:\delta_n<\Delta \quad \text{and} \quad \delta_{n-1}-\delta_n>c_n-c_{n-1}\}\,.$$

Summarizing the results of this section, the Neyman-Pearson theory is the theory of dichotomous testing subject to constant regrets. For fixed e, the optimum test (viz. that which minimizes expected regret) is a likelihood ratio test, with critical ratio given by the product of the prior odds by the quotient of the constant regrets r and s incurred by the two kinds of error. A pre-determined sample size n^0 can be so chosen that (e_{n^0}, d_0) constitutes an optimum insurance policy in the sense of offering maximum security per unit cost among all those policies whose security surpasses a specified tolerance level.

IV. ENTROPY

We turn now to the development of a theory of disinterested information appropriate for those contexts in which utilities do not enter, quantitatively or qualitatively, into the choice of an optimum experiment.

Let $\tilde{x}$ be a random variable. Then, depending on whether $\tilde{x}$ is discrete, with $p_i = P(x = x_i)$, or continuous, with density $f(x)$, we define the *entropy* of $\tilde{x}$ (or of its probability law) by

$$H(p_1, \ldots, p_m) = - \Sigma_i p_i \log p_i \tag{19'}$$

or

$$H(\tilde{x}) = - \int f(x) \log f(x) \, dx . \tag{19''}$$

Summation or integration, unless otherwise specified, is always over the entire range of values of $\tilde{x}$. Logarithms may be taken to any base $b > 1$. For expository convenience, we confine the following discussion to the discrete case. The proofs of theorems given here extend readily to the continuous case, *mutatis mutandis*, as the reader may verify.

When logarithms are taken to the base two, the unit of the entropy scale is called a 'bit' – short for 'binary digit'. A one-bit reduction in H is brought about by eliminating half of an even number of equi-probable alternatives.

A more illuminating motivation for (1) can be given by noting that *$H = \log m$ is the minimum number of binary digits needed to code answers to m yes/no questions.* E.g., to locate a single square on a checkerboard (containing 64 squares), a minimum number of $6 = \log_2 64$ questions is required, and the minimum is attained just in case each question splits the remaining alternatives in half. Thus, the first question reduces uncertainty to 32 alternatives, the second to 16 alternatives, and so on.

A binary sequence of six digits, like '001100', could be used to code the answers, so that each sequence would correspond biuniquely to a square on the board. Seen in this light, a coordinate system is a way of coding positions, and often, a more efficient coding can be brought about by the simple device of relocating the origin – witness the Copernican revolution.

Given m equi-probable alternatives, each has probability $p=1/m$; hence, for this case $H=-\Sigma p \log p=-mp \log p=-\log p$. In other words, the average uncertainty associated with each alternative is $h_i=-\log p$. Consider next the case of unequally probable alternatives. For example, let E_i be the event that a pair of six-sided dice turn up a sum of i spots, $i=2, \ldots, 12$. Suppose we restrict the outcome space to the event E_7. Then our uncertainty is reduced to six equi-probable alternatives – viz. the ordered pairs (1, 6), (6, 1), (2, 5), (5, 2), (3, 4), (4, 3) – and is measured by $h_7=\log 6=-\log P(E_7)$. And, in general, when restricted to E_i, our uncertainty is measured by $-\log p_i, p_i=P(E_i)$. Consequently, our uncertainty regarding the partition $E_2, \ldots, E_{12}$ is given by the weighted average of the h_1, viz.,

$$H = \Sigma_i h_i P(E_i) = -\Sigma_i p_i \log p_i$$

that is, by (19′).

For our example we have $p_i=(i-1)/36$ for $i=2, \ldots, 7$, and $p_i=[12-(i-1)]/36$ for $i=8, \ldots, 12$, whence $H=3.71$ bits. The reader can verify that a minimum of three or four yes/no questions is required to ascertain which of the alternatives $E_2, \ldots, E_{12}$ obtains.

By contrast, our uncertainty regarding the finest partition of the outcome space, viz. the set of all thirty-six ordered pairs (i, j), is given by $\log_2 36=5.17$ bits, implying a reduction of $5.17-3.71=1.46$ bits when we pass to the coarser partition of the E_i $(i=2, \ldots, 12)$. That is, our uncertainty is reduced when we lump the outcomes, and, as a result, a shorter sequence of binary digits would be required to encode the outcome of N trials when classified as E_i than when classified as (i, j). In what follows we continue to take logarithms to the base two.

That entropy measures uncertainty is further substantiated by noting the following properties.

(i) $H(p_1, \ldots, p_m') = H(p_1, \ldots, p_m, 0)$, the agent's uncertainty is wholly determined by those alternatives to which he ascribes a non-zero probability.

(ii) When all the p_i are equal, $H(p_1, \ldots, p_m)$ is increasing in m, the number of equi-probable alternatives.

(iii) $H(p_1, \ldots, p_m) = 0$, a minimum, when some $p_i = 1$.

(iv) $H(p_1, \ldots, p_m) = \log m$, a maximum, when each $p_i = 1/m$.

(v) Any averaging of the p_i (i.e. any flattening of the distribution) increases H.

Properties (iii)–(v) assume, of course, that m is fixed. Less fundamental are the following properties:

(vi) H is non-negative.

(vii) $H(p_i, \ldots, p_m)$ is invariant under every permutation of the indices $1, \ldots, m$.

(viii) $H(p_i, \ldots, p_m)$ is continuous in its arguments.

The informational measures that interest us presuppose the concepts of joint entropy and conditional entropy. The *joint entropy* of $\tilde{x}$ and $\tilde{y}$ is defined by

$$H(\tilde{x}, \tilde{y}) = -\Sigma_i^m \Sigma_j^n p_{ij} \log p_{ij}, \quad p_{ij} = P(\tilde{x} = x_i, \tilde{y} = y_j) \tag{20}$$

and measures the joint uncertainty (or informational overlap) of $\tilde{x}$ and $\tilde{y}$. (Tildes will always distinguish the joint entropy $H(\tilde{x}, \tilde{y})$ from the entropy $H(p, q)$ of a single random variable whose two possible values have probabilities p and q.) The *entropy* $H_i(\tilde{y}) = H_{x_i}(\tilde{y})$ of $\tilde{y}$ *conditional on* x_i is given by

$$H_i(\tilde{y}) = -\Sigma_j p_i(j) \log p_i(j), \quad p_i(j) = P(\tilde{y} = y_j / \tilde{x} = x_i) \tag{21}$$

and measures the uncertainty in $\tilde{y}$ when it is known that $\tilde{x} = x_i$. $H_i(\tilde{y})$ is a random variable defined on the range of $\tilde{x}$, and its expected value

$$H_{\tilde{x}}(\tilde{y}) = \Sigma_i p_i H_i(\tilde{y}) = -\Sigma_i p_i \Sigma_j p_j(j) \log p_i(j) \tag{22}$$

is called the *entropy of* $\tilde{y}$ *conditional on* $\tilde{x}$. It measures the average uncertainty in $\tilde{y}$ given $\tilde{x}$ (i.e. given that some unspecified one of the states $x_i, \ldots, x_m$ obtains). Since $\Sigma_j p_i(j) = 1$

$$\begin{aligned} -H(\tilde{x}, \tilde{y}) &= \Sigma_{ij} p_{ij} \log p_{ij} \\ &= \Sigma_i p_i \Sigma_j p_i(j) \log p_i(j) + \Sigma_j p_i(j)\, \Sigma_i p_i \log p_i \\ &= -H_{\tilde{x}}(\tilde{y}) - H(\tilde{x}) \end{aligned}$$

writing $p_{ij} = p_i(j) p_i$, whence

$$H(\tilde{x}, \tilde{y}) = H_{\tilde{x}}(\tilde{y}) + H(\tilde{x}). \tag{23'}$$

I.e., the uncertainty in both $\tilde{x}$ and $\tilde{y}$ is obtained by adding the uncertainty in $\tilde{x}$ to the uncertainty in $\tilde{y}$ given $\tilde{x}$. Equation (23′) is just the additive analogue of the multiplicative relation

$$P(x, y) = P(y/x)\,P(x)$$

utilized in its proof. By writing the joint probability p_{ij} in the alternative form $p_j(i)q_j$, where $q_j = P(\tilde{y} = y_j)$, we obtain

$$H(\tilde{x}, \tilde{y}) = H_y(\tilde{x}) + H(\tilde{y}) \tag{23''}$$

as an alternative form of (23′). Together (23′) and (23″) express the symmetry of the joint entropy.

Exploiting the convexity of $g(t) = t \log t$, i.e., that $\Sigma_i p_i g(t_i) \geqslant g(\Sigma_i p_i t_i)$, by the substitution $t_i = p_i(j)$, we obtain

$$\Sigma_i p_i p_i(j) \log p_i(j) \geqslant \Sigma_i p_i p_i(j) \log(\Sigma_i p_i p_i(j)) = q_j \log q_j$$

whence

$$\Sigma_i p_i \Sigma_j p_i(j) \log p_i(j) \geqslant \Sigma_j q_j \log q_j$$

or

$$-H_{\tilde{x}}(\tilde{y}) \geqslant -H(\tilde{y}),$$

which is equivalent to *Shannon's inequality*:

$$H_{\tilde{x}}(\tilde{y}) \leqslant H(\tilde{y}). \tag{24}$$

From the proof it is obvious that equality holds just in case $\tilde{x}$ and $\tilde{y}$ are independent. Loosely expressed, (24) asserts that entropy (average uncertainty) never increases with additional specification of the state of nature.

On the basis of (23) and (24), we are now in a position to round out our list of the fundamental properties of entropy with the following additions:

(ix) $H(\tilde{x},\tilde{y}) = H_{\tilde{x}}(\tilde{y}) + H(\tilde{x}) = H_{\tilde{y}}(\tilde{x}) + H(\tilde{y})$.
(x) $H(\tilde{x}) - H_{\tilde{y}}(\tilde{x}) = H(\tilde{y}) - H_{\tilde{x}}(\tilde{y})$ (from (ix)).
(xi) $H_{\tilde{x}}(\tilde{y}) \leqslant H(\tilde{y})$, with equality iff $\tilde{x}$ and $\tilde{y}$ are independent.
(xii) $H(\tilde{x},\tilde{y}) \leqslant H(\tilde{x}) + H(\tilde{y})$, with equality as in (xi) (from (ix) and (xi)).

It can be shown[2], moreover, that properties (i), (iv), (viii) and (ix) uniquely characterize H (up to a positive constant). An alternative characterization can be given in terms of (ii), (viii), and (ix).[3]

By (iv), the maximum value of H for a given number m of alternatives is $\log m$. It is often convenient to describe one's uncertainty as the proportion of the maximal uncertainty obtainable with the same number of alternatives:

$$R = H/H_{\max} = H/\log m\,.$$

The quantity R is called the *relative entropy*, and the complementary quantity $1-R$ is called the *redundancy* (for reasons which will appear).

V. EXPECTED SAMPLE INFORMATION

The quantity $T(\tilde{x};\tilde{y})=H(\tilde{x})-H_{\tilde{y}}(\tilde{x})$ is called the *information transmitted from $\tilde{x}$ to $\tilde{y}$*. In view of property (x) of the last section, the terminology is somewhat misleading: $T(\tilde{x};\tilde{y})=T(\tilde{y};\tilde{x})$ and no direction of informational flow is implied. T measures the *association* between $\tilde{x}$ and $\tilde{y}$; it is indifferent to which of the two is taken as source. Thus, $T(\tilde{x};\tilde{y})=0$ iff $\tilde{x}$ and $\tilde{y}$ are independent. The symmetry of T is more evident when it is expressed in the equivalent form

$$T(\tilde{x};\tilde{y}) = H(\tilde{x}) + H(\tilde{y}) - H(\tilde{x},\tilde{y})\,. \tag{25}$$

It is now an easy matter to substantiate the claim that $T(\tilde{z};\tilde{\omega})$ measures the amount by which an observation on $\tilde{z}$ can be expected to reduce the agent's uncertainty about $\tilde{\omega}$.

Let $p'_i=\pi(\tilde{\omega}=\omega_i)$, $i=1,\ldots,m$ be the agent's prior probabilities, and $p''_j(i)=\pi(\tilde{\omega}=\omega_i/\tilde{z}=z_j)$ his posterior probabilities when $\tilde{z}=z_j$. Then the difference

$$H(\tilde{\omega}) - H_j(\tilde{\omega}) = H(p'_1,\ldots,p'_m) - H[p''_1(i),\ldots,p''_m(i)]$$

measures the amount by which observation of z_j reduces prior uncertainty regarding $\tilde{\omega}$. Consequently, its expected value, $\Sigma_j q_j[H(\tilde{\omega})-H_j(\tilde{\omega})]=$ $=H(\tilde{\omega})-\Sigma_j q_j H_j(\tilde{\omega})=H(\tilde{\omega})-H_{\tilde{z}}(\tilde{\omega})$ measures the *average* amount by which the experiment e (which comprises a single observation on $\tilde{z}$) reduces uncertainty about $\tilde{\omega}$.

We call the former quantity

$$I(e, z_j) = H(\tilde{\omega}) - H_j(\tilde{\omega}) \tag{26}$$

the *conditional sample information* (the *CSI*) in analogy to the *CVSI*.

Its expected value,

$$I^*(e) = E_{z/e}\, I(e, \tilde{z}) = T(\tilde{z}; \tilde{\omega}) \tag{27}$$

is called the *expected sample information* yielded by the experiment e (the *ESI*) in analogy to the *EVSI*.

That the *ESI*, unlike the *CSI*, is always non-negative follows from Shannon's inequality. Conversely, our definition of the *ESI* lends precision to the intuitive interpretation of that inequality: that one's average uncertainty is never increased by taking additional observations. In these terms, one can also say that an experiment is *relevant* when its *ESI* is strictly positive.

Shannon's inequality can be generalized in two directions. First, the ω_i need not exhaust the whole parameter space – all that is required is that they be exclusive alternatives. Secondly, $H(\tilde{\omega}) \geqslant H_{\tilde{z}}(\tilde{\omega}) \geqslant H_{\tilde{y}\tilde{z}}(\tilde{\omega}) \geqslant$ $\geqslant H_{\tilde{x}\tilde{y}\tilde{z}}(\tilde{\omega})$ etc., where $H_{\tilde{x}\tilde{y}}(\tilde{\omega}) = \Sigma_{i,j} P(x_i, y_j)\, \Sigma_k\, P(\omega_k / x_i, y_j)$ is the uncertainty about $\tilde{\omega}$ when both $\tilde{x}$ and $\tilde{y}$ are observed, etc. Now write $T(\tilde{x}, \tilde{y}; \tilde{\omega}) = H(\tilde{\omega}) - H_{\tilde{x}\tilde{y}}(\tilde{\omega})$ for the information which $\tilde{x}$ and $\tilde{y}$ jointly transmit about $\tilde{\omega}$, and let $e = (e_1, e_2)$ be the *composite experiment* which consists in observing both $\tilde{x}$ and $\tilde{y}$. Thus the outcome space of the composite experiment has typical element $x_i y_j$, and the generalized Shannon inequality implies that it transmits more information about $\tilde{\omega}$, on the average, than either of its components.

Interesting cases arise, according as the component experimental variables are independent or strictly correlated. More generally, one of the variables might be a sufficient statistic for the other. For the former case, we show that the *ESI* is additive; for the latter, that the *ESI* is preserved iff the experimental variable is replaced by a sufficient statistic. These results comprise the following two theorems.

THEOROM 1: *The ESI is additive in independent experiments; i.e., if $\tilde{x}$ and $\tilde{y}$ are independent, then*

$$T(\tilde{x}, \tilde{y}; \tilde{\omega}) = T(\tilde{x}; \tilde{\omega}) + T(\tilde{y}; \tilde{\omega}). \tag{28}$$

THEOREM 2: *No transformation of the outcome space can increase the ESI, and the ESI is preserved by such a transformation iff it is a sufficient statistic. In symbols,*

$$T(t(\tilde{z}); \tilde{\omega}) \leqslant T(\tilde{z}; \tilde{\omega}) \tag{29}$$

with equality iff $t(z)$ is sufficient.

Proof of theorem 1: Using the symmetry of T, the proof reduces to showing that $T(\tilde{\omega}; \tilde{x}, \tilde{y}) = H(\tilde{x}, \tilde{y}) - H(_{\tilde{\omega}}\tilde{x}, \tilde{y}) = T(\tilde{\omega}; \tilde{x}) + T(\tilde{\omega}; \tilde{y})$ when $\tilde{x}$ and $\tilde{y}$ are independent. The theorem is therefore proved if property (xii) can be extended to conditional entropy, so that $H_{\tilde{\omega}}(\tilde{x}, \tilde{y}) = H_{\omega}(\tilde{x}) + H_{\tilde{\omega}}(\tilde{y})$ also holds when $\tilde{x}$ and $\tilde{y}$ are independent. But this extension is perfectly obvious from its multiplicative analogue, the extension of $P(x,y) = P(x)P(y)$ to $P(x,y/\omega) = P(x/\omega)P(y/\omega)$ when $\tilde{x}$ and $\tilde{y}$ are independent.

Proof of theorem 2: By the generalized Shannon inequality we know that $T(z, t(z); \omega) \geqslant T(t(z); \omega)$. But $T(z, t(z); \omega) = T(z; \omega)$, since observing z and $t(z)$ together is clearly equivalent to observing $\tilde{z}$ alone. If t is a sufficient statistic, then $\pi(\omega/z) = \pi(\omega/t(z))$, by Bayes' theorem and the definition of sufficiency, so in this case equality holds in (29). When t is not sufficient, the posterior distributions are distinct, hence the conditional entropies $H_j(\tilde{\omega})$ and $H_{t(z_j)}(\tilde{\omega})$ are distinct for some j, and strict inequality must hold in (29).

Combining these theorems, we see that the maximal *ESI* for composite experiments is achieved when the component experiments are independent; the minimum *ESI* is obtained when one of the component experimental random variables is a sufficient statistic for the others. The inequality (29) must not be confused with our earlier finding that lumping events of a partition *decreases* the entropy. The present result pertains to uncertainty regarding a parameter relative to a reduced and an unreduced sample space.

We have noted that T provides a measure of association and could be computed for any contingency table instead of χ^2. If $\hat{T}(\tilde{x}; \tilde{y})$ is an estimate of $T(\tilde{x}; \tilde{y})$ – obtained by replacing the probabilities p_{ij} by their estimates n_{ij}/n – then the approximate equality

$$\chi^2 = 1.3863\, n\hat{T}(\tilde{x}; \tilde{y}) \tag{30}$$

holds. Thus, a null hypothesis of non-association could be tested by applying the χ^2 tables to $1.3863\, n\hat{T}(\tilde{x}; \tilde{y})$ for the same degrees of freedom $(m-1)(n-1)$ appropriate for the corresponding χ^2 test. Qua measure of association, T has the advantage of being expressible in meaningful units (bits) and requires only nominal scaling. It is also additive (Theorem 1) unlike χ^2. For further discussion of multivariate informational analysis, consult Attneave (1959), ch. 3, McGill (1954), Garner and McGill (1956), and Miller and Madow (1954).

For dichotomous $\tilde{x}$ and $\tilde{y}$, (30) implies that $\hat{T}(\tilde{x};\tilde{y})$ is increasing in the absolute value of the determinant of the joint probability matrix with typical entry $p_{ij}=P(x_i, y_j)$. In particular, $\hat{T}(\tilde{x};\tilde{y})$ is minimal when $\tilde{x}$ and $\tilde{y}$ are independent and the determinant is zero, and maximal when the joint matrix is diagonal (for a suitable permutation of the indices). We apply this case of the result to the solution of a sampling problem.

Example (on the relation between two traits). We assume that the proportions with which the two traits H and S occur in a given population are known to be h and s, but that the proportion $\tilde{\omega}=P(H,S)$ of population elements having both traits is unknown. The probabilities of the four possible outcomes HS, $H\bar{S}$, $\bar{H}S$, $\overline{HS}$ under the hypothesis $\omega_1:\tilde{\omega}=hs$ of independence, and under the hypothesis $\omega_2:\tilde{\omega}=\delta\neq hs$ of dependence, are as in Table I.

TABLE I

	ω_1		ω_2	
	S	$\bar{S}$	S	$\bar{S}$
H	hs	$h(1-s)$	δ	$h-\delta$
$\bar{H}$	$(1-h)s$	$(1-s)(1-h)$	$s-\delta$	$1-h-s+\delta$

Now let $e_S, e_H, e_{\bar{S}}, e_{\bar{H}}$ be the experiments of drawing a random sample of size n from S, H, $\bar{S}$, $\bar{H}$, respectively, and let $p=\pi(\omega_1)$ and $q=1-p==\pi(\omega_2)$ be the prior probabilities. Considering first the case $n=1$, the joint matrices of e_H and e_S are found (using the above table) to be as in Table II.

TABLE II

e_S	H	$\bar{H}$	e_H	S	$\bar{S}$
ω_1	ph	$p(1-h)$	ω_1	ps	$p(1-s)$
ω_2	$q\delta/s$	$q(1-\delta/s)$	ω_2	$q\delta/h$	$q(1-\delta/h)$

The determinants are det $e_S=pq(hs-\delta)/s$ and det $e_H=pq(hs-\delta)/h$. Assume for definiteness that $hs-\delta$ is positive, so that these values are also the absolute values of the determinants. One similarly computes det $e_S = pq(hs-\delta)/(1-s)$ and det $e_{\bar{H}} = pq(hs-\delta)/(1-h)$ with identical absolute values. Since all four determinants have the same numerators, the one with the smallest denominator $\mu = \min(h, s, 1-h, 1-s)$ is the largest: that is, the maximal *ESI* is obtained by sampling the rarest of the four traits. The same result accrues if $hs-\delta$ is negative.

Since the *ESI* obtained for any mixed sample (i.e. sampling from each of the traits with a specified probability) is a weighted average of the *ESI*'s of the component pure samples, it is clear that no mixed sample has a higher *ESI* than that of the optimum pure sample. In particular, sampling the rarest trait is more informative than sampling the population at large. In addition, the restriction to $n=1$ represents no loss of generality in virtue of Theorem 1: for random samples the *ESI* is additive. Finally, note that the result is independent of the prior probabilities $\pi(\omega_i)$ and the choice of δ (the particular dependency asserted).

The case $\delta = h$ (i.e. the hypothesis that H is a subset of S) has been widely discussed in connection with Hempel's notorious paradox of confirmation. Thus, our present result generalizes the Bayesian solution to Hempel's paradox (cf. Suppes, 1966) and brings out the continuity between generalizations ('All H are S') and their probabilistic counterparts ('H and S are associated'). If it is most informative to sample infected patients ($\bar{S}$) in testing the efficacy of a vaccine, then it should be most informative to sample $\bar{S}$ in testing the stronger hypothesis 'All vaccinated patients (H) are uninfected (S)' – given that $\bar{S}$ is the rarest of the four traits. As regards Hempel's illustrative example ('All ravens are black'), our intuition that it is most informative to sample ravens is explained by the fact that ravenhood is the rarest of the four traits H, S, $\bar{H}$, $\bar{S}$ in any reasonably chosen population.

The sampling rule in question has been obtained under rather different assumptions. These extensions depend on the fact that a sample drawn from the rarest trait is sufficient for the others. To see this, suppose that $h \leqslant s \leqslant \frac{1}{2}$. Let $\tilde{x}$ take the values 1 or 0 according as an individual drawn at random from H is S or $\bar{S}$. Define $\tilde{x}'$ similarly for a sample drawn from S. Then $P(\tilde{x}=1/\omega_1) = P(S/H, \omega_1) = s$, $P(x=1/\omega_2) = P(S/H, \omega_2) = \delta/h$, while $P(\tilde{x}'=1/\omega_1) = P(H/S, \omega_1) = h$ and $P(\tilde{x}'=1/\omega_2) = P(H/S, \omega_2) = \delta/s$. Now let $\tilde{u}$ be uniformly distributed on [0,1] independently of $\tilde{x}$, and define $\tilde{y} = h(\tilde{x}, \tilde{u}) = 1$ if $\tilde{x}=1$ and $\tilde{u} \leqslant h/s$, and $y=0$ otherwise.

Then $P(\tilde{y}=1/\omega_1)=P(\tilde{x}=1/\omega_1)\,P(u\leqslant h/s)=s(h/s)=h$, and $P(\tilde{y}=1/\omega_2)=$ $=P(\tilde{x}=1/\omega_2)\,P(u\leqslant h/s)=(\delta/h)\;(h/s)=\delta/s$, so that $\tilde{y}$ has the same distribution as $\tilde{x}'$. That is, the distribution of $\tilde{x}'$ can be obtained from that of $\tilde{x}$ together with a table of random digits, so that $\tilde{x}$ is sufficient for $\tilde{x}'$. If $h=\min\,(h, s, 1-s)$, the result can be extended to show that sampling from H is sufficient for sampling from $\bar{H}$ and $\bar{S}$.

It now follows that sampling the rarest trait is most informative in the sense of Neyman-Pearson theory (see Lehmann, 1959, pp. 75–77). Namely, let $\beta(\alpha)$ and $\beta'(\alpha)$ be the power of the most powerful level α test based on the experiments e and e'. Then e is *more informative* than e' *in the sense of Neyman-Pearson theory* if $\beta'(\alpha)\geqslant\beta(\alpha)$ for all α. It is then easily shown that if the associated random variables $\tilde{x}$ and $\tilde{x}'$ are such that $\tilde{x}$ is sufficient for $\tilde{x}'$, them e is more informative than e' in the sense of the definition.

For the same reason, sampling the rarest trait maximizes the *EVSI* in any decision-making context. For to say that $\tilde{x}$ is sufficient for $\tilde{x}'$ is to say that the two induce identical posterior distributions over the parameter space, and consequently, the associated experiments lead to identical sets of Bayes acts.

For those who are chary of prior probabilities, the *EVSI* can be replaced by the following definition of 'more informative than' (Blackwell and Girshick, 1954, ch. 12). We assume for simplicity that the outcome and parameter spaces are of the same finite size, and identify the terminal actions with their associated loss vectors $a=(a_1,\ldots, a_n)$, where a_i is the loss when $\tilde{\omega}=\omega_i$. Then the components $d_j=(d_{j1},\ldots, d_{jn})$ of the derived acts can also be written as vectors, where d_j is the act selected when the jth outcome is observed and d_{ji} is the loss associated with this act when $\tilde{\omega}=\omega_i$. Finally, if $p_i(j)=P\,(z_j/\omega_i)$,

$$\rho_i = \Sigma_j p_{ij} d_{ji}$$

is the risk associated with the derivated act $d=(d_1,\ldots, d_n)$ when $\tilde{\omega}=\omega_i$. Thus the risk vector is the diagonal of the matrix product PD, $P=(p_{ij})$, $D=(d_{ij})$. We define e to be *more informative* than e' *in the sense of Blackwell and Girshick* iff any risk vector obtainable with e' is also obtainable with e (i.e. iff the class of decision rules obtainable with e comprises a complete class with respect to the derived acts obtainable with either e or e').

Under this definition the least informative e is one for which the associated $P=(p_{ij})$ has identical rows (so that $\tilde{z}$ and $\tilde{\omega}$ are independent), while the most informative experiment is one for which P is the $n \times n$ identity matrix. Again it can be shown that observing $\tilde{x}$ is more informative than observing $\tilde{x}'$ if $\tilde{x}$ is sufficient for $\tilde{x}'$. (This condition is one that every reasonable definition of 'more informative than' should satisfy.)

We have already commented upon the relation between the Neyman-Pearson and Bayesian approaches to experimental design in decision-making contexts. The connection between the *EVSI* and the Blackwell-Girshick definition is also quite clear: viz., it can be shown that for any space of terminal actions, e is more informative than e' in their sense iff the *EVSI* of e exceeds that of e' when the prior distribution of $\tilde{\omega}$ is uniform. Of course, both the Neyman-Pearson and Blackwell-Girshick definitions, unlike their Bayesian counterpart, admit of incomparabilities – pairs of experiments neither of which is more informative than the other. Depending on one's attitude toward prior probabilities, this will appear as an advantage or a disadvantage.

We have yet to discuss the relationship between the *EVSI* and the *ESI*. In this direction, we state a fairly strong negative result: viz., *the EVSI is not in general an increasing function of the ESI.*[4]

Example: Let the utility matrix be $U = \begin{pmatrix} .2 & 0 \\ 0 & 1 \end{pmatrix}$ so that a_2 is initially preferred iff $\pi(\omega_1) \geqslant \frac{5}{6}$. Let e and e' (with associated r.v.'s $\tilde{z}$ and $\tilde{z}'$) have joint matrices P and P':

$$14P = \begin{pmatrix} 6 & 5 \\ 1 & 2 \end{pmatrix} \quad 14P' = \begin{pmatrix} 2 & 4 \\ 6 & 2 \end{pmatrix},$$

where $p_{ij} = P(\omega_i, z_j)$ and $p'_{ij} = P(\omega_i, z'_j)$. Then $|\det P'| = 14^{-2} \cdot 20 > 14^{-2} \cdot 7 = |\det P|$, whence $T(\tilde{z}'; \tilde{\omega}) > T(\tilde{z}; \tilde{\omega})$. However, since both maximal column entries of

$$70 \times UP' = \begin{pmatrix} 2 & 4 \\ 30 & 10 \end{pmatrix}$$

occupy the same row, $v_t(e') = 0$. On the other hand, z_i is effective, since $\pi(\omega_1/z_1) > \frac{5}{6}$, so that $v_t(e) > 0$. Hence, the *EVSI* of e exceeds that of e', even though the *ESI* of the latter exceeds that of the former.

The monotonivity breaks down because non-effective outcomes (viz. z for which $a_z = a'$) may increase the *ESI* without increasing the *EVSI*. The conditions on the utility matrix under which the *EVSI* is increasing in the *ESI* is a matter deserving of further consideration, but we shall not pursue it here.

VI. EXPLANATORY POWER

Instead, we turn to the important concept of explanatory power. We have suggested that the sensitivity of an experiment, or the weight of the evidence it provides, be identified with its *ESI*, the average amount of information it transmits about the unknown parameters of an assumed model. Thus, an experiment is sensitive if the likelihood functions of its most probable outcomes are peaked. The extreme case of perfect information is that in which each possible experimental outcome univocally determines a correct hypothesis.

Analogously, a theory may be considered informative if it is a sensitive discriminator of possible experimental outcomes, the extreme case being that in which parametric hypotheses uniquely determine the occurrence of a specified experimental outcome. Just as the *ESI* $T(\tilde{z};\tilde{\omega})$ of an experiment measures its sensitivity, so the reciprocal quantity $T(\tilde{\omega};\tilde{z})$ might be taken as an appropriate measure of the *explanatory power* of the theory whose undetermined parameter is $\tilde{\omega}$. Roughly speaking, then, a theory maximizes explanatory power in the sense of this measure if the most probable hypotheses regarding its parameters impose the strongest possible constraints on the possible outcomes of the relevant experiment.

This explication accounts for the increased explanatory power of the non-trivial quantitative extensions of a qualitative theory. One can construe the outcomes which are discriminable by the quantitative theory as being lumped into equivalence classes the differences between whose individual members lie beneath the discriminatory threshold of the qualitative theory. The data distribution to which the qualitative theory gives rise should be relatively flat, if not constant, over each such equivalence class. Hence, the qualitative theory averages the conditional probabilities $P(z_j/\omega_i)$ of the quantitative theory over the outcomes which lie in the same equivalence class. But any such averaging increases the conditional entropies $H_i(\tilde{z})$, and so increases their weighted average $H_{\tilde{\omega}}(\tilde{z})$, thereby decreasing the explanatory power as measured by $T(\tilde{\omega};\tilde{z})$. One might extend this analysis, as in Adams (1966), to the informativeness of scales. Thus, a theory which justifies interval scaling, say, exceeds the explanatory power of a theory which justifies only nominal scaling, etc. A closely related application is to the informativeness of a typology.

Unfortunately, $T(\tilde{\omega};\tilde{z})$ is not an altogether appropriate measure when we wish to compare the explanatory power of one model *vis-à-vis* a given experiment with that of another model with respect to a different experiment. For one model may score lower than another merely because less initial uncertainty surrounds the outcomes of its experiment. Thus, $T(\tilde{\omega};\tilde{z}) = H(\tilde{z}) - H_{\tilde{\omega}}(\tilde{z})$ results from two separable constraints: (i) the *distributional constraint* $H_{\max} - H(\tilde{z})$, reflecting the a priori predictability of the outcome, and (ii) the *correlational constraint* $H_{\max} - H_{\tilde{\omega}}(\tilde{z})$, reflecting the predictability of the outcome by the model per se. (Here $H_{\max} = \log n$, where n is the number of possible experimental outcomes.)

From our definitions, the transmitted information is seen to equal the correlational constraint less the distributional constraint,

$$T(\tilde{\omega};\tilde{z}) = [H_{\max} - H_{\tilde{\omega}}(\tilde{z})] - [H_{\max} - H(\tilde{z})],$$

whence the transmitted information is decreasing in the distributional constraint. But the latter may depend on such features as the partition of the outcome space (i.e. the number of outcomes) selected, the sampling rule, etc., which have nothing to do with the model's explanatory power per se. Explanatory power is therefore more meaningfully measured by the magnitude of the transmitted information relative to that of the initial uncertainty, viz., by

$$T(\tilde{\omega};\tilde{z})/H(\tilde{z}).$$

For decisive experiments, $H_{\tilde{\omega}}(\tilde{z}) = 0$, and $T(\tilde{\omega};\tilde{z})/H(\tilde{z}) = 1$, a maximum. When $\tilde{\omega}$ and $\tilde{z}$ are independent, $H_{\tilde{\omega}}(\tilde{z}) = H(\tilde{z})$ and $T(\tilde{\omega};\tilde{z})/H(\tilde{z}) = 0$, a minimum. The quantity $1 - T(\tilde{\omega};\tilde{z})/H(\tilde{z})$ might be called the *equivocation* of the associated model. Distinct models can meaningfully be compared in terms of their equivocations with respect to different experiments. We remark that the notion of experimental sensitivity can be refined in an analogous way, using $T(\tilde{z};\tilde{\omega})/H(\tilde{\omega})$ in place of $T(\tilde{\omega};\tilde{z})/H(\tilde{z})$.

VII. THEORETICAL SIMPLICITY

The literature on simplicity[5] is rather chaotic: a multitude of notions have been canvassed and disparate rationales supplied. One proposal, that of Popper (1935), makes a serious effort to supply a reasonably compelling rationale for preferring hypotheses which are simplest in the sense delineated. Roughly, Popper has identified the simplicity of hypotheses with their strength or specificity – with the paucity of Carnapian state descriptions in which they hold. Thus, an hypothesis which excludes nothing is clearly of no value. On the other hand, hypotheses and models which impose strong constraints on the state of nature have high explanatory power in the sense of Section VI. In addition, such hypotheses are held to be more open to experimental falsification. Falsifiability seems to Popper to provide another reason – indeed, the primary reason – for preferring strong to weak hypotheses when both are equally in accord with the data. A Baconian or eliminativist rationale also underlies Popper's preference for strong hypotheses: such hypotheses should lead one most rapidly to the discovery of true hypotheses. Let us examine these contentions in detail.

To begin with, while Popper's sense of 'simplicity' certainly has a strong connection with the explanatory power of a model, its relation to the problem of deciding what questions the scientist should put to nature next that he may uncover her secrets seems tenuous. The appropriate strategy for 'twenty questions' is not to ask maximally specific questions, but questions which split the remaining alternatives in half. Such questions, as we saw with reference to the checkerboard example of Section IV, *maximize average informative intake*. And that would appear to constitute the most compelling desideratum in the context of discovery. (When the possible states of nature are unequally probable, the fission strategy generalizes to splitting the sum of the probabilities of the remaining alternatives.) At any rate, such relatively transparent contexts point to the necessity of distinguishing the role of hypotheses in *questioning procedures* from their role in *explanation*.

To be sure, the accessibility of an hypothesis to experimental check is an important desideratum, but this feature is no simple function of the specificity of the hypothesis. To see this, consider the well-known recreational problem of locating that one of twelve balls whose weight differs from the others, and determining whether this 'odd ball' is heavier or lighter, using an ordinary pan balance. An hypothesis asserting that the odd ball is one of four, if false, would certainly be refuted by weighing four of the remaining balls against the other four remaining balls, for these would not balance. On the other hand, the maximally specific hypothesis asserting that one particular ball is the odd ball could not assuredly be refuted by a single weighing, be it ever so false, for an odd number of balls remain, and a one against one weighing could show at most that one of the two was the odd ball but not which one.

The notion of falsifiability is itself by no means clear. In experimental contexts the most natural meaning of 'falsifiability' is the probability of refuting the pertinent hypothesis conditional on the falsehood of that hypothesis, or the average number of observations required to do so. (This construal is closely related to the concept of *power* in Neyman-Pearson theory.) It is fairly clear from the above example that the falsifiability of an hypothesis is no simple function of its strength, but depends as well on the structure of the contemplated experiment. Indeed, the notion seems to have little meaning when divorced from a particular space of performable experiments or observational procedures.

Worse still, the notion of content or strength blurs when extended beyond the rather limiting confines of Carnapian state descriptions. For example, while both the conjectures of Fermat and Goldbach are essentially of the form 'for all n, $P(n)$', the former might be thought stronger on the grounds that its '$P(n)$' asserts a property of all triples of natural numbers, while that of the Goldbach conjecture attributes to n a property – that of being odd or the sum of two primes – which is statable without recourse to quantifiers. But by virtue of the greater logical complexity of the property which it projects, Fermat's conjecture is less amenable to refutation, for, unlike that of the Goldbach theorem, the property of the Fermat theorem is apparently non-recursive. At the very least, pairs of this sort pose a serious problem for those who would maintain that the relation of comparative strength admits of no incomparables.

These difficulties notwithstanding, there is a temptation to suppose that an hypothesis which is clearly weaker than another by virtue of inclusion among its logical consequences ought to be discarded and is, in any case, less well confirmed when both hypotheses are equally in accord with the data. To appraise this contention, we adopt the measure $Dc(H/E) = P(H/E) - P(H)$ for the degree to which E confirms H.[6] Then by Bayes' theorem, $Dc(H/E) = P(H)[P(E/H)/P(E) - 1]$. Consequently, if H' implies H, and H implies E, our definition leads us to conclude that H' is less well confirmed than the weaker hypothesis H. Obviously a necessary condition that the stronger hypothesis be better confirmed is that it imparts to E a higher conditional probability, and to render this condition sufficient one requires that the inequality

$$\frac{P(E/H') - P(E)}{P(E/H) - P(E)} > \frac{P(H)}{P(H')}$$

hold. Nor is this result counterintuitive. Many would agree, e.g., that the hypothesis 'All emeralds are grue' is less well confirmed by current observation of green emeralds than the weaker hypothesis 'All emeralds examined prior to time t are green'.

Talk of falsifiability presupposes an acceptivist or eliminativist approach to hypothesis testing like that of Neyman-Pearson theory. However, even those who work within this framework are compelled to admit that rejection of hypotheses is never final nor so decisive as the simple logical scheme of modus tollens would suggest. A more realistic account, and one that could be squared with a degrees-of-belief approach, would consider instead how sharply a contemplated experiment could be expected to reduce one's uncertainty regarding an hypothesis. The cost of achieving a tolerable reduction would then be balanced against the expected value of finding out whether the pertinent hypothesis was true or false. Seen in this light, a questioning procedure is essentially a sequential experiment. Accordingly, it generates a tree, where the number of branches emanating from each node represents the number of possible answers to the associated question. The value of information transmitted by each branch can then be computed and weighted by the probability that the branch in question will be traversed. The resulting weighted average measures the value of information which the associated questioning procedure transmits (on the average). Thus, the theory of the *EVSI* extends intact to questioning procedures.

The dual problems of measuring simplicity and providing a rationale are solved at one stroke, where hypotheses are concerned, by passing to informational measures and subsuming the problem what hypotheses to test next under the more general problem of experimental design. Given its ability to clarify this problem, it is not unreasonable to expect information theory to illuminate the notion of simplicity or economy as it pertains to models.

Prominence has been accorded the notion of simplicity in curve-fitting. However, the considerations of tractability that guide the choice of an empirical curve, qua compact summary of the data, are rather different from those which inform the selection of a curve *family*, qua model or explanation of the data. In the latter context, one wants the parameters of the model or family to be theoretically interpretable, for even if the underlying theory is rather qualitative, it will often suggest appropriate relations between the parameters. If the number of equations relating them is large compared to the number of parameters, then, usually, some of them will be overdetermined. The requirement that different parameter estimates – obtained from different equations – be concordant, indirectly imposes additional constraints on the data and checks on the model. The theory is thereby rendered applicable far beyond the domains over which its theoretical parameters are experimentally determinable.

Intuitively, the number of parameters should mirror the number of degrees of freedom in the data. Mere paucity of the model's parameters is a poor index of economy, for like the efficiency of a code, that of a model depends on the compression actually achieved compared to the compression that could possibly be achieved. (Of course, no lower bounds can be specified for models or codes generally.)

Increasing economy is nicely illustrated in the development of planetary astronomy. The Ptolemaic model was vitiated less by sheer excess of parameters per se, than by the uninterpretability of its parameters in physical terms. The introduction of secondary motions effectively divorced ancient astronomy from ancient physics and cosmology. For the latter posited a mechanism of interlocking spherical crystalline shells to drive the planets in their orbits, and this model was difficult to reconcile with secondary motions. By the simple device of relocating the origin of the system, Copernicus was able to radically reduce the number of parameters needed to specify the orbits. Kepler's refinement of the geocentric model further enhanced its economy, not only by relating the rate at which a planet moved to its distance from the sun, but also by relating the motions of different planets. This paved the way to Newton's formulation of the gravitation law (a law which Newton seems to have deduced from Kepler's laws), which enabled the theory to move beyond a kinematical level and relate the masses of the planets to their motions. At each stage in the evolution of the theory, redundancies implicit in the data were made explicit. Often they were revealed by virtue of a simplifying transformation, like Copernicus's inversion. Such transformations improve overall efficiency of coding when the saving they effect exceeds the number of information units required for their description.

APPENDIX: OTHER INFORMATION MEASURES

The Neyman-Pearson and Blackwell-Girshick definitions of 'more informative than' for decision making contexts encountered in Section V were 'objective' in the sense that they do not involve the prior measure $\pi(\omega)$. We sketch here the relations between the *ESI* and several 'objective' measures of disinterested information.

Consider the context of dichotomous testing, where H_1 and H_2 are the rival hypotheses, and assume that the experimental random variable $\tilde{z}$ is discrete. Write $P_i(z)=P(z/H_i)$, $i=1, 2$, and $L(z)=P_1(z)/P_2(z)$ for the likelihood ratio. Then

$$J(1, 2; \tilde{z}) = E(\log L/H_1) = \Sigma_i P_1(z_i) \log L(z_i) \tag{31}$$

the conditional expectation (against the distribution of $\tilde{z}$ given by H_1) of the log likelihood ratio, is called the *discrimination information* for H_1 against H_2 provided by the experiment whose random variable is $\tilde{z}$.

The properties of discrimination, barring symmetry, parallel those of transmitted information (the *ESI*): (*i*) J is non-negative, (ii) J is additive in independent experiments, and (iii) no transformation of the outcome space increases J, and the only transformations that preserve it are sufficient statistics.

When $P_2(z_i)=1/n$, $i=1,\ldots, n$, so that the data distribution is uniform conditional on H_2,

$$\begin{aligned} J(1, 2; \tilde{z}) &= \Sigma_i P_1(z_i) \log P_1(z_i) + \log n \\ &= H_{\max} - H[p_1(z_1), \ldots, P_1(z_n)] \end{aligned} \tag{32}$$

the distributional constraint imposed by H_1.

If $\tilde{z}^{(n)}$ comprises n independent observations on $\tilde{z}$,

$$P(L(\tilde{z}^{(n)}) > \gamma/H_1) > 1 - \delta, \tag{33}$$

where γ is arbitrarily large, δ arbitrarily small, and n is taken sufficiently large – provided that $\pi(H_1) \neq 0$. Hence,

$$P(J(1,2;\tilde{z}^{(n)}) > \gamma/H_1) > 1 - \delta, \tag{34}$$

when n is sufficiently large. As an alternative to the standpoint of Neyman-Pearson theory, from which the design problem is viewed as that of selecting e so as to reduce error probabilities to a tolerable level, one can regard the problem instead as that of selecting an e which provides a sufficient amount of discrimination information at minimal cost.

Suppose that the data distribution is a one-parameter family, $P_i(z) = P_i(z/\tilde{\omega})$. We consider the case where H_1 asserts $\tilde{\omega} = \omega$ and H_2 that $\tilde{\omega} = \omega + \Delta\omega$, where $\Delta\omega$ is small, writing $J(1,2;\tilde{z}) = J(\omega, \omega + \Delta\omega; \tilde{z})$. Obviously, $J(\omega, \omega + \Delta\omega; \tilde{z})$ approaches zero as $\Delta\omega$ approaches zero. Indeed, if the data distribution is continuous in $\tilde{\omega}$, the discrimination information which $\tilde{z}$ supplies varies inversely with $\Delta\omega$, so that more data are required to bring about convergence of opinion when initial estimates are not greatly divergent. More generally, it can be shown (Savage, 1954, p. 236) that

$$\lim_{\Delta\omega \to 0} \frac{J(\omega, \omega + \Delta\omega; \tilde{z})}{\Delta\omega^2} = \tfrac{1}{2} I_F(\omega; \tilde{z}), \tag{35}$$

where

$$I_F(\omega; \tilde{z}) = E_{z/e}\left(\frac{\partial^2 L(\tilde{z}/\omega)}{\partial\omega^2}\right) \tag{36}$$

and $L(z/\omega) = \log P(z/\omega)$.

I_F is *Fisher's information measure*. It is asymptotically equal to the inverse of the sampling variance of an efficient estimator of $\tilde{\omega}$, and consequently plays a fundamental role in Fisher's theory of estimation (Fisher, 1922, 1934). Like discrimination information, of which it is a special case, Fisherian information shares the fundamental properties (i)–(iii) of non-negativity, additivity in independent experiments, and preservation under sufficient statistics, with the *ESI*. Fisher's measure differs from the latter in dispensing with $\pi(\omega)$. However, insofar as the comparison of experiments turns primarily on the likelihood ratio, the two measures will often select the same e. Nevertheless, the suppression of $\pi(\omega)$ is of importance when the latter is peaked. For if there is a concentration of prior probability mass at ω_1, say, then a much larger sample will be required to bring about a concentration of posterior probability mass at a different value ω_2, given that $\tilde{\omega}=\omega_2$, than would be required if no prior probability mass were concentrated at any value of $\tilde{\omega}$. Of the two measures, only the *ESI* is sensitive to the lawlikeness of the rival hypotheses considered. The lawlikeness of simple alternative hypotheses of the form $\tilde{\omega}=\omega_2$ might be compared by estimating the sample size required to concentrate $100\alpha\%$ of the posterior density in a δ neighborhood of ω_2 when $\tilde{\omega}=\omega_2$ is the true hypothesis (using fixed values of α and δ). Of course, the proposed measure is relative to a particular distribution family.

The literature on information has been somewhat confused by the usage of communications theorists, who often use the terms 'entropy' and 'information' interchangeably. Statisticians have occasionally felt called upon to deny that entropy is a suitable measure of information for the purposes of statistics. We agree that entropy is not in general a suitable measure – for one thing, sufficiency reductions of the data always result in a reduction of entropy – but we would argue that transmitted information or reduction of entropy has many of the features – e.g. (i)–(iii) – one should expect of an adequate measure. The conditions under which entropy itself measures the information yield of an experiment are easily specified. If the experiment is decisive in the strong sense that $H_j(\tilde{\omega})=0$ for each outcome z_j, then $T(\tilde{z};\tilde{\omega})=H(\tilde{\omega})$. This is precisely the case communications theorists have in mind when they identify the entropy of a source as 'information'. They are merely equating the entropy of a source with the amount of information that would result were a message to be selected by the source and *definitively identified.*

City College of New York

BIBLIOGRAPHY

Adams, E. W.: 1966, 'On the Nature and Purpose of Measurement', *Synthese* **16**, 125–168.

Attneave, F. J.: 1954, 'Some Informational Aspects of Visual Perception', *Psych. Review* **61**, 183–193.

Attneave, F. J.: 1959, *Applications of Information Theory to Psychology*, Holt, Rinehart, and Winston, New York.

Blackwell, D. and Girshick, M. A.: 1954, *Theory of Games and Statistical Decisions*, Wiley, New York.

Fisher, R. A.: 1922, 'On the Foundations of Theoretical Statistics', *Phil. Trans. Roy. Soc., London* **A222**, 309–368.
Fisher, R. A.: 1934, 'Probability, Likelihood, and Quantity of Information in the Logic or Uncertain Inference', *Proc. Roy. Soc., London* **A146**, 1–8.
Foster, M. H. and Martin, M. L. (eds.): 1966, *Probability, Confirmation and Simplicity*, Odyssey Press, New York.
Garner, W. R.: 1962, *Uncertainty and Structure as Psychological Concepts*, Wiley, New York.
Garner, W. R. and McGill, W. J.: 1956, 'Relation between Uncertainty, Variance, and Correlation Analyses', *Psychometrika* **21**, 219–228.
Good, I. J.: 1968, 'Corroboration, Explanation, Evolving Probabilities, Simplicity and a Sharpened Razor', *Brit. J. Phil. Sci.* **19**, 123–143.
Hintikka, K. J. J. and Pietarinen, J.: 1966, 'Semantic Information and Inductive Logic', in Hintikka and Suppes (1966).
Hintikka, K. J. J. and Suppes, P. (eds.): 1966, *Aspects of Inductive Logic*, North-Holland Publ. Co., Amsterdam.
Khinchin, A. I.: 1957, *Mathematical Foundations of Information Theory*, translated from the Russian by R. A. Silverman and M. D. Friedman, Dover, New York.
Kullback, S.: 1959, *Information Theory and Statistics*, Wiley, New York.
Kullback, S. and Leibler, R. A.: 1951, 'On Information and Sufficiency', *Ann. Math. Stat.* **22**, 79–86.
Lehmann, E.: 1959, *Testing Statistical Hypotheses*, Wiley, New York.
Lindley, D. V.: 1956, 'On a Measure of Information provided by an Experiment', *Ann. Math. Stat.* **29**, 986–1005.
McGill, W. J.: 1954, 'Multivariate Information Transmission', *Psychometrika* **19**, 97–116.
Miller, G. A.: 1956, 'The Magic Number Seven, Plus or Minus Two', *Psych. Review* **63**, 81–97.
Miller, G. A. and Madow, W. G.: 1954, *On the Maximum Likelihood Estimate of the Shannon-Wiener Measure of Information*, Air Force Cambridge Research Center: Technical Report, 54–75.
Popper, K. R.: 1935, *Logik der Forschung*, Julius Springer, Wien. Engl. transl. with notes and appendices under the title *The Logic of Scientific Discovery*, Hutchinson, London, 1959.
Raiffa, H. and Schlaifer, R.: 1961, *Applied Statistical Decision Theory*, Harvard Business School, Boston.
Savage, L. J.: 1954, *The Foundations of Statistics*, Wiley, New York.
Shannon, C. E. and Weaver, W.: 1949, *The Mathematical Theory of Communication*, Univ. of Illinois Press, Urbana, Ill.
Sneed, J.: 1967, 'Entropy, Information and Decision', *Synthese* **17**, 392–407.
Suppes, P.: 1966, 'A Bayesian Approach to the Paradoxes of Confirmation', in Hintikka and Suppes (1966).

REFERENCES

[1] Notation: a tilde distinguishes a random variable – qua function – from its generic value. In conjunction with an expectation sign a tilde indicates the random variable being expected, while subscripts 'ω', 'ω/z', 'z/e', etc., appended to an expectation sign indicate the measures $\pi(\omega)$, $\pi(\omega/z)$, $P_{z/e}$, etc., with respect to which the expectation is

taken. Finally, a single prime indicates expectation against a prior distribution; a double prime, expectation against a posterior distribution. The notation is that of Raiffa and Schlaifer (1961) and is adhered to throughout. The present section, up to the discussion of Neyman-Pearson theory, is little more than a summary of their more detailed exposition.

[2] Khinchin (1957), p. 9f.

[3] Shannon and Weaver (1949), p. 49f.

[4] An appropriate counterexample appeared for the first time in Sneed (1967).

[5] Specimen articles (by Ackermann, Barker, Goodman and Rudner) may be found in *Philosophy of Science* **28**; cf. also Foster and Martin (1966).

[6] For an interesting motivation of this measure which turns on regarding information as an 'epistemic utility', see Hintikka and Pietarinen (1966), p. 108. Popper himself has proposed an analogous measure.

PART II

INFORMATION AND SOME PROBLEMS OF THE SCIENTIFIC METHOD

RISTO HILPINEN

ON THE INFORMATION PROVIDED BY OBSERVATIONS*

I. AYER'S PROBLEM

In his article 'The Conception of Probability as a Logical Relation' [1] A. J. Ayer has criticized the logical interpretation of inductive probability on which J. M. Keynes [22], Rudolf Carnap [6], and many other writers on the philosophy of induction have based their conceptions of inductive inference. Ayer's criticism is concerned with the principle of total evidence and with the obvious fact that it is often reasonable to collect new evidence when we are studying the credibility of some hypothesis. According to Ayer, it is impossible to understand this simple fact and justify the principle of total evidence, if the concept of inductive probability is interpreted in the way suggested by Keynes, Jeffreys, Carnap, and other proponents of the logical interpretation. Ayer's argument runs as follows:

According to the logical interpretation, probability is a logical relation between sentences. All elementary probability statements '$P(h \mid e)=r$', where h and e are two sentences of a given language L and r is a real number in the closed interval (0, 1), are analytic, i.e. necessarily true or necessarily false. Suppose now that h is some hypothesis in which an investigator X is interested, and let e represent the 'evidence' (e.g. results of observations) which is used to determine the credibility of h or 'justified degree of belief' in h. The value of $P(h \mid e)$ depends on the choice of e: probabilities vary with evidence. How should X, then, determine the credibility of h in the situation at hand? According to the *principle of total evidence*, "in the application of inductive logic to a given knowledge situation, the total evidence available must be taken as basis for determining the degree of confirmation" (Carnap [6], p. 211). Hence if $P(h \mid e)$ is used to express the credibility of h in the situation in question, the sentence e has to describe the total evidence available in this situation (or at least all such evidence as is not irrelevant to h[1]). But $P(h \mid e)=r$ is an analytic sentence regardless of whether e represents the total evidence available or not, and therefore the probability based on total

J. Hintikka and P. Suppes (eds.), Information and Inference, 97–122.

evidence cannot be 'better' or 'more correct' than a probability which is based on a part of the total evidence only. If probability statements are analytic, the principle of total evidence cannot, in Ayer's opinion, be justified in any reasonable way and it becomes impossible to understand this principle at all. For the same reason Ayer thinks that it is impossible to explicate the rationality of collecting new evidence in a theory of induction based on the logical interpretation of probability. "Once we have assembled some trustworthy data by these means, there can be no reason, on Keynes's system, why we should trouble to carry our investigations any further. The addition of more evidence can, indeed, yield a higher or lower probability for the statement in which we are interested. But unless we have made some logical mistake, this probability cannot be said to be more, or less, correct than the one that was yielded by the evidence with which we started." (Ayer [1], p. 14.)

In fact, Ayer could still have sharpened his criticism at least as far as it is directed against Carnap's probability theory. On Carnap's theory, it is possible to determine the probability of a hypothesis with respect to tautological evidence also (the *a priori* probability of the hypothesis). Tautological evidence is 'empty' evidence: having tautological evidence only amounts to having no evidence at all. In addition to Carnap's theory, the existence of *a priori* probabilities is assumed also in all other Bayesian systems of inductive logic. If Ayer's criticism were correct, all observation making would, according to these theories, be totally unnecessary.

The logical concept of probability or 'degree of confirmation' which Ayer criticizes has been characterized by Carnap ([5], p. 514) as follows: "... to decide to what degree *h* is confirmed by *e* ... all that we need is a logical analysis of the meanings of the two sentences." This conception can, however, be shown to be inadequate on grounds which are wholly independent of Ayer's criticism. In a given language *L* it is possible to define several (in point of fact, infinitely many) probability-measures, or, in Carnap's terminology, *c*-functions. In [7] Carnap has defined a class of probability functions, called the λ-continuum, for first-order languages which contain monadic predicates only. In this class each *c*-function is characterized by one particular value of a parameter $\lambda (0 \leqq \lambda \leqq \infty)$. By giving different values to λ it is possible to weight, in different ways, a 'logical factor', based on certain aprioristic symmetry considerations concerning the structure of L, and an 'empirical factor', based on observed relative frequencies. In [15] Jaakko Hintikka has generalized Carnap's λ-continuum and constructed a two-parametric system in which inductive probabilities are defined with the help of two parameters, λ (which corresponds to Carnap's λ) and α. Carnap's λ-continuum is obtained in Hintikka's system as a special case in which $\alpha=\infty$. In systems of this kind, explicit probability statements of the form '$P(h \mid e)=r$' presuppose a choice of particular values for these parameters. But what is the meaning of this choice, i.e. how should these parameters be interpreted? It turns out that we can give them a natural *subjectivistic* interpretation: the values of λ and α can be taken to indicate certain beliefs or assumptions of the investigator concerning the universe he is investigating; especially its irregularity or the 'amount of disorder' in the universe (Hintikka [16]). On the other hand, the 'adequacy' of an inductive method, characterized by a particular c-function, depends on how correct these assumptions are; hence the parameters in question can be given an *objectivistic* interpretation, too (Carnap [7], pp. 68–73; Hintikka [16]). Hence the degree of confirmation of h with respect to e depends, not only on "the meanings of the two sentences", but also on certain extralogical assumptions concerning the universe of which we are speaking. This is not, so it seems to me, in accord with the strictly logical interpretation of inductive probability.

The inadequacy of the strictly logical interpretation of inductive probability becomes more evident when probability-measures are defined in languages which are logically richer than the language-systems considered by Carnap [7] and Hintikka [15]. If L contains, for instance, binary predicates, we have to consider several 'logical factors', and to describe all the intuitively plausible inductive methods we need more than two free parameters. In this case the logical factors alone can be weighted in different ways, and they may have an opposite role in inductive inference: weighting one ('logical') factor may, for instance, favor one kind of inductive generalization, weighting another, another kind of generalization. Here the choice of an inductive method is thus a more complicated matter than in the case of unary languages, and our extralogical assumptions concerning the universe of which we are speaking are more important (Hilpinen [14]).[2]

Ayer's problem is not, however, restricted to the logical interpretation of inductive probability only. It can also be raised, as I. J. Good ([11], p. 319) has pointed out, in connection with the 'subjectivistic' (or 'personalistic') interpretation. The questions of the rationality of making observations and of the justification of the principle of total evidence can be raised, it seems to me, in any theory of induction which is based on the concept of inductive probability regardless of the precise interpretation of this concept.

II. THE PRINCIPLE OF TOTAL EVIDENCE AND THE UTILITY OF OBSERVATIONS

Ayer's problem has already been solved – at least in part. Although Ayer presents the questions about the rationality of observations and about the justification of the principle of total evidence as two ways of formulating the same difficulty, inherent in the logical interpretation of probability, these questions should be distinguished from each other, and they have, in fact, been answered in different ways. Ayer's failure to distinguish the two questions from each other is perhaps in part due to the fact that Carnap's formulation of the principle of total evidence (see above p. 97) is not quite precise. The expression 'available evidence' can be understood in two different ways. First, 'the total evidence available' can be understood as referring to such evidence as the investigator in some sense already has at his disposal; in this sense the total evidence consists of sentences that the investigator knows (or believes) to be true and accepts as evidence. On the other hand, 'total evidence available' can also be taken to refer to those evidential data that the investigator can get to his disposal, if he wants to, or chooses to. If the evidential statement *e* describes observational results relevant to the hypothesis, 'total evidence', according to the first interpretation, refers to all the observations already made, and according to the second interpretation, it refers to such observations as the investigator can possibly make or is going to make, or which he perhaps ought to make and take into account.

If the concept of total evidence is interpreted in the way mentioned first, it has a precise and unambiguous meaning in each 'knowledge situation' or context of application of inductive logic, and thus the principle of total evidence, too, has a precise meaning. But the second interpretation does not give any precise meaning to the concept nor to the principle of total evidence. Carnap ([6], p. 211) has laid down the following rule concerning the application of inductive logic to a knowledge situation T:

(1) If inductive logic supplies a result of the form $P(h \mid e) = r$ and if e expresses the total knowledge of X at the time T, that is to say, his total knowledge of the results of his observations, then X is justified at this time to believe h to the degree r.

In (1) Carnap obviously gives the concept of total evidence the first interpretation. But in [1] Ayer seems to interpret the expression 'total evidence available' (at least in some cases) in the second way; therefore it is not surprising that he finds the concept vague and fails to distinguish the two questions mentioned above from each other.[3] In the sequel, I shall interpret the concept of total evidence in the first-mentioned, precise way, and distinguish the problem of the rationality of making observations from the problem of total evidence.

Carnap has emphasized that the principle of total evidence is not a principle of inductive logic proper, but belongs to what Carnap calls the 'methodology' of induction (see e.g. Carnap [6], p. 204; cf. also Hempel [12], p. 453). By inductive logic Carnap understands the theory of inductive probability (i.e. theory of c-functions). All questions concerning the application of this theory (e.g. those connected with the concept of total evidence) belong to the methodology of induction. The principle of total evidence refers to some 'knowledge situation' and it is concerned with the credibility of h for X (i.e. "the degree to which X is justified to believe h") in this situation. Let $P_T(h)$ be the credibility of h at T,[4] and let e_T be the totality of evidence which X accepts at T (i.e. a description of the results of the observations X has made). P_T is a probability measure by means of which it is possible to express the credibility of any sentence of L at T. According to (1),

$$P_T(h) = P(h \mid e_T), \quad (2)$$

where P is the probability measure defined in the system of inductive logic used by X. According to (2), $P_T(e_T) = 1$. This probability is not determined with the help of inductive logic; it simply expresses the fact that X accepts the sentence e_T as his evidence at T, i.e. e_T describes observational results known by X at T. What observational results X accepts at T as evidence determines uniquely the probabilities P_T of all the other sentences of L by (2).[5] Hence the measure P_T takes 'automatically' into account all the evidence accepted at T (see Suppes [30], p. 50), and the principle of total evidence (when interpreted in the present way) raises no theoretical difficulties; in fact, it is superfluous as a separate principle of the methodology of induction, because it follows from the requirement that P_T has to satisfy the axioms of probability. 'Knowledge of the results of observations' is in the present context equivalent to acceptance of the results in question as evidence, in other words, assigning to them the probability P_T, or credibility, 1. If $P_T(i)<1$, X does not accept i as evidence at T and hence does not 'know' the results described in i; in this case the acceptance of i would presuppose new observations. But this takes us to Ayer's second problem, the problem of justifying the rationality of making observations.

In discussions concerning the 'inductive inconsistencies' generated by 'statistical' or inductive syllogisms (see e.g. Hempel [12], pp. 40ff.) it is sometimes assumed that we can both know that i is true (in the sense that we could accept i as evidence without making new observations) and at the same time assign i a credibility less than 1 (i.e. *not* know the truth of i). But this assumption becomes self-contradictory, when 'knowledge of the results of observations' is interpreted in probabilistic terms, i.e. in terms of the measure P_T. Hence no separate principle of total evidence is needed.

A detailed argument of the kind sketched above has been presented by Patrick Suppes ([30], especially pp. 57–62). It does not, however, remove all the difficulties connected with the concept of total evidence. According to Suppes, there is another problem of total evidence, "that of characterizing what part of the welter of potential information impinging an organism is to be accepted as evidence" ([30], p. 62). But in so far as this problem is related to Ayer's questions at all, it seems to be related to the second problem, that of explicating the rationality of observations.

Ayer says that probability statements cannot be 'guides in life' if they are analytic ([1], p. 18).[6] There is undoubtedly some truth in this statement. But even if statements about the conditional probabilities defined in inductive logic were analytic, sentences of the form $P_T(h)=r$ on which we base our practical decisions in the situation T are not analytic, but have factual content. $P_T(h)=r$ can be taken to be equivalent to 'the total evidence accepted at T is e_T, and $P(h \mid e_T)=r$'. Hence the content of '$P_T(h)=r$' depends on the content of e_T. '$P_T(h)=r$' can be analyzed into two components, one of which is (according to the logical interpretation of inductive probability) an analytic sentence, the other a factual statement. Hence this 'paradox of the logical probability' is due to a failure to distinguish the probabilities $P_T(h)$ and $P(h \mid e_T)$ from each other.

Above it was assumed that the sentence e_T which X accepts as evidence at T describes the results of observations already made. When the observations are made and their credibility is established, no theoretical difficulties concerning the concept of total evidence will arise. But so far nothing has been said of Ayer's second problem, the rationality of collecting new evidence.

I. J. Good [11] has shown that the rationality of collecting new evidence can be based on the well-known Bayesian decision-principle, the principle of maximizing expected utility. If the decision maker follows Bayes's principle and if the cost of making new observations is not taken into account, collecting new evidence is always rational. In this case the expectation of the utility provided by the 'best' alternative course of action ('best' in the sense of Bayes's principle), based on observations already made, cannot be greater than the corresponding expectation based on new observations, combined with the observations already made. Good's argument runs as follows:

Suppose that $a_1, a_2, \ldots, a_j, \ldots, a_s$ are the alternative courses of action open to the decision maker X. These alternatives are mutually exclusive and jointly exhaustive; X has to choose one and only one alternative. Let $h_1, h_2, \ldots, h_i, \ldots, h_r$ be mutually exclusive and jointly exhaustive hypotheses which describe the alternative 'states of nature' in the decision situation, and let the utility that results from a_j under circumstances h_i be u_{ij}. The decision maker has accepted the sentence e as evidence and considers making new observations the possible outcomes of which are $e_1, e_2, \ldots, e_k, \ldots, e_t$. According to Bayes's decision rule, the expected value of the utility provided by the best alternative is with respect to e

$$\text{(3)} \qquad \max_j \sum_i P(h_i \mid e)\, u_{ij}.$$

If X makes new observations whose outcome is e_k, the corresponding expectation relative to the total evidence now available to X is

$$(4)\qquad \max_j \sum_i P(h_i \mid e \,\&\, e_k)\, u_{ij}.$$

The probability of the result e_k is $P(e_k \mid e)$; hence making new observations and taking them into account yields the expectation

$$(5)\qquad \sum_k P(e_k \mid e) \max_j \sum_i P(h_i \mid e \,\&\, e_k)\, u_{ij} = \sum_k \max_j \sum_i P(h_i \mid e)\, P(e_k \mid h_i \,\&\, e)\, u_{ij}.$$

Good shows that (3) cannot be larger than (5). (3) and (5) are equal only if the alternative recommended by Bayes's principle is the same irrespective of which of the results e_k actually occurs, otherwise (5) is larger than (3). Hence collecting new evidence is rational, according to Bayes's principle, if the cost of making observations is negligible.

Good's argument seems to me a perfectly satisfactory solution to Ayer's second question. Ayer is evidently looking for some 'pragmatic' justification for making new observations, and Good's solution – reference to the principle of maximizing expected utility – is, indeed, such a justification.[7]

In Good's argument the cost of making observations is not taken into account. Actually, observations do cost something; a part of this cost is the time it takes to make observations. There are obviously cases in which it does not pay to make observations, because the cost of making them outweighs the increase in utility otherwise expected. Collecting new evidence is not *always* rational.

III. THE INFORMATION CARRIED BY OBSERVATIONS

The results described above give a reasonably satisfactory solution to Ayer's problems. It seems to me, however, that there is more to be said about the principle of total evidence and the rationality of making observations. Irrespective of whether the principle of total evidence is necessary as a separate principle of the methodology of induction or not, total evidence gives in any case – so it seems to me – a 'better' basis, in some sense, for the evaluation of the credibility of hypotheses than any part of it (cf. Ayer [1], p. 14). This superiority cannot, however, be explicated by referring to the principle of maximizing expected utility. Suppose that, for example, e & e_k describes the total evidence available at T. Then (4) is the expectation of utility based on the total evidence, and (3) is the corresponding expectation based on the part e of the total evidence. But (4) is not necessarily larger than (3); (3) can be greater than (4), although (5) is always as large as (3). The superiority of the total evidence accepted at T in comparison to its parts cannot be explicated by pragmatic considerations of this kind. Instead of a justification of this kind, I have in my mind a very simple and intuitively obvious way of explicating this 'superiority' of total evidence.

Why do investigators make observations? Perhaps the simplest (though not a very informative) answer is as follows: Observations give information, and scientists make observations because one of the aims of inquiry is to obtain information. The investigator usually is not very much interested in the observations in themselves; but, for instance, in certain hypotheses, and he makes observations because he assumes that they provide information concerning the hypotheses in question. Now obviously a large number of observations carries more information than a small number of observations (in some sense of the word 'information'), other things being equal.[8] Collecting new evidence cannot reduce the amount of information available to the investigator, but it may increase it. The total evidence accepted carries more information than its parts, and therefore collecting new evidence, too, is rational, if the aim of the investigator is to acquire information concerning the hypotheses in which he is interested.

This way of describing the superiority of total evidence (in comparison to its parts) is not a 'pragmatic' justification. It does not point out any practical reasons for accepting new evidence, and probably Ayer would reject it as an unsatisfactory answer to his questions. It seems to me, however, that the concept of information carried by observations concerning a hypothesis, or a set of hypotheses, which was used above, is an interesting concept which should have an adequate *explicatum* in inductive logic. An explication of this kind would make it possible to describe certain essential aspects of scientific method by means of the concepts of inductive logic, e.g. the requirement that our opinions of the credibility of hypotheses should, in order to be rational, have sufficient objective warrant. In Section IV I shall propose a simple *explicatum* for this concept of information.

Ayer's problems are by no means altogether new problems. Similar questions have been raised by others before Ayer in the philosophy of induction. Ayer himself observes that J. M. Keynes ([22], chapter VI) has considered a question similar to the problems discussed in [1]. According to Keynes, the *weight* which an inductive argument has with respect to a hypothesis increases with the number of observational data: "As the relevant evidence at our disposal increases, the magnitude of the probability of the argument may either decrease or increase, according as the new knowledge strengthens the unfavorable or the favorable evidence, but *something* seems to have increased in either case – we have a more substantial basis upon which to base our conclusion" (Keynes [22], p. 71). Keynes expresses the weight of the evidence e with respect to h by $V(h \mid e)$. According to Keynes,

$$(6) \qquad V(h \mid e \,\&\, j) \geqq V(h \mid e)$$

is always true. Above we have assumed that the concept of information provided by e concerning h satisfies a similar condition.

Before Keynes, A. Meinong ([27], pp. 70–71) has distinguished two different aspects or 'dimensions' in a probabilistic argument, its 'intensity' ('Intensität') and its 'reliability' ('Sicherheit'). The intensity of an argument is, according to Meinong, what is usually called probability ("Vermutungsgrad zwischen der gewissen Bejahung und der gewissen Verneinigung", [27], p. 71), whereas the reliability is comparable to, e.g., Keynes's concept of weight (cf. also the discussion in Nitsche [28]).[9] C. D. Broad [4] and Janina Hosiasson [19] have discussed the question as to why we prefer probabilities relative to many data. This question is, according to G. H. von Wright, "one of the fundamental questions in the philosophy of probability" ([33], p. 349). Ayer's problems can be viewed as special cases of this problem. The answer depends on what is meant by this 'preference'; in this paper I will not discuss the ways in which Broad, Hosiasson and von Wright have tried to tackle this problem.

IV THE MEASURE $Q(e|h)$ OF THE INFORMATION PROVIDED BY OBSERVATIONS

I shall propose below an explicatum $Q(e \mid h)$ for the concept of information carried by the evidence e about a hypothesis h. The interpretation of this measure will be discussed in detail in Section V and it will be compared with other measures of information defined in current literature. $Q(e \mid h)$ is defined here as a quantitative concept; it is a function whose arguments are the sentences of a language L and whose range of values is a suitably selected real interval. $Q(e \mid h)$ is assumed to have a value only if h is not logically true; from the intuitive point of view this restriction is unimportant because the information carried by empirical evidence (e.g. results of observations) is relevant to factual hypotheses only. $Q(e \mid h)$ is defined by a set of axioms presented below; the choice of these axioms is justified by the intuitive plausibility of the theorems entailed by them.

The first axiom,

(Q1) If $\vdash e \equiv i$ and $\vdash h \equiv k$,
$Q(e \mid h) = Q(i \mid h)$ and $Q(e \mid h) = Q(e \mid k)$,

is obvious: logically equivalent sentences have the same meaning and the same content; the same information can be formulated in different, logically equivalent ways. The following conditions, too, are intuitively plausible:

(7) If $\vdash e$, $Q(e \mid h) = \min$.

A logically true or tautological statement conveys no information. If all the sentences we can cite as 'evidence' for some hypothesis are logical truths, we actually have no evidence at all. Hence, in this case, the function Q assumes its minimum.

(8) If $\vdash e \supset h$, $Q(e \mid h) = \max$.

If our evidence is strong enough to imply the hypothesis, it conveys all the information yielded by h. Hence $Q(e \mid h)$ assumes in this case its maximum. (7) and (8) imply

(9) If $\vdash e \supset h$ but not $\vdash i \supset k$, $Q(e \mid h) \geqq Q(i \mid k)$.

The choice of the extreme values of $Q(e \mid h)$ is a matter of convention. Perhaps the most natural choice is to take 0 as the minimum and 1 as the maximum. This is expressed by our second axiom

(Q2) $0 \leqq Q(e \mid h) \leqq 1$.

(7) and (8) can now be expressed in the form:

(10) If $\vdash e$, $Q(e \mid h) = 0$;
(Q3) If $\vdash e \supset h$, $Q(e \mid h) = 1$.

(10) follows from (Q2), (Q3) and the axiom (Q4) presented below.

The measure $Q(e \mid h)$ is assumed to be *additive* in the following sense:

(11) $Q(e \,\&\, i \mid h) = Q(e \mid h) + Q(i \mid h)$, if e and i convey no common information about h.

The information shared by e and i is obviously expressed by the disjunction $e \vee i$: e and i may say something more than the disjunction, but both give at least the information carried by $e \vee i$. If $e \vee i$ is logically true, e and i convey no information in common, and if $e \vee i \vee h$ is a logical truth, e and i convey no common information about the subject-matter of h. Hence (11) can be replaced by the axiom

(Q4) If $\vdash e \vee i \vee h$, $Q(e \,\&\, i \mid h) = Q(e \mid h) + Q(i \mid h)$,

which implies (together with (Q1) and (Q3))

(12) $Q(e \,\&\, i \mid h) = Q(e \mid h) + Q(i \mid h) - Q(e \vee i \mid h)$.

In Section III the measure of the information which e gives about the subject-matter of h was assumed to satisfy the condition

(13) $Q(e \,\&\, i \mid h) \geqq Q(e \mid h)$.

This condition is similar to the Condition (6) imposed by Keynes on the concept of the weight of inductive arguments. (13) can also be formulated as follows:

(14) If $\vdash e \supset i$, $Q(e \mid h) \geqq Q(i \mid h)$,

or as

(15) $Q(e \vee i \mid h) \leqq Q(e \mid h)$.

(15) is obviously implied by (Q2) and the axiom

(Q5) $Q(e \vee i \mid h) = Q(e \mid h)\, Q(i \mid h \vee e)$.

(Q1)–(Q5) also imply e.g. the following theorems concerning $Q(e \mid h)$:

(16) $Q(\sim e \mid h) = 1 - Q(e \mid h)$ (from (Q4) by (Q3));

(17) If $\vdash h \supset e$, $Q(e \vee i \mid h) = Q(e \mid h)\, Q(i \mid e)$. ((Q5) by (Q1)).

(Q1)–(Q5) and the theorems implied by them are all the conditions imposed here on the concept $Q(e \mid h)$.

$Q(e \mid h)$ has no value if h is a logical truth. This restriction is plausible, because speaking of the information carried by observations concerning logical truths seems to have no intuitive meaning. Now it may sound equally meaningless to speak of information concerning a logically false hypothesis f, but $Q(e \mid f)$ can nevertheless be given a meaningful interpretation. If 'h' is replaced in the axioms (Q1)–(Q5) by 'f' (which represents a logically false proposition), we obtain the theorems

(F1) If $\vdash e \equiv i$, $Q(e \mid f) = Q(i \mid f)$;
(F2) $0 \leqq Q(e \mid f) \leqq 1$;
(F3) If $\vdash \sim e$, $Q(e \mid f) = 1$;
(F4) If $\vdash e \vee i$, $Q(e \,\&\, i \mid f) = Q(e \mid f) + Q(i \mid f)$;
(F5) $Q(e \vee i \mid f) = Q(e \mid f)\, Q(i \mid e)$.

The *explicatum cont* of the amount of information carried by a sentence which has been widely discussed in current literature on the philosophy of induction satisfies conditions similar to (F1)–(F4) (see e.g. Carnap and Bar-Hillel [10], p. 238; Hintikka [17]; Hempel and Oppenheim [13], pp. 171–172). The measure $1 - cont(e)$ (and hence $1 - Q(e \mid f)$, too) can be shown to satisfy the axioms of probability calculus (see Hintikka [17]), and conversely the definition

(18)
$$Q(e \mid f) = cont(e) = 1 - P(e),$$

where P is a probability measure, gives $Q(e \mid f)$ the properties (F1)–(F4). Hence it is tempting to identify $Q(e \mid f)$ and $cont(e)$ with $1 - P(e)$ and define this measure of information by (18). *Cont*(e) or $Q(e \mid f)$ is a measure of the *absolute* or *total* amount of information carried by e or of the total *content* of e, whereas $Q(e \mid h)$ is a measure of *relative* information (i.e. of the information which e gives about the subject-matter of h).

(F5) can also be expressed in the form

(19)
$$Q(e \mid h) = \frac{Q(e \vee h \mid f)}{Q(h \mid f)}$$

Hence $Q(e \mid h)$ is definable by (F5) in terms of the amount of absolute information. Because $Q(e \mid f)$ is equivalent to *cont*(*e*), this definition can also be written as

$$(20) \qquad Q(e \mid h) = \frac{cont(e \vee h)}{cont(h)}.$$

As I pointed out above (see p. 108), the sentence $e \vee h$ expresses the information shared by *e* and *h*. According to (20), the measure $Q(e \mid h)$ expresses the degree to which the information carried by *h* is shared by *h* and *e*, i.e. $Q(e \mid h)$ says how much *e* conveys of the informative content of *h*. Hintikka [17] has called measures of this kind (in analogy to statistical communication theory) measures of *transmitted information*. The measure $Q(e \mid h)$ of transmitted information can be described intuitively in the following way: it expresses how much information the knowledge of the truth of *e* gives us about the state of affairs described by *h*, if *h* is assumed to be true.

According to (12), (19) and (20) can be expressed in the form

$$(21) \qquad Q(e \mid h) = \frac{Q(h \mid f) + Q(e \mid f) - Q(h \,\&\, e \mid f)}{Q(h \mid f)} = \frac{cont(h) + cont(e) - cont(h \,\&\, e)}{cont(h)}$$

and, according to (18) and (20), $Q(e \mid h)$ can be defined in terms of the probability measure *P* as follows:

$$(22) \qquad Q(e \mid h) = \frac{1 - P(h \vee e)}{1 - P(h)} = 1 - P(e)\left[\frac{1 - P(h \mid e)}{1 - P(h)}\right] = P(\sim e \mid \sim h).$$

(22) gives the concept $Q(e \mid h)$ a simple interpretation in terms of probability. The amount of information provided by *e* about the hypothesis *h* increases when the probability of *e* relative to $\sim h$ decreases, that is to say, *e* gives much information about *h*, if the observational results described by *e* are very *improbable* unless *h* is true. (22) also shows the dependence of $Q(e \mid h)$ on the *a priori* probability of *e*, and hence on the total content of *e*, and on the absolute and conditional content of *h*.

As was indicated earlier, $Q(e \mid h)$ is in certain respects similar to Keynes's concept of the weight of inductive arguments. It satisfies some of the requirements imposed by Keynes on the concept of weight, but not all of them. According to Keynes ([22], p. 73), $V(h \mid e)$ satisfies the condition

$$V(h \mid e) = V(\sim h \mid e). \tag{23}$$

Keynes justifies this requirement by saying that "an argument is always as near proving or disproving a proposition, as it is to disproving or proving its contradictory" ([22], p. 73). Keynes seems to assume that $V(h \mid e)$ assumes its maximum if $P(h \mid e)=1$ *or* if $P(h \mid e)=0$. This assumption is plausible if $V(h \mid e)$ is interpreted as the amount of information concerning the *truth* of h. Because the truth-conditions of a hypothesis determine uniquely the truth-conditions of its negation, information concerning the truth of a hypothesis is obviously information concerning the truth of its negation, too. But $Q(e \mid h)$ does not satisfy the condition (23). $Q(e \mid h)$ can be said to be a measure of information concerning the state of affairs described by h, or the *content* of h, whereas $V(h \mid e)$ corresponds more closely to the amount of information concerning the truth of h. This concept can, however, be given an *explicatum* $Q.E(e \mid h)$ the formal properties of which are similar to those of $Q(e \mid h)$. $Q.E(e \mid h)$ is a special case of a more general concept $Q.E(e \mid H)$, the amount of information concerning the truth of the alternative (that is to say, mutually exclusive and jointly exhaustive) hypotheses in the set $H=\{h_1, h_2, \ldots, h_i, \ldots, h_r\}$. The concept $Q.E(e \mid H)$ is called here simply the amount of information carried by e about a set of hypotheses H. $Q.E(e \mid h)=Q.E(e \mid H)$, if $H=\{h, \sim h\}$. $Q(e \mid h)$ is definable in terms of the amount of absolute information carried by h and e, and $Q.E(e \mid H)$ is defined in the same way in terms of the amount of information given by e and the expected amount of information carried by the hypotheses $h_i \in H$. The prior expectation of the content of the members of H, or the *a priori* content-entropy in H,[10] is

$$E_i(cont(h_i)) = \sum_i P(h_i)\,(1 - P(h_i)) \tag{24}$$

and the expected value of the content of h_i & e is

$$E_i(cont(h_i \,\&\, e)) = \sum_i P(h_i \mid e)\,(1 - P(h_i \,\&\, e)). \tag{25}$$

The amount of information provided by e about the set H can now be defined with the help of the *cont*-measure in the same way as $Q(e \mid h)$:

$$\text{(26)} \qquad Q.E(e \mid H) = \frac{E_i(cont(h_i)) + cont(e) - E_i(cont(h_i \,\&\, e))}{E_i(cont(h_i))}.$$

According to (18), (24), (25) and (26), $Q.E(e \mid H)$ can be defined in terms of the probability measure P as follows:

$$\text{(27)} \qquad Q.E(e \mid H) = 1 - P(e)\left[\frac{1 - \sum_i (P(h_i \mid e))^2}{1 - \sum_i (P(h_i))^2}\right].$$

$Q.E(e \mid h)$ is a special case of (27) in which $H = \{h, \sim h\}$:

$$\text{(28)} \qquad Q.E(e \mid h) = 1 - P(e)\left[\frac{P(h \mid e)(1 - P(h \mid e))}{P(h)(1 - P(h))}\right].$$

(28) satisfies Keynes's requirement (23); it is symmetrical with respect to h and $\sim h$.

If the information carried by the evidence is measured by (27) or (28) instead of $Q(e \mid h)$, the question turns up whether these measures satisfy the Condition (6), i.e. whether the 'superiority' of total evidence can be explicated in terms of these measures. The answer is in the affirmative, that is to say,

$$\text{(29)} \qquad Q.E(e \,\&\, j \mid H) \geqq Q.E(e \mid H)$$

holds for any e, j and H. The proof is simple: (29) holds only if

$$\text{(30)} \quad Q.E(e \,\&\, j \mid H) - Q.E(e \mid H) = \frac{\sum_i (P(h_i \,\&\, e)(1 - P(h_i \mid e)) - P(h_i \,\&\, e \,\&\, j)(1 - P(h_i \mid e \,\&\, j)))}{\sum_i P(h_i)(1 - P(h_i))} \geqq 0,$$

and every term in the sum in the numerator of (30) can easily be shown to be non-negative. Hence (30) is always non-negative, and (29) is satisfied. Hence $Q.E(e \mid h)$, too, satisfies (29), because it is a special case of $Q.E(e \mid H)$.

According to Keynes ([22], p. 76), "weight cannot be explained in terms of probability". The weight and the probability of an argument are independent properties. This is true of the measures $Q(e \mid h)$ and $Q.E(e \mid h)$ in the sense that they do not depend on the *a posteriori* probability of h alone, but on other factors, too. The measures in question can nevertheless be defined in terms of a probability measure P in the way specified in (22) and (28). Apparently Keynes did not think that a definition of this kind is possible. $Q.E(e \mid h)$ differs from $V(h \mid e)$ in other respects, too; these differences will be discussed briefly in Section V. As measures of the information carried by observations they have, however, many important common properties, as we have seen.

Hempel and Oppenheim ([14], part IV:9) have used a measure equivalent to $Q(h \mid e)$ to express the 'systematic power' of a theory h with respect to the evidence e. In our terminology, 'systematic power', defined in this way, is equivalent to the amount of information conveyed by the theory about the results of observations.

V. $Q(e \mid h)$ AND OTHER MEASURES OF TRANSMITTED INFORMATION

The measure $Q(e \mid h)$ was called above a measure of *transmitted* information. I will now discuss some other measures of transmitted information defined in recent literature on the theory of induction and compare these measures with $Q(e \mid h)$. (See especially Hintikka [17].)[11] Suppose that $I(e)$ expresses the amount of total (or absolute) information carried by e. The amount of *relative* information

(31) $$I(h \mid e) = I(h \,\&\, e) - I(e)$$

is the amount of the information which h adds to the information carried by e, that is to say, the amount of information carried by h in excess of the information supplied by e. The amount of *transmitted* information, or the information e conveys concerning the subject-matter of h, is defined by

(32) $$I(eTh) = I(h) - I(h \mid e),$$

and the corresponding normalized measure by

(33) $$I_h(eTh) = I(eTh)/I(h).$$

(32) can be taken to express the amount of information shared by h and e, and the normalized measure (33) indicates the degree to which the infor-

mation carried by h is shared by e (Törnebohm [31], p. 84).[12] $I(e\mathrm{T}h)$ is symmetrical, i.e. $I(e\mathrm{T}h)=I(h\mathrm{T}e)$, whereas the normalized measure is not symmetrical; $I_h(e\mathrm{T}h)=I_e(h\mathrm{T}e)$ is not always true.

Definitions (31)–(33) suffice to show the ambiguity of the presystematic concept of information. But the expression 'information' is ambiguous in another way, too. $I(e)$ has been given two different explicata in inductive logic. One of these explicata is the logarithmic measure of information[13]

$$(34) \qquad inf(e) = -\log P(e),$$

and another the measure *cont* discussed above (see (20)).[14] If the amount of information is expressed in terms of the logarithmic measure *inf*, the measures (32) and (33) of transmitted information are

$$(35) \qquad inf(e\mathrm{T}h) = \log P(h \mid e) - \log P(h)$$

and

$$(36) \qquad inf_h(e\mathrm{T}h) = \frac{\log P(h) - \log P(h \mid e)}{\log P(h)},$$

and if the measure *cont* is used, (32) becomes

$$(37) \qquad cont(e\mathrm{T}h) = 1 - P(h \vee e).$$

The corresponding normalized measure is the measure $Q(e \mid h)$ (22) which has been discussed in detail in Section IV.

The logarithmic measure (35) of transmitted information expresses how much e changes the (prior) probability of h. This change may be either positive or negative. If the probability of h is increased, i.e. $P(h \mid e) > P(h)$, (35) is positive; if $P(h \mid e) < P(h)$, (35) is negative; if $P(h \mid e) = P(h)$, $inf(e\mathrm{T}h)=0$. Hence $inf(e\mathrm{T}h)$ shows whether e is positively or negatively relevant to h or irrelevant to h (in the sense of Carnap [6], chapter VI; see p. 347). Håkan Törnebohm ([31], [32]) has used the normalized measure (36) as a measure of the 'degree of confirmation'[15] of h by e. This measure distinguishes between positive and negative evidence in the same way as (35).

In Section IV we distinguished, in the case of the measure Q of transmitted information, between information concerning a hypothesis h and information concerning a set of hypotheses $H=\{h_1, h_2, \ldots, h_i, \ldots, h_r\}$. This distinction can, of course, be made irrespective of what kind of *explicatum* is used as the measure of the amount of information. The information carried by H is defined as the expected amount of the information carried by the hypotheses in H. The prior expectation of the information carried by the hypotheses $h_i \in H$ is

$$(38) \qquad E_i I(h_i) = \sum_i P(h_i)\, I(h_i)$$

and the expectation of the amount of information provided by H relative to e is

$$(39) \qquad E_i I(h_i \mid e) = \sum_i P(h_i \mid e)\, I(h_i \,\&\, e) - I(e).$$

Transmitted information, or the information carried by e concerning the set H is defined by

$$(40) \qquad I.E(e\mathsf{T}H) = E_i I(h_i) - E_i I(h_i \mid e),$$ [16]

and the corresponding normalized measure is

$$(41) \qquad I_H.E(e\mathsf{T}H) = I.E(e\mathsf{T}H)/E_i I(h_i).$$

If the amount of information is measured with *cont*, (40) becomes

$$(42) \qquad \begin{aligned} cont.E(e\mathsf{T}H) &= E_i\, cont(h_i) - E_i\, cont(h_i \mid e) \\ &= 1 - \sum_i (P(h_i))^2 - P(e)\left(1 - \sum_i (P(h_i \mid e))^2\right); \end{aligned}$$

the normalized measure $cont_H.E(e\mathsf{T}H)$ is identical with the measure $Q.E(e\mathsf{T}H)$ (27) discussed above (see pp. 111–112).

If the amount of information is defined in terms of the *explicatum inf*, the amount of information carried by e about H is

$$(43) \qquad inf.E(e\mathsf{T}H) = E_i\, inf(h_i) - E_i\, inf(h_i \mid e),$$

where

$$(44) \qquad E_i\, inf(h_i) = -\sum_i P(h_i) \log P(h_i)$$

is called the *entropy* of the (prior) probability distribution $(P(h_1), P(h_2), \ldots, P(h_r))$, and

$$(45) \qquad E_i\, inf(h_i \mid e) = -\sum_i P(h_i \mid e) \log P(h_i \mid e)$$

is the corresponding conditional entropy in the set H. (43) has been used by D. V. Lindley [26] to express the amount of information provided by the result e of an experiment about the set of hypotheses $H = \{h_1,$

$h_2, \ldots, h_r\}$.[17] (43) can be either positive or negative, but its expectation

$$(46) \qquad \sum_k \sum_i P(e_k)\, P(h_i \mid e_k)\, inf(e_k \mathrm{T} h_i) = E_i\, inf(h_i) - \sum_k P(e_k)\, (E_i\, inf(h_i \mid e_k)),$$

where $e_1, e_2, \ldots, e_k, \ldots, e_t$ are the possible results of the experiment, is always non-negative (see e.g. Shannon [29], pp. 21–22).[18]

(35), (36) and (43) do not satisfy the conditions (7) and (11)–(15), and the measures (35) and (43) do not satisfy (8) either. No logarithmic measure of information satisfies the axioms (Q4) and (Q5) imposed on $Q(e \mid h)$. Hence the logarithmic concept of transmitted information is quite different from the concept $Q(e \mid h)$ which we have in mind when saying that the total evidence available is at least as informative as any part of it. Especially the condition (13) (and hence (14) and (15), too) is essential for the latter concept of information. Many students of inductive logic have distinguished the measures *inf* and *cont* from each other by calling *inf* a measure of 'surprise value' or of 'unexpectedness' and *cont* a measure of 'substantive information' or '(factual) content' (see Bar-Hillel [3], pp. 307–308; Hintikka [17]). The differences between the concepts of transmitted information based on *inf* and those based on *cont* are in accord with this interpretation. $inf(h)$ measures the prior surprise value or unexpectedness of the truth of h. If e increases the probability of h, it reduces the unexpectedness of h and gives 'positive' information about it (in the present sense); in this case $inf(e\mathrm{T}h)$ is positive. If the probability of h is decreased by e, its surprise value is increased and $inf(e\mathrm{T}h)$ is negative. If the amount of information carried by a sentence means its unexpectedness (or the surprise value of its truth), information transmission means, in the present 'semantical' sense, the reduction of this unexpectedness.

In the same way, (43) indicates the effect of the evidence e on our uncertainty about which of the hypotheses $h_i \in H$ is the true alternative. An even probability distribution indicates considerable uncertainty about the true alternative h_i; in this case the entropy of the distribution is large. If e reduces this entropy (that is to say, if the conditional entropy (45) is smaller than the initial entropy (44)), it reduces our uncertainty about the true alternative and conveys ('positive') information about H; in this case (43) is positive. If e increases the entropy in H, (43) is negative. New observations can either decrease or increase our uncertainty about the true hypothesis h_i. But the expectation (46) of (43) is nevertheless always non-negative; (46) is 0 only if the hypotheses h_i and the results e_k are probabilistically independent, otherwise it is positive. Hence new evidence can always be expected to reduce our uncertainty concerning the true hypothesis (or at least not to increase it). If certainty of the true alternative in H is the 'utility' which an investigator attempts to maximize (i.e. his aim is to minimize the entropy in the set H), making observations is rational; it is, indeed, always rational, if the cost of collecting new evidence is not taken into account. Actually, making observations costs something, and therefore investigators cannot continue indefinitely making new observations. But although new observations can always be expected to reduce the entropy of the distribution we are interested in, the observations actually made may nevertheless increase it. Therefore the 'superiority' of the total evidence in comparison with its parts cannot be explicated in terms of the logarithmic concepts of transmitted information defined above.

The measures $cont(e \top h)$, $cont_h(e \top h)$ (i.e. $Q(e \mid h)$), $cont.E(e \top H)$ and $cont_H.E(e \top H)$ (that is to say, $Q.E(e \mid H)$) of the 'substantive' information conveyed by e or the 'factual content' of e are always non-negative. All these measures satisfy the conditions (13)–(15) imposed on $Q(e \mid h)$. Irrespective of how new evidence affects the unexpectedness of h, or the uncertainty concerning the alternatives $h_i \in H$, it usually increases, but never reduces, the factual information concerning h (or H) available to the investigator. The word 'information' used in Sections III–IV refers to this substantive information or factual content. $cont(e \top h)$ does not satisfy (Q3), but its maximum (for a constant h) depends on the amount of content of h; therefore $cont(e \top h)$ does not satisfy the condition (9) either. When the number of observations increases, the value of $Q(e \mid h)$ approaches 1, whereas $cont(e \top h)$ approaches $cont(h)$. $cont(e \top h)$ is symmetric, in othcr words, $cont(e \top h) = cont(h \top e)$ holds for any e and h, but $Q(e \mid h)$ is not symmetric. The former measure expresses the 'absolute' amount of information carried by e concerning h, whereas $Q(e \mid h)$ expresses the relative amount of transmitted information. If, for instance, $\vdash e \supset h$ but not $\vdash h \supset e$, it sounds plausible to say that e gives more information about h than h gives about e; in this case we have in mind the relative amount of transmitted information. Similarly, (9) holds for the relative amount of transmitted information only. The difference between *cont*.$E(e\mathsf{T}H)$ and the corresponding normalized measure $Q.E(e\mathsf{T}H)$ can be described in the same way. These measures of transmitted information represent different ways of expressing the amount of the objective facts on which the evaluation of the credibility of h is based when e describes the totality of the evidence accepted. Correspondingly, $cont(e_T)$ can be taken to be a measure of the objective basis of the credence-measure P_T used at T. All these measures can be defined in terms of the same measure-function on L as the measures of probability and of uncertainty discussed above.[19]

If we "prefer probabilities relative to many data", a 'good' evidence should presumably be informative in both of the two main senses discussed above. It should have sufficient factual content and it should also reduce our uncertainty concerning the truth of the hypotheses in which we are interested. The relation between these two requirements depends on the measure-function used as the basis for the measures of information and on the set of hypotheses H in which we are interested.[20]

As was pointed out in Sections III and IV, Keynes's concept of the weight of inductive arguments is in many respects similar to the measure $Q.E(e \mid h)$ of transmitted information. According to Keynes, "we may say that the weight of the probability is increased, as the field of possibility is contracted" ([22], p. 77); this is precisely what the measures of transmitted information based on *cont* do express. On the other hand, Keynes also says that the weight of an argument expresses how near the argument is to proving or disproving a proposition ([22], p. 73; cf. also the paragraph cited above, p. 111). Therefore he assumes that $V(h \mid e) = V(k \mid e)$, if $\vdash e \supset (h \equiv k)$, for "in proving or disproving one, we are necessarily proving or disproving the other" (p. 73). But this characterization does not apply to the concept $Q.E(e \mid h)$; it does not satisfy the condition mentioned above.[21] $Q.E(e \mid h)$ cannot be said to express "how near to proving or disproving h we are", if this expression is understood in the way in which Keynes is using it. It might perhaps be suggested that there is the same kind of ambiguity in Keynes's concept of weight as in the presystematic concept of information.

University of Helsinki

BIBLIOGRAPHY

[1] Ayer, Alfred J., 'The Conception of Probability as a Logical Relation', in *Observation and Interpretation* (ed. by S. Körner), Butterworth, London, 1958, pp. 12–17.

[2] Ayer, A. J., Bohm, D. *et al.*, 'Discussion' (of Ayer [1]), in *Observation and Interpretation* (ed. by S. Körner), Butterworth, London, 1958, pp. 18–30.

[3] Bar-Hillel, Yehoshua, 'Semantic Information and Its Measures', in *Language and Information* (ed. by Y. Bar-Hillel), Addison-Wesley, Reading, Mass., 1964, pp. 298–312.

[4] Broad, C. D., 'Critical Notice on J. M. Keynes, *A Treatise on Probability*', *Mind* **31** (1922) 72–85.

[5] Carnap, Rudolf, 'The Two Concepts of Probability', *Philosophy and Phenomenological Research* **5** (1945) 513–532.

[6] Carnap, Rudolf, *Logical Foundations of Probability*, University of Chicago Press, Chicago, 1950.

[7] Carnap, Rudolf, *The Continuum of Inductive Methods*, University of Chicago Press, Chicago, 1952.

[8] Carnap, Rudolf, 'The Aim of Inductive Logic', in *Logic, Methodology and Philosophy of Science* (ed. by E. Nagel, P. Suppes, and A. Tarski), Stanford University Press, Stanford, 1962, pp. 303–318.

[9] Carnap, Rudolf, 'Inductive Logic and Intuition', *The Problem of Inductive Logic* (ed. by I. Lakatos), North-Holland Publ. Comp., Amsterdam, 1968, pp. 257–267.

[10] Carnap, Rudolf and Bar-Hillel, Yehoshua, 'An Outline of the Theory of Semantic Information', in *Language and Information* (by Y. Bar-Hillel), Addison-Wesley, Reading, Mass., 1964, pp. 221–274.

[11] Good, I. J., 'On the Principle of Total Evidence'. *The British Journal for the Philosophy of Science* **17** (1967) 319–321.

[12] Hempel, Carl G., 'Inductive Inconsistencies', *Synthese* **12** (1960) 439–469.

[13] Hempel, Carl G. and Oppenheim, Paul, 'Studies in the Logic of Explanation', *Philosophy of Science* **15** (1948) 131–175.

[14] Hilpinen, Risto, 'On Inductive Generalization in Binary First-Order Languages' (unpublished).

[15] Hintikka, Jaakko, 'A Two-Dimensional Continuum of Inductive Methods', in *Aspects of Inductive Logic* (ed. by J. Hintikka and P. Suppes), North-Holland Publ. Comp., Amsterdam, 1966, pp. 113–132.

[16] Hintikka, Jaakko, 'On Semantic Information', present volume. Also in *Logic, Physical Reality and History, Proceedings of the International Colloquium at the University of Denver* (ed. by W. Yourgrau), The Plenum Press, New York (forthcoming).

[17] Hintikka, Jaakko, 'The Varieties of Information and Scientific Explanation', in *Logic, Methodology, and Philosophy of Science III, Proceedings of the 1967 International Congress* (ed. by B. v. Rootselaar and J. F. Staal), North-Holland Publ. Comp., Amsterdam 1968, pp. 311–331.

[18] Hintikka, Jaakko and Hilpinen, Risto, 'Knowledge, Acceptance, and Inductive Logic', in *Aspects of Inductive Logic* (ed. by J. Hintikka and P. Suppes), North-Holland Publ. Comp., Amsterdam, 1966, pp. 1–20.

[19] Hosiasson, Janina, 'Why Do We Prefer Probabilities Relative to Many Data', *Mind* **40** (1931) 23–32.

[20] Jeffrey, Richard C., *The Logic of Decision*, McGraw-Hill, New York, 1965.
[21] Kemeny, John G., 'Fair Bets and Inductive Probabilities', *Journal of Symbolic Logic* **20** (1955) 263–273.
[22] Keynes, J. M., *A Treatise on Probability*, Macmillan, London, 1921.
[23] Khinchin, A. I., *Mathematical Foundations of Information Theory*, Dover Publications, New York, 1957.
[24] Lenz, John W., 'Carnap on Defining "Degree of Confirmation"', *Philosophy of Science* **23** (1956) 230–236.
[25] Levi, Isaac, *Gambling with Truth*, Alfred A. Knopf, New York, 1967.
[26] Lindley, D. V., 'On a Measure of the Information Provided by an Experiment', *Annals of Mathematical Statistics* **27** (1956) 986–1005.
[27] Meinong, A., 'Kries, Johannes, v.: Die Principien der Wahrscheinlichkeits-Rechnung', *Göttingsche Gelehrte Anzeigen* (1890) 56–75.
[28] Nitsche, Ad., 'Die Dimensionen der Wahrscheinlichkeit und die Evidenz der Ungewissheit', *Vierteljahresschrift für wissenschaftliche Philosophie* **16** (1892) 20–35.
[29] Shannon, Claude E., 'The Mathematical Theory of Communication', in *The Mathematical Theory of Communication* (ed. by C. E. Shannon and W. Weaver), University of Illinois Press, Urbana, 1949, pp. 3–91.
[30] Suppes, Patrick, 'Probabilistic Inference and the Concept of Total Evidence', in *Aspects of Inductive Logic* (ed. by J. Hintikka and P. Suppes), North-Holland Publ. Comp., Amsterdam, 1966, pp. 49–65.
[31] Törnebohm, Håkan, 'Two Measures of Evidential Strength', in *Aspects of Inductive Logic* (ed. by J. Hintikka and P. Suppes), North-Holland Publ. Comp., Amsterdam, 1966, pp. 81–95.
[32] Törnebohm, Håkan, 'On the Confirmation of Hypotheses about Regions of Existence', *Synthese* **18** (1968) 28–45.
[33] von Wright, G. H., 'Broad on Induction and Probability', in *The Philosophy of C. D. Broad* (ed. by P. A. Schilpp), Tudor, New York, 1949, pp. 313–352.

REFERENCES

* This study has been facilitated by a Finnish State Fellowship *(Valtion apuraha nuorille tieteenharjoittajille)*.

[1] See Carnap [6], p. 211. For the concepts of relevance and irrelevance, see Carnap [6], ch. VI.

[2] In spite of these objections to the logical *interpretation* of inductive probability, the probability measures defined in inductive logic (e.g. [7] and [16]) can, of course, be called 'logical' probabilities. Carnap's conception of the interpretation of inductive probability seems to have changed after the publication of [6]. In [6], p. 299, "the choice of an *m*-function is regarded as a purely logical question". According to [7], the choice of an inductive method depends on "performance, economy, aesthetic satisfaction" (p. 55). In [8] Carnap seems to have shifted towards the subjectivistic conception (see especially p. 315).

[3] Perhaps the word 'available' is one source of confusion here.

[4] P_T is relative to X, but, for the sake of brevity, explicit reference to X is omitted here. In [8] P_T is called by Carnap a (rational) credence function, and $P_T(h)$ is called the credence of h for X at T.

[5] The model of the application of inductive logic accepted here is called the *conditionalization* model. According to this model, 'learning from experience' (or rational change in belief) takes place through the conditionalization of the measure P_T to the evidence accepted. The limitations of this model have been discussed e.g. by Jeffrey [20], ch. 11, and Suppes [30], pp. 60–65. In many situations, especially in scientific inquiry, this model seems to me, however, fairly realistic.

The confirmation function P is called *regular*, if $P(i \mid e) = 1$ if and only if i is logically implied by e. The concept of regularity is equivalent to 'strict coherence' or 'strict fairness' (see e.g. Kemeny [21]). If P is strictly coherent, $P_T(i) = 1$ only if i is accepted as evidence at T or logically implied by a sentence accepted as evidence at T. The regularity (or strict coherence) of P can also be defined as follows: $P(i) = 0$, if and only if i is logically false. Carnap imposes the requirement of strict coherence on all credence functions Cr or, in our terminology, probability measures P_T [8], p. 308; [9]. According to Carnap, P_T is regular if $P_T(i) = 0$ only if i is (logically) impossible ([8], p. 308; [9], p. 262). But this requirement is incompatible with the conditionalization model; if X has, in the sense of the conditionalization model, 'learnt something from experience', i.e. accepted evidence, $P_T(e_T) = 1$, although $\sim e_T$ is not logically false (here e_T represents the evidence accepted at T). The requirement of strict coherence should be imposed on the 'initial' credence function P only; all reasonable coherence requirements concerning P_T can be defined in terms of the coherence of P.

[6] This 'paradox of the logical interpretation' has also been pointed out by John W. Lenz in [24], p. 232.

[7] Suggestions for this kind of justification of the rationality of collecting new evidence have been made before Good by other authors. For instance, according to [2], p. 23, footnote 1, U. Öpik produced a similar argument after the discussion on Ayer's paper [1] in the Colston Symposium (1957). In [22], p. 77, Keynes says: "We may argue that, when our knowledge is slight but capable of increase, the course of action, which will, relative to such knowledge, probably produce the greatest amount of good, will often consist in the acquisition of more knowledge."

[8] The word 'information' is here used in its loose, presystematic sense, not in the technical sense in which it is used in information theory.

[9] According to Meinong [27], p. 70, "Vermutungen sind um so weniger Wert, je mehr sie auf Unwissheit basieren, bei Gleichsetzung von Vermutungen aber hat man da, wo diese Gleichsetzung durch unser Wissen gefordert, nicht durch unser Nicht-Wissen bloss gestatted wird, den Idealfall vor sich".

[10] If the measure *cont* used in (24) is replaced with the logarithmic measure of information, we obtain the entropy-expression (44) (see below p. 115). Hence the term 'content-entropy'.

[11] In addition to the measures defined below, Hintikka has defined in [17] some other interesting measures of transmitted information which are not discussed here.

[12] Törnebohm also calls the *explicatum* (36) for (33) the degree of 'information overlap' and 'degree of covering' ([31], p. 84). These interpretations are plausible if the amount of information is explicated in terms of *cont*, e.g. in the case of the measure $Q(e|h)$, but they fail in the case of *inf*, if (32) and (33) are negative. See [31], p. 84.

[13] The logarithms are usually assumed to be to the base 2; the choice of the base is obviously a matter of convention.

[14] The crucial difference between these measures concerns additivity; the *inf*-measures of h_1 and h_2 are additive, if the sentences are probabilistically independent, that is, $P(h_1 \,\&\, h_2) = P(h_1)\, P(h_2)$.

[15] This concept is, of course, different from the concept of degree of confirmation used by Carnap. In this paper, the expression 'degree of confirmation' is used in the Carnapian way, i.e. as a synonym of 'inductive probability'.
[16] Carnap and Bar-Hillel call this measure 'the amount of specification of H through e'. See [10], p. 266.
[17] Lindley does not define (43) in the way in which it is defined here. Lindley's measure is, however, equivalent to (43) if the number of alternative hypotheses $h_i \in H$ is finite.
[18] This is expressed by Shannon's 'fundamental inequality'. Cf. also Khinchin [23], pp. 5–6.
[19] In recent literature on inductive logic it is usually assumed that *inf* and *cont* are defined in terms of the same measure function on L as the degree of confirmation. Such a definition is presupposed here, also, in the Sections IV and V. This assumption is obvious in the case of *inf*, if this measure is interpreted in the way described above. In the case of *cont* it has, however, been questioned by Isaac Levi ([25], especially pp. 164–165).
[20] In many cases the increase of the amount of evidence, i.e. the increase of $Q(e \mid h)$, will necessarily reduce the entropy in H. One interesting case of this kind has been discussed by Hintikka and Hilpinen [18]. Given a suitable probability measure on a monadic first-order language L, the degree of confirmation of one constituent C_w (constituents are the strongest, that is to say, most informative, generalizations in L) approaches 1 when the number of observations increases. Hence the entropy in the set of the constituents of L approaches 0, when the amount of evidence increases.
[21] The condition in question is satisfied by $Q.E(e \mid h)$ only if $cont(h) = cont(k)$.

JUHANI PIETARINEN

QUANTITATIVE TOOLS FOR EVALUATING SCIENTIFIC SYSTEMATIZATIONS*

1. THE SYSTEMATIC POWER OF A SCIENTIFIC HYPOTHESIS

One of the basic functions of a system is to provide information on certain facts or states of affairs about which we are uncertain or agnostic. And to provide information is to reduce uncertainty or agnosticism. For instance, knowing the catalogue system of the University Library in Helsinki helps one to find answers to such questions as whether or not Popper's *Logik der Forschung* is there and, if it is, where among the multitude of books it can be found.

The same feature is also essential to those systems whose aim is to account for or to foretell empirical phenomena. To fulfil their task, such explanatory or predictive systems must provide information about those aspects of the world which perplex us. Such scientific systematizations are centred around empirical hypotheses, usually general laws or theories, but sometimes singular descriptive statements.

For instance, one explanation of the fact that a certain object X floats on water would be that X is a wooden object, and that all wooden objects float on water. Granted the generalization, no additional information is needed to understand why X (which has been verified to be a wooden object) floats.

Take another example. How is it possible that there are large boulders of the crystalline bedrock of Northern Europe in the sedimentary rock districts of Central and Eastern Europe? The most common scientific explanation is that in the Quaternary period the European Continental glacier moved from Scandinavia to Central and Eastern Europe and carried the boulders with it. Here again the hypothesis, this time a singular statement, provides information sufficient to enable one to understand the surprising deposits.

The information offered by explanatory hypotheses does not always release us from all our agnosticism or uncertainty concerning the fact to be explained, however. For instance, suppose we have observed that

J. Hintikka and P. Suppes (eds.), Information and Inference, 123–147.

a subject X chooses in a two-choice learning experiment the alternative A with a relative frequency asymptotically equal to r. A learning theory T may account for this by stating that, in this kind of experimental situation, the asymptotic choice probability of X is π. Here the explanatory hypothesis offers grounds for expecting the result in question, not conclusively, however, but to an extent depending on the closeness of r and π. This feature is present also in the following explanation of the rain-making ceremonies of the Hopi Indians: these ceremonies give the members of the society an occasion to engage in a common activity, and thus have the function of reinforcing their group identity. Although this hypothesis offers information for the understanding of the puzzling phenomenon, we are still left more or less uncertain about the ceremonies: why do the Hopis perform these rather than one of the many other possible activities fulfilling the same function?

The characterization of scientific systematizations as information-providing arguments at once suggests a method for quantifying the systematic power of an empirical hypothesis. All we have to do is to try to find suitable measures of uncertainty, and then to compare the initial uncertainty concerning the facts to be systematized with the uncertainty after embedding the hypothesis into the explanatory or predictive (or postdictive, or some other similar) system. In the present paper we shall explore this method.

One general remark may be in order here. It has often been thought that the powerfulness of a system includes something that could be called its simplicity or conceptual strength or generality, and that this component should be taken into account in measuring systematic power. Thus, if we have two hypotheses equally capable of relieving our agnosticism concerning some state of affairs, the simpler or stronger or more general one should have greater systematic power. Why? Because it can systematize a larger number of facts than the other. But this criticism disintegrates if we distinguish systematic power in a relative or local sense from this notion in some other sense.[1] The local sense of systematic power, the sense that interests us here, is concerned with the explanatory or predictive capacity of hypotheses with respect to certain determinate data, and, it appears, this notion is not essentially connected with such concepts as the simplicity or logical strength of theories.

Is this local sense of systematic power of genuine interest? The power of many theories lies not in their capacity for providing information concerning some restricted number of facts, however large, but in their capacity to produce an unlimited number of facts. Is not such 'fact-generating' power of theories their really important aspect? The answer is, of course, that there are numerous activities in science which aim at the explanation and prediction of some particular facts. A situation of this kind is met even at the early stages of such theorizing as has as its final outcome a powerful theory in the 'fact-generating' sense. For the starting-point of all theorizing is that "some surprising phenomenon P is observed [but] P would be explicable as a matter of course if [some hypothesis] H were true".[2]

2. DEFINITION OF A MEASURE OF SYSTEMATIC POWER

Let h and d be sentences of a well-defined scientific language L such that d is neither logically true nor logically inconsistent. Assume further an uncertainty measure U defined on the sentences of L. Let $U(d|h)$ denote the measure of uncertainty or agnosticism associated with d when the truth of h is taken for granted, and let $U(d)$ be the initial uncertainty of d (i.e., $U(d)=U(d|t)$, where t is a tautological hypothesis).[3]

Then the measure $S(h, d)$ of the systematic power of h with respect to d will have to satisfy the following requirements:

(R1) $\quad S(h, d) = F[U(d), U(d|h)]$;

(R2) $\quad S(h, d) \gtreqless 0 \quad \text{iff} \quad U(d|h) \lesseqgtr U(d)$;

(R3) $\quad S(h, d) = \max S \quad \text{iff} \quad U(d|h) = \min U$;

(R4) $\quad \max S = 1$;

(R5) $\quad S(h, d) \geqslant S(k, d) \quad \text{iff} \quad U(d|h) \leqslant U(d|k)$.

Here $F(x, y)$ is some (so far unspecified) real-valued function of two arguments.

We are looking for a definition as simple as these five requirements permit. It is reasonable to require that $U(s) > \min U$ for any sentence s that is not a truth of logic, and that $\min U = 0$. By (R5) F is monotone decreasing in its second argument. For simplicity, we shall assume

(R6) $\quad F(x, y)$ is a linear function of y.

This enables us to put $F(x, y)=yB(x)+C(x)$. Furthermore, we have by (R2) $F(x, x)=xB(x)+C(x)=0$. Hence $B(x)=-C(x)/x$, and $F(x, y)=$ $=-yC(x)/x+C(x)=C(x)\,(x-y)/x$. By (R3) and (R4) and by the fact that $\min U=0$ we have that $C(x)=1$. Thus we obtain the following measure for the systematic power of h with respect to d:

$$(1) \qquad S(h, d)=\frac{U(d)-U(d \mid h)}{U(d)}.$$ [4]

This measure is additive as follows:

$$S(h \,\&\, k, d)=S(h, d)+S(k, d) \quad \text{iff}$$
$$U(d \mid h)+U(d \mid k)-U(d \mid h \,\&\, k)=U(d).$$

The definition (1) does not impose any restrictions on the minimum value of the S-measure. One possible condition on the minimum is the following:

$$(\text{R7}) \qquad S(h, d)=\min S \quad \text{iff} \quad U(d \mid h)=\max U.$$

On the other hand, it may be argued that the maximal amount of uncertainty that can remain about d when h is given is exactly the initial uncertainty of d. Accordingly, we might require that

$$(\text{R7*}) \qquad S(h, d)=\min S \quad \text{iff} \quad U(d \mid h)=U(d).$$

These two conditions on the minimum of the S-measure are both plausible; yet they are incompatible. This shows that our presystematic concept of uncertainty is likely to be ambiguous and in need of explication.

3. SPECIFIC CONDITIONS OF ADEQUACY FOR SYSTEMATIC POWER

What can be said of the systematic power of h with respect to d, interpreted as the amount of uncertainty h is capable of removing from d, if we try to lay bare what is contained in the concept of the amount of uncertainty? At least two requirements are necessary. (It should be remembered that d is neither logically true nor logically false.) First, if d follows logically from h, h cannot leave any uncertainty concerning d:

$$(\text{S1}) \qquad \text{If} \vdash h \supset d, \quad \text{then} \quad S(h, d)=1.$$

Secondly, if h is a logical truth, it cannot provide any information on d:

(S2) If $\vdash h$, then $S(h, d) = 0$.

These two conditions seem to be perfectly clear. But in trying to determine the minimum for $S(h, d)$, two alternative conditions can be proposed. One is this:

(S3) If $\vdash h \supset \sim d$, $S(h, d) = \min S$.

For if it is a logical truth that $\sim d$ follows from h, then the certainty with which we expect d to be true is minimal and the uncertainty correspondingly maximal. But the uncertainty may refer to another 'intuition': although we cannot expect d to be the case on the basis of h if $\sim d$ is logically implied by h, h and d may still have something in common which reduces our agnosticism concerning the state of affairs described by d. For instance, if we know h, we also know $h \vee d$, and this is a factual sentence. But we also know $h \vee d$ if we know d. In this case, the proper requirement for the minimum is as follows:

(S3*) If $\vdash \sim h \supset d$, $S(h, d) = \min S$.

Conditions (S3) and (R7) correspond to each other in that they both accord with the concept of the uncertainty of a sentence as the unexpectedness of the state of affairs described by the sentence. Conditions (S3*) and (R7*), in contrast, interpret uncertainty as lack of knowledge of the state of affairs described by a sentence.

4. MEASURES OF UNCERTAINTY

On the basis of what has been said in the preceding sections, the measure $U(d|h)$ of the uncertainty of d given h must meet the following conditions:

(U1) $U(d|h) \geqslant 0$;

(U2) if $\vdash h \supset d$, $U(d|h) = 0$;

(U3) if $\vdash h$, $U(d|h) = U(d)$.

Moreover, two alternative conditions can be stated. By (S3) and (R7),

(U4) if $\vdash h \supset \sim d$, $U(d|h) = \max U$,

and by (S3*) and (R7*),

(U4*) if $\vdash \sim h \supset d$, $U(d|h) = U(d)$.

It is usual to quantify uncertainty or information by means of probability measures, perhaps because these are rather familiar and thoroughly investigated.[5] This strategy is followed here, too.

If p is a probability measure (call it a measure of inductive probability) defined on the sentences of L such that it satisfies the usual conditions of adequacy,[6] two simple definitions of $U(d|h)$ which satisfy the requirements (U1)–(U4) are the following:

(i) $\text{unc}_1(d|h) = -\log p(d|h)$;

(ii) $\text{unc}_2(d|h) = 1 - p(d|h)$.

These may be called measures of relative uncertainty or information of d with respect to h. The corresponding measures of absolute uncertainty or information of d are obtained by taking h to be a tautological sentence t:

(iii) $\text{unc}_1(d) = \text{unc}_1(d|t) = -\log p(d|t) = -\log p(d)$;

(iv) $\text{unc}_2(d) = \text{unc}_2(d|t) = 1 - p(d|t) = 1 - p(d)$.

Of these measures, (iv) is the content measure proposed, e.g., by Popper, by Hempel and Oppenheim, and by Carnap and Bar-Hillel;[7] (ii) is a definition of what Hintikka calls conditional content of d with respect to h;[8] and (i) and (iii), respectively, are the relative and absolute information measures called inf by Carnap and Bar-Hillel. (The logarithms are to the base two in the definitions (i) and (iii).)

There are other possibilities for measuring the relative uncertainty of d with respect to h. The following two measures satisfy the requirements (U1)–(U3) and (U4*):

(v)
$$\begin{aligned}\text{unc}_3(d|h) &= p(h \vee d) - p(d)\\ &= p(h) - p(h \,\&\, d)\\ &= p(h)[1 - p(d|h)];\end{aligned}$$

(vi)
$$\begin{aligned}\text{unc}_4(d|h) &= \log[p(h \vee d)/p(d)]\\ &= \log\left\{1 + \frac{p(h)}{p(d)}[1 - p(d|h)]\right\}.\end{aligned}$$

Replacing h by a tautological sentence t we obtain

(vii) $\mathrm{unc}_3(d) = \mathrm{unc}_3(d|t) = 1 - p(d)$;

(viii) $\mathrm{unc}_4(d) = \mathrm{unc}_4(d|t) = -\log p(d)$.

Thus (vii) is identical with (iv), and (viii) correspondingly with (iii). Of these measures, unc_3 has appeared earlier in the literature;[9] it has been termed by Hintikka the incremental content of d with respect to h. The fourth measure, unc_4, is a logarithmic transformation of unc_3.

The intuitive content of the four measures of uncertainty has perhaps not yet become apparent. Some further light on this matter is thrown by the following discussion where we consider the use of these measures in the expression (1) for systematic power.

5. FOUR MEASURES OF SYSTEMATIC POWER

As has been already mentioned, one very common sense of explanation (or prediction) of the occurrence of some phenomenon is that the explaining hypothesis offers information for the anticipation of this occurrence.[10] One way of quantifying this kind of systematic power of a hypothesis is to use the following measure:[11]

(2) $$\mathrm{syst}_1(h, d) = \frac{\mathrm{unc}_1(d) - \mathrm{unc}_1(d|h)}{\mathrm{unc}_1(d)} = \frac{\log p(d) - \log p(d|h)}{\log p(d)}.$$

It can be verified that

(i) $-\infty \leqslant \mathrm{syst}_1(h, d) \leqslant 1$;

(ii) if (a) $p(h \,\&\, k) = p(h)\,p(k)$ and (b) $p(h \,\&\, k|d) = p(h|d)\,p(k|d)$, $\mathrm{syst}_1(h \,\&\, k, d) = \mathrm{syst}_1(h, d) + \mathrm{syst}_1(k, d)$;

(iii) if (a) $p(h \,\&\, k) = p(h)\,p(k)$, (b) $p(h \,\&\, k|d) = p(h|d)\,p(k|d)$, and (c) $p(k \,\&\, d) = p(k)\,p(d)$, then $\mathrm{syst}_1(h \,\&\, k, d) = \mathrm{syst}_1(h, d)$;

(iv) if $p(h \,\&\, d) = p(h)\,p(d)$, $\mathrm{syst}_1(h, d) = 0$;

(v) $\mathrm{syst}_1(h, d) \leqslant 0$ iff $p(h|d) \leqslant p(h)$.

Compare this with a second measure:

$$\text{(3)} \qquad \text{syst}_2(h, d) = \frac{\text{unc}_2(d) - \text{unc}_2(d \mid h)}{\text{unc}_2(d)} = \frac{p(d \mid h) - p(d)}{1 - p(d)},$$

which behaves as follows:

(i) $-p(d)/[1 - p(d)] \leqslant \text{syst}_2(h, d) \leqslant 1$;

(ii) if $\vdash h \vee k$, $\text{syst}_2(h \,\&\, k, d) = [p(h)/p(h \,\&\, k)]\, \text{syst}_2(h, d) + [p(k)/p(h \,\&\, k)]\, \text{syst}_2(k, d)$;

(iii) if (a) $\vdash h \vee k$ and (b) $p(k \,\&\, d) = p(k)\, p(d)$, $\text{syst}_2(h \,\&\, k, d) = [p(h)/p(h \,\&\, k)]\, \text{syst}_2(h, d)$;

(iv) if $p(h \,\&\, d) = p(h)\, p(k)$, $\text{syst}_2(h, d) = 0$;

(v) $\text{syst}_2(h, d) \leqslant 0$ iff $p(h \mid d) \leqslant p(h)$.

The only essential differences appear in the conditions for additivity (ii) and irrelevance (iii). Although in the case of syst_2 the additivity condition is not as clear-cut as one could hope,[12] it shows in effect that the independence of alternative hypotheses from the point of view of systematization is guaranteed if the hypotheses do not have any content in common. In contrast, the former measure requires that the hypotheses must be probabilistically or, as it is often said, inductively independent. This kind of difference is also seen in the irrelevance condition (iii). The difference may be significant. The ordinary multivariate procedures which are very popular in the behavioral sciences take into account, like syst_1, only the probabilistic dependencies between the explaining variables. On the other hand, measures like syst_2 offer a possibility for evaluating the explanatory or predictive capacity of combined hypotheses whose shared content is of more interest than their inductive dependence or independence.

If, on the other hand, we are interested in the common content rather than the inductive dependence or independence between hypotheses and data, there are again two measures at hand for judging the adequacy of systematization. The first,

$$\text{(4)}\qquad \text{syst}_3(h, d) = \frac{\text{unc}_3(d) - \text{unc}_3(d \mid h)}{\text{unc}_3(d)}$$
$$= \frac{1 - p(h \vee d)}{1 - p(d)}$$
$$= 1 - \frac{p(h)[1 - p(d \mid h)]}{1 - p(d)}$$
$$= p(\sim h \mid \sim d),$$

which has been proposed by Hempel and Oppenheim in their classic paper on the logic of explanation,[13] has the following properties:

(i) $0 \leqslant \text{syst}_3(h, d) \leqslant 1$;

(ii) if $\vdash h \vee k$, $\text{syst}_3(h \,\&\, k, d) = \text{syst}_3(h, d) + \text{syst}_3(k, d)$;

(iii) if $\vdash k \vee d$, $\text{syst}_3(h \,\&\, k, d) = \text{syst}_3(h, d)$;

(iv) if $\vdash h \vee d$, $\text{syst}_3(h, d) = 0$;

(v) $\text{syst}_3(\sim h, d) = 1 - \text{syst}_3(h, d)$.

This measure could be called the descriptive power of h with respect to d: it indicates how much knowledge h conveys about d, or to what extent h is capable of describing the state of affairs expressed by d.

The second measure of this kind,

$$\text{(5)}\qquad \text{syst}_4(h, d) = \frac{\text{unc}_4(d) - \text{unc}_4(d \mid h)}{\text{unc}_4(d)}$$
$$= \frac{\log p(h \vee d)}{\log p(d)}$$
$$= \frac{\log \{p(d) + p(h)[1 - p(d \mid h)]\}}{\log p(d)}$$

behaves in the same respect in the following way:

(i) $0 \leqslant \text{syst}_4(h, d) \leqslant 1$;

(ii) if $p[(h \,\&\, k) \vee d] = p(h \vee d)\, p(k \vee d)$, $\text{syst}_4(h \,\&\, k, d) = \text{syst}_4(h, d) + \text{syst}_4(k, d)$;

(iii) if (a) $p[(h \& k) \vee d] = p(h \vee d)\, p(k \vee d)$ and (b) $\vdash k \vee d$, $\text{syst}_4(h \& k, d) = \text{syst}_4(h, d)$;

(iv) if $\vdash h \vee d$, $\text{syst}_4(h, d) = 0$.

Here, again, the two measures differ from each other essentially with respect to their additivity and irrelevance properties, and in somewhat the same way as before. A sufficient condition for two hypotheses to be independent with respect to the power of systematizing given data d is, for syst_3, that the hypotheses have no content in common, but, for syst_4, that the content shared by one of the hypotheses with d is inductively independent of the content shared by the other hypothesis with d. This same feature also distinguishes syst_4 from syst_3 with respect to the irrelevance condition (iii).

One characteristic difference between syst_1 and syst_2, on the one hand, and syst_3 and syst_4, on the other, has immediate bearings on discussions about scientific systematizations. The former measures are in accordance with the claim frequently made in philosophical discussion that a theory explains or predicts certain data the better the more probable it makes them.[14] But the requirement that the best theory is that which 'maximizes likelihood' fails for the second two measures. The concept of explanation seems thus to be most naturally connected with the idea of 'reducing unexpectedness', rather than with that of 'relieving agnosticism' in some substantial sense. Nonetheless, the latter concept seems to be much more appropriate in situations like the explanation of boulder-deposits or of the Hopis' ceremonies cited in the beginning.[15]

Another fact that concerns the behaviour of the first of our four measures of systematic power is of particular interest. In general,

$$\text{syst}_1(h \& k, d) = \text{syst}_1(h, d) + \text{syst}_2(k, d) - R(h, k, d),$$

where

$$R(h, k, d) = \frac{[\text{unc}_1(h) - \text{unc}_1(h \mid k)] - [\text{unc}_1(h \mid d) - \text{unc}_1(h \mid k \& d)]}{\text{unc}_1(d)}.$$

The term $R(h, k, d)$ can be negative, and, consequently, the systematic power of two hypotheses taken together may exceed the sum of the systematic powers of the individual hypotheses. The situation is here quite analogous to the analysis of variance with a positive interaction term[61].

Interestingly enough, a sufficient condition for $R(h, k, d)$ not to be positive is that $p(h \,\&\, k) = p(h)\, p(k)$, that is, that h and k are inductively independent. The exactly corresponding condition occurs in ordinary analysis of variance, and this explains why the interaction effect is there as though the interaction term never were non-negative.

6. SYSTEMATIC POWER, TESTABILITY, AND SIMPLICITY

In an essay entitled 'Truth, Rationality, and the Growth of Knowledge',[17] Karl Popper gives a criterion of what he calls the relative potential satisfactoriness, or potential progressiveness, of scientific theories. This criterion "characterizes as preferable the theory which tells us more; that is to say, the theory which contains the greater amount of empirical information or *content*; which is logically stronger; which has the greater explanatory and predictive power; and which can therefore be *more severely tested* by comparing predicted facts with observations". Popper claims further that "all these properties ... can be shown to amount to one and the same thing: to a higher degree of empirical *content* or of testability".[18] In other words, Popper wants to identify essentially the concepts of informative content, logical strength, testability, and systematic (or explanatory and predictive) power of empirical hypotheses. This thesis is apt to arouse some questions.

Let us start with the notion of testability. Intuitively, the stronger the assertions a theory makes about the world, that is, the more possibilities for the 'real state of affairs' it forbids, the more liable it is to be refuted by experience. Thus, if we consider experimental facts as a test of a hypothesis h, the severity of this test is the greater the more unknown or uncertain these facts are in themselves, and the more obvious they seem to be in the light of h. This idea of testability seems to be the very same idea used here in defining the notion of the systematic power of hypotheses. But our measures of systematic power do not accord with Popper's intention. To see this, let $C(h|b)$ measure the content that h is capable of adding to our body of knowledge b, and $S(h, d|b)$ the systematic power of h, given b, with respect to some fact d not belonging to b. The following assertion would have to be accepted by Popper:

$$\text{(I)} \qquad S(h, d|b) \geqslant S(k, d|b) \quad \text{iff} \quad C(h|b) \geqslant C(k|b).$$

But if the systematic power is measured by any one of the four explicata presented above and the content by any of the expressions proposed by Popper (they are $1-p(h|b)$ or our unc_2, $1/p(h|b)$, and $-\log p(h|b)$ or our unc_1), or by our unc_3 or unc_4, this assertion is not valid in general; only the special case where k is a logical consequence of h satisfies (I).

The same is true of another thesis included in the above quotation from Popper, namely that

$$\text{(II)} \qquad S(h, d|b) \geqslant S(k, d|b) \quad \text{iff} \quad M(h) \geqslant M(k),$$

where M is a measure of the logical strength of the sentences in L. For instance, $1-p(h)$ and $-\log p(h)$ are proper measures of the logical strength of h. Again, however, (II) is generally valid only in cases where h logically implies k.[19]

To make valid assertions out of (I) and (II), the quantity S must be replaced by some other measure. By which one? The answer is not quite clear, but Popper seems to be after a notion of testability or systematic power which is not so strictly local as our concept of systematic power. The precise measures he has formulated,[20] if used in (I) and (II), succeed in satisfying them not only when k follows logically from h but also when one can logically derive from h a test statement which is initially less probable than any statements derivable from k. But, apart from the deductive case, these measures do not satisfy (I) and (II).[21] It seems that even Popper's own measures do not serve his intentions. One reason for this is clear: he does not carefully distinguish the local testability (or explanatory power) of theories from the global sense of this concept.

The fact that (II) cannot be accepted as a valid condition further illustrates the relation between the notions of systematic power (in its local sense) and simplicity. The simplicity of a theory has sometimes been identified with its logical strength.[22] To what extent this gives an adequate understanding of the motley notion of simplicity is not considered here. It suggests, however, that the simplicity of scientific hypotheses is something which is not associated with the capacity of these hypotheses to systematize facts of experience (a point made at the beginning of this paper). Thus simplicity, along with logical strength and informative content, is a desideratum different from systematic power, and it should be given an independent position in decisions where the choice between theories is concerned.

7. SOME STATISTICAL MEASURES OF SYSTEMATIC POWER

An interesting analogue of our expression (1) for systematic power appears in the statistical methodology of the behavioral sciences.

Consider two discrete (quantitative) random variables X and Y with respective values $x_1, x_2, \ldots$ and $y_1, y_2, \ldots$.[23] The mean of Y is defined by

$$E(X) = \sum_{y} \Pr[Y = y] \cdot y,$$

and the variancc by

$$\sigma^2(X) = E[X - E(X)]^2 = E(X^2) - [E(X)]^2.$$

The conditional variance of Y when the distribution of X is known is

$$\sigma^2(Y|X) = \sigma^2(Y)[1 - \rho^2(X, Y)], \tag{12}$$

where $\rho(X, Y)$ is the coefficient of linear correlation between X and Y. This expression shows the variability in Y not accounted for by linear regression. Thus,

$$\rho^2(X, Y) = \frac{\sigma^2(Y) - \sigma^2(Y|X)}{\sigma^2(Y)} = 1 - \frac{\sigma^2(Y|X)}{\sigma^2(Y)} \tag{13}$$

shows the relative reduction in the variance of Y accomplished by the use of a linear prediction. The analogy between (13) and (1) is obvious; here, the 'uncertainty' is measured by variance rather than by information.

However, in some situations it is possible to determine a statistical measure of systematic power which is based on a concept of information. For instance, let X and Y be continuous variables with a normal distribution function. If the equation for the Shannonian information

$$C(Y) = -\int_{-\infty}^{+\infty} f(y) \log f(y)\, dy \tag{14}$$

is substituted in the density function

$$f(y) = \frac{1}{\sigma(Y)\sqrt{2\pi}} \exp\left(-\frac{1}{2}\frac{(y - E(Y))^2}{\sigma^2(Y)}\right)$$

for the normal distribution, one obtains

$$C(Y) = \log \sigma(Y) + \tfrac{1}{2} \log 2\pi e.$$[24]

An expression for conditional information is obtained from (12)

$$C(Y|X) = \log \sigma(Y) \sqrt{1 - \rho^2(X, Y)} + \tfrac{1}{2} \log 2\pi e.$$

Thus the quantity

$$\text{(15)} \qquad \frac{C(Y) - C(Y|X)}{C(Y)} = 1 - \frac{\log \sigma(Y) \sqrt{1 - \rho^2(X, Y)} + \tfrac{1}{2} \log 2\pi e}{\log \sigma(Y) + \tfrac{1}{2} \log 2\pi e} = 1 - \frac{\log \sigma(Y|X) + k}{\log \sigma(Y) + k}$$

measures the systematic power of X with respect to Y. Again the analogy between (15) and (1) (and between them and (13)) is obvious.

One could try to replace the Shannonian measure (14) by a function corresponding to the content measure

$$\int_{-\infty}^{+\infty} f(y) [1 - f(y)] \, dy,$$

but this is of no use in the case of normal distribution: the information value will always be zero. This seems to be one reason why the content measure of semantic information has no generally accepted analogue in statistics.

Our main purpose in mentioning these statistical measures of systematic power has been to emphasize the basic intuitive feature common to all of them. They are all concerned with effects to reduce the uncertainty connected with the phenomenon to be explained or predicted. A more penetrating analysis of the central problems connected with these measures cannot be attempted here. One of the most intriguing of these further problems is the estimation of the population mean and variances as well as Shannonian information from observations concerning a finite sample of individuals. Some questions concerning the estimation of parameters are discussed elsewhere – in a somewhat different context but still in a form relevant to the estimation of distribution parameters.[25]

There is another point to be made here. In the behavioral sciences, we often speak of 'explaining a variable by means of other variables'. The basic idea is to reduce the variance of the variable (or a set of variables) to be explained by means of a set of other variables. A hypothesis about the probability distribution of a variable in a population is a probabilistic generalization. Here is a very clear and impressive example of an explanation of statistical generalizations which does not follow the ordinary deductive pattern. This shows a claim put forward by Nagel to be wrong. Nagel says: "The expectation is ... not unreasonable that the formal structure of explanations of statistical generalizations in the social sciences is also deductive..."; "This expectation is indeed fully confirmed, and in consequence, nothing further needs to be said concerning the overall pattern exhibited by explanations for statistical social generalizations".[26] The explanation of variables by variables is not in accordance with Nagel's expectation.

8. KINDS OF SYSTEMATIZATIONS

A scientific systematization is called deterministic or probabilistic according to whether the laws in its premises are deterministic or probabilistic. A probabilistic hypothesis is distinguished from a deterministic one essentially by the fact that it contains probability functors.[27] Typical probabilistic generalizations describe the population distribution of a variable or of a set of variables.

On the other hand, a distinction has to be made between deductive and inductive systematization. Thus we have four kinds of scientific systematization: deductive-deterministic, deductive-probabilistic, inductive-deterministic and inductive-probabilistic, all of which are exemplified in various branches of science. Undoubtedly the most common type of systematization in the behavioral and social sciences is the inductive-probabilistic variety.[28]

9. THE EMPIRICAL ADEQUACY OF SYSTEMATIZATIONS

The evaluation of the empirical adequacy of a scientific systematization presupposes an answer to two questions: (i) What is the systematic power of the hypotheses occurring in premises? (ii) To what extent are these hypotheses empirically confirmed? The former question does not, of course, cause any trouble in the case of deductive systematization. Only in evaluating inductive systematizations do certain typical problems arise.

One question is centered around the likelihoods $p(d|h)$. If h is a probabilistic generalization (i.e., a statement about the population probability distribution), one special condition may be stated: the inductive probability of the description d of a sample frequency distribution, given a probabilistic hypothesis h, must be equal to the empirical probability which is specified by the distribution functor stated in h. Thus, if h is 'For each individual in the population, the probability of its having the property Q if it has the property P is π' and d is 'a has Q' (and the background information includes the statement that a has P), then $p(d|h)=\pi$.[29] This condition remains valid if d is a probabilistic generalization.

A corollary concerning relative sample frequencies is that the inductive probability of a sample frequency distribution has a maximum if the population distribution stated by h specifies probabilities which are equal, or as nearly equal as possible, to the relative sample frequencies.[30]

Another question concerns the prior probabilities $p(h)$ that are needed in the case of syst_3 and syst_4: how to bet rationally on general assertions, especially if they are probabilistic?

The same question is relevant when we have to specify the posterior inductive probabilities $p(h|e)$ which indicate the degree of confirmation of h with respect to e, the totality of the available evidence for h. One possibility of approaching this problem has been shown by de Finetti.[31] The implications of his answer are not discussed here; we just point to one requirement concerning posterior inductive probabilities: when the number of observations included in e increases, the posterior inductive probability of that hypothesis which describes the true state of affairs should converge to one.

The reason for seeking evidential support for a systematization is, of course, the desire to avoid systematizations which are completely *ad hoc*. To characterize an explanation offered by some hypothesis as *ad hoc* implies that the only evidence in support of it is the explanandum phenomenon, and usually this gives poor grounds for claiming empirical soundness for the explanation.[32] Thus, in the case of explanation, the total evidence e must contain at least some information b in addition to d. But if d is not a known fact but a fact to be predicted, it will not be included in the total evidence e. (In what follows, we shall write the posterior probability as $p(h|e)$ understanding that e may consist either (i) of two parts, of the data d and of the confirmatory evidence b, or (ii) of b alone. Similarly, systematic power is denoted generally by $S(h, d|b)$, where b may be a sentence such that b & d is equivalent to e, or b alone may be equivalent to e.)

10. ON THE ACCEPTANCE OF SCIENTIFIC SYSTEMATIZATIONS

When should we accept a scientific systematization, that is, some explanation or prediction? In order to be able to answer this question, we must qualify it. Do we mean by acceptance that a systematization is satisfactory or empirically adequate, or that it is the most adequate among a set of alternatives or, as it is perhaps most reasonable to think, that it is the most adequate among satisfactory alternatives? We shall now try to formulate some rules of acceptance as answers to these questions, characterizing the problem as a decision-making situation.

A systematization concerning some given data d is regarded as satisfactory if and only if it is confirmed highly enough and has sufficient systematic power; and we should accept a hypothesis in so far as it is satisfactory:

RULE 1: Accept a hypothesis h if and only if

$$p(h|e) \geqslant 1-\varepsilon \text{ and } S(h, d|b) \geqslant 1-\varepsilon'.$$

It is reasonable to take ε to be a fraction greater than 0 and less than .50, and ε' a fraction less than 1.

The acceptance class specified by Rule 1 may include several hypotheses. To find in some sense the best hypothesis requires another kind of consideration.

One possible procedure is as follows. Form an index that reflects the two objectives in question, i.e., the degree of confirmation and the systematic power, and base your decision on this index. One measure that can be used for this purpose is simply the product of the objectives. The acceptance rule would then be

RULE 2: Accept a hypothesis h in so far as there is no alternative k such that

$$p(h|e)\, S(h, d|b) < p(k|e)\, S(k, d|b).$$

Putting somewhat differently, h should be accepted only if there exist no alternative hypothesis k such that

$$\frac{p(e \mid h)}{p(e \mid k)} < \frac{p(k) S(k, d \mid b)}{p(h) S(h, d \mid b)}.$$

That is, what is important in deciding between rival hypotheses is their likelihood with respect to the totality of the available evidence weighted against, as Lindley puts it, the seriousness of rejection.[33] In general, a hypothesis is the more serious the higher its prior confirmation and the more one can lose in rejecting it; here, the loss function is determined by the systematic power of hypotheses.

Rule 2 may be said to exemplify decision-making under certainty. All that is needed for the decision is known: there is no risk involved in determining the posterior probability and the systematic power of each of the alternative hypotheses.

It may be argued, however, that this kind of thinking does not reflect the real aims of science. The question is not of finding *probable* hypotheses with high systematic power; rather, what is aimed at are *true* hypotheses capable of systematizing certain data. But it is not known which of the alternatives, if any, is the true one; therefore, some uncertainty or risk is involved in the acceptance situation.[34]

Under risk, the decision is based on the expected gain of accepting a hypothesis h:

$$EG(h) = p(h \mid e)\, G(h) + p(\sim h \mid e)\, G(\sim h).$$

If h is accepted and it is true, we gain its systematic power, but if h is false and $\sim h$ true, we lose the systematic power of $\sim h$. The expected gain of accepting h is consequently

$$(6) \qquad EG(h) = p(h \mid e)\, S(h, d \mid b) - p(\sim h \mid e)\, S(\sim h, d \mid b).$$

Thus our third rule of acceptance is as follows:

RULE 3: Accept a hypothesis h as long as there is no alternative k such that $EG(h) < EG(k)$.

The most straightforward recommendation for a rule of acceptance is obtained when the systematic power is measured by $syst_3$. It can be shown that (6) is maximized by a hypothesis h which maximizes the difference

$$p(h \mid d \,\&\, b) - p(h \mid b).$$

If h is a weak hypothesis, i.e., a disjunction of mutually exclusive strong hypotheses h_i, the difference can be expressed as the sum

$$\sum_i [p(h_i|d \& b) - p(h_i|b)].$$

This sum is maximal when the addenda are all strong hypotheses h_i such that $p(h_i|d \& b) - p(h_i|b) > 0$, or, equivalently, such that $p(d|h_i \& b) > p(d|b)$. In other words, the condition imposed by Rule 3 is satisfied by the disjunction of all those strong components that increase the likelihood of the data.[35] It does not make any difference whether we think the question is of explanation or of prediction.

Precisely the same result is obtained when the measure syst_2 is used and prediction is understood to be in question, but in the case of explanation the situation is much more complicated. A more detailed consideration of the special rules of acceptance obtained by using different measures of systematic power in (6) is omitted here. Instead, two general observations concerning Rules 2 and 3 are worth making.

In the first place, these two rules do not lead in general to the acceptance of the same hypothesis (or to the same acceptance class of hypotheses). Thus the characterization of the acceptance situation as decision-making under certainty or under risk is not just a matter of taste; the arguments which lead to this characterization are of primary importance and have far-reaching implications.

In the second place, it may happen that the hypothesis (or there may be several) accepted according to these rules is not satisfactory. It therefore seems necessary to combine these rules with Rule 1. Moreover, to enable one to decide between several alternatives of maximal expected gain, the criterion of simplicity (or strength) is added. The following Rule 5 is suggested by these reflections:

RULE 5: Accept a hypothesis h if and only if (i) $p(h|e) \geqslant 1 - \varepsilon$ and $S(h, d|b) \geqslant 1 - \varepsilon'$, and (ii) there is no alternative k such that $EG(h) < EG(k)$. If there are several hypotheses in the acceptance class determined by (i) and (ii), accept the simplest (strongest) one.

Alternatively, the condition expressed by Rule 2 can be substituted for (ii).

One advantage of the five rules presented above is that they are quite general with respect to the hypotheses under consideration: they need not be mutually exclusive nor form an exhaustive set. We can compare strong and weak hypotheses with each other. One important qualification has to be made, however: if these hypotheses are probabilistic, we have to impose certain restrictions, for the following reason.[36]

Consider the following two arguments:

(10) If *a* obtains a high score on the verbal scale of WAIS (Wechsler's Adult Intelligence Scale), it is very certain that he will pass his graduate examination.

(11) If *a* obtains a low score on McClelland's test for the achievement motivation, it is very certain that he will fail to pass his graduate examination.

In both of these arguments two probabilistic generalizations are used; let these be h_1='$Pr[Y=y_1|X_1=x_1]=\pi$', where Y denotes the variable 'to take one's graduate examination' with the values y_1='pass' and y_2='fail to pass', and X_1 denotes the variable 'score on VS of WAIS' with the values 'high' (x_1) and 'low' (x_2), and h_2='$Pr[Y=y_2|X_2=x_2]=\phi$' where X_2 is the variable 'score on McClelland's test'; Pr denotes the empirical probability functor. Let d be the sentence '$Y(a)=y_1$'. The phrase 'is very certain' is to be understood in (10) that the uncertainty $U(d|h_1)$ is very low, and correspondingly in (11) that $U(\sim d|h_2)$ is very low, such that $S(h_1, d)\geqslant 1-\varepsilon'$ and $S(h_2, \sim d)\geqslant 1-\varepsilon'$. Suppose, furthermore, that both h_1 and h_2 have sufficient empirical support. Then condition (i) of Rule 5 is satisfied by both of the hypotheses h_1 and h_2; that is to say, h_1 offers a satisfactory systematization for d and h_2 for $\sim d$.

And further, it is perfectly possible that the other condition (ii) stated in Rule 5 is also fulfilled by h_1 and h_2; i.e., it is not the case that $EG(h_1)<$ $<EG(k)$ for some alternative k available for systematizing d, and not $EG(h_2)<EG(k')$ for some k' available for systematizing $\sim d$. Therefore, using Rule 5, a systematization can be accepted both for d and for its negation $\sim d$, in effect, for some occurrence of a phenomenon as well as for the non-occurrence of this. This is undesirable.

However, it is not difficult to see how this kind of ambiguity can be avoided: we have to take into account the joint effect on Y of both experimental variables X_1 and X_2; i.e., hypotheses of three-variate type have to be formulated instead of two-variate ones. These hypotheses specify different conditional distributions of the form $Pr(Y|X_1X_2)$, and no ambiguity of the above kind can arise when the systematization under consideration is restricted to the two variables X_1 and X_2. In general, if the aim is to systematize certain results d by means of a set $X=\{X_1, X_2, ..., X_q\}$ of variables, one must take into account the joint effect of all of the variables in X. This requirement calls for the use of systematizations of multivariate type, a requirement often put forward by practicing statistical methodologists.

National Research Council for the Humanities, Helsinki, Finland

REFERENCES

* Many suggestions and comments by Prof. Jaakko Hintikka, Mr. David Miller, Dr. Risto Hilpinen, Dr. Raimo Tuomela, and Mr. Kimmo Linnilä have been of great value in preparing this paper.

[1] Cf. the distinction between local and global theorizing in Jaakko Hintikka's paper 'The Varieties of Information and Scientific Explanation', in *Logic, Methodology and Philosophy of Science III, Proceedings of the 1967 International Congress* (ed. by B. van Rootselaar and J. F. Staal), Amsterdam 1968, pp. 151–71. This paper contains many suggestive ideas concerning the use of measures of information in scientific systematizations.

[2] This line of thought appears in Peirce's retroductive inference, as presented in N. R. Hanson, *Patterns of Discovery*, Cambridge 1958, p. 86. Similarly, Karl Popper writes: "What is the general problem situation in which the scientist finds himself? He has before him a scientific problem: he wants to find a new theory capable of explaining certain experimental facts; facts which the earlier theories successfully explained; others which they could not explain; and some by which they were actually falsified" (Popper, *Conjectures and Refutations*, New York 1962, p. 241).

[3] To be accurate, some phrase like 'with respect to everything else known' should be added here. That is, if the 'background knowledge' is b, the measures should read $U(d|h\&b)$ and $U(d|b)$. The background knowledge may contain other hypotheses accepted at a given time as well as descriptions of observational results different from d; for instance, the antecedent conditions for inferring d from h should be included in b. Following the tradition, this background knowledge is usually left implicit in the expressions under consideration.

[4] If the background information b is written explicitly, expression (1) obtains the following form:

$$S(h, d|b) = \frac{U(d|b) - U(d|h\&b)}{U(d|b)}.$$

[5] For instance, Karl Popper (see *The Logic of Scientific Discovery*, London 1959, Appendix *IX), Carl G. Hempel and Paul Oppenheim ('Studies in the Logic of Explanation', *Philosophy of Science* **15** (1948) 135–75), Rudolf Carnap and Yehoshua Bar-Hillel ('An Outline of a Theory of Semantic Information', *Technical Report No. 247 of the Research Laboratory of Electronics, MIT*, 1952; reprinted in Y. Bar-Hillel,

Language and Information, Reading, Mass., 1964, pp. 221–74), and J. G. Kemeny ('A Logical Measure Function', *Journal of Symbolic Logic* **18** (1953) 289–308) proceed in this way. One notable exception to this tradition is Isaac Levi (see Levi, *Gambling with Truth*, New York 1967, and especially his paper 'Information and Inference', *Synthese* **17** (1967) 369–91).

[6] By the 'usual conditions of adequacy' we mean in the first place such restrictions on the measure p as are implied by defining p as a fair-betting ratio; see e.g. Kemeny's essay 'Carnap's Theory of Probability and Induction', in *The Philosophy of Rudolf Carnap* (ed. by P. A. Schilpp) La Salle, Ill., 1963, pp. 711–38. What else should be required of p in order for it to offer appropriate tools for defining measures of uncertainty is left open to a large extent. One group of measure functions (such as give for general sentences a zero probability in an infinite domain) is argued to be inadequate for this purpose by Hintikka and Pietarinen ('Semantic Information and Inductive Logic', in *Aspects of Inductive Logic* (ed. by K. J. Hintikka and P. Suppes), Amsterdam 1966, pp. 96–112). Certain general difficulties and open questions concerning the inductive probabilities should perhaps be mentioned here. The main difficulties are the following (see Kemeny, *loc. cit.*, and also Carnap's 'Replies and Expositions', in the same volume): (i) how to extend the methods of determining inductive probabilities for sentences from such simple languages as the monadic predicate calculus to languages with more than one family of predicates of first and higher order; and (ii) how to find satisfactory inductive probabilities for general propositions. Kemeny (as well as Carnap) points out that the extension mentioned under (i) does not cause new problems in principle, though it does mean vast and difficult mathematical work. The problems under (ii), on the other hand, raise new questions. One answer has been offered by Hintikka (see his 'Two-Dimensional Continuum of Inductive Methods', in *Aspects of Inductive Logic*, pp. 113–32) for a monadic first-order language. Carnap in his 'Replies' (p. 977) mentions that he also has a – so far unpublished – solution to the problem.

[7] See note 5 for the references.

[8] In 'The Varieties of Information and Scientific Explanation'. In his *Conjectures and Refutations*, p. 390, Popper seems to have the same measure in mind; similarly, and more explicitly, in 'Theories, Experience and Probabilistic Intuitions', in *The Problem of Inductive Logic* (ed. by Imre Lakatos), Amsterdam 1968, p. 287.

[9] E.g., in Carnap and Bar-Hillel, *op. cit.*

[10] This sense of explanation is illustrated by what Hempel regards as a general condition of adequacy for any rationally acceptable explanation of a particular event. To quote Hempel, "any rationally acceptable answer to the question 'Why did event X occur?' must offer information which shows that X was to be expected – if not definitely, as in the case of D-N explanation, then at least with reasonable probability. Thus, the explanatory information must provide good grounds for believing that X did in fact occur; otherwise, that information would give us no adequate reason for saying: 'That explains it – that does show why X occurred.' "(C. G. Hempel, *Aspects of Scientific Explanation*, New York 1965, pp. 367–8).

[11] The idea of using the logarithmic measure of transmitted information as the basis for defining expressions for the explanatory power is not new. It is discussed by Popper in *Logic of Scientific Discovery*, p. 403; similarly, I. J. Good argues that this measure is "an explication for 'explanatory power' but not for corroboration" (see I. J. Good, 'Weight of Evidence, Corroboration, Explanatory Power, Information and the Utility of Experiments', *Journal of the Royal Statistical Society*, B, **123** (1960) 319–31).

[12] This remark also concerns the measure of explanatory power

$$E(h, d) = \frac{p(d|h) - p(d)}{p(d|h) + p(d)}$$

proposed by Popper (e.g., p. 400 in *Logic of Scientific Discovery*).
[13] For the reference, see note 5.
[14] Perhaps this requirement is made most explicitly by Joseph Hanna on p. 13 of his paper 'A New Approach to the Formulation and Testing of Learning Models', *Synthese* **16** (1966) 344–380. Hanna's ideas come very close to the approach presented here: he relies entirely, however, on statistical concepts of probability and uncertainty. (For a further comparison of Hanna's approach and the one sketched here, see J. Pietarinen and R. Tuomela, 'An Information Theoretic Approach to the Evaluation of Behavioral Theories', *Reports from the Institute of Social Psychology, Univ. of Helsinki*, No. 2, 1968.) In other standard references to the characterization of what we shall call inductive systematization and what is variously called statistical (e.g. by W. C. Salmon in 'The Status of Prior Probabilities in Statistical Explanation', *Philosophy of Science* **32** (1965) 137–46), or probabilistic (e.g. by Nagel in *Structure of Science*), or inductive (by Hempel in *Aspects of Scientific Explanation*) explanation or prediction is based on the idea that the explanans makes the explanandum highly probable.
[15] A particularly interesting field of application for the measures $syst_3$ and $syst_4$ is offered by historical research. It is proper for historians to ask how much the evidence material we have to hand has common content with such and such narrative.
[16] That there is a one-to-one correspondence between the structure of statistical informational analysis with that of the usual analysis of variance has been shown by Garner and McGill in their paper 'Relation between Uncertainty, Variance, and Correlational Analysis', *Psychometrica* **21** (1956) 219–28. Since the unc_1-measure is quite analogous to the statistical (Shannonian) measure of information, it is not surprising to find terms similar to those in the variance analysis in our context too.
[17] In *Conjectures and Refutations*, pp. 215–50.
[18] *Loc. cit.*, p. 217. The same ideas can be found in many of the earlier publications of Popper, esp. in *Logic of Scientific Discovery*, as is indicated by Popper himself on the page cited.
[19] The literature on scientific explanation has an argument (relying on Popper's ideas) which is relevant here. H. E. Kyburg's theorem put forward in his discussion 'On Salmon's Paper', *Philosophy of Science* **32** (1965) 147–51, p. 148, says that the explanatory powers of two theories are equal if and only if the prior probabilities of these theories are equal. This argument is built on the premise that explanatory power is a monotone increasing function of the logical strength of theories, that is, on the idea that the explanatory power of some theory is a monotone increasing function of a measure of the possibilities which are excluded by the theory. But it is not valid if the premise is so qualified that it corresponds to the intuition behind our concept of explanatory power: the explanatory power of a theory with respect to an explanandum is a monotone increasing function of a measure of possibilities excluded by the theory *from the possibilities allowed by the explanandum.*
[20] See e.g. *Conjectures and Refutations*, pp. 390–1, and *Logic of Scientific Discovery*, pp. 400–3.
[21] Consider, for example his measure

$$\text{(T)} \qquad \frac{p(d|h\&b) - p(d|b)}{p(d|h\&b) + p(d|b)}.$$

If now from h we can deduce a fact d, and from k a fact f such that $p(d) \leqslant p(f)$, h gives a higher value to (T) than k. It is then not difficult to show that $p(h \mid b) \leqslant p(k \mid b)$ and that $p(h) \leqslant p(k)$; hence (I) and (II) are valid. But if h and k make the test statements only more or less probable the corresponding proof cannot be stated.

[22] By Popper and e.g. by J. G. Kemeny in 'Two Measures of Complexity', *Journal of Philosophy* **52** (1955) 131–75.

[23] For the notion of random variable, see e.g. W. Feller, *An Introduction to Probability Theory and its Applications*, Vol. I, 2nd ed., New York 1957, Chapter IX.

[24] See C. E. Shannon and W. Weaver, *The Mathematical Theory of Communication*, Urbana, Ill., 1949, p. 56.

[25] See Pietarinen and Tuomela, *op. cit.*

[26] *Structure of Science*, p. 139.

[27] Often the empirical (physical) probabilities which occur in probabilistic hypotheses are given a relative frequency interpretation. Occasionally empirical interpretation other than the statistical one is preferable, however. For instance, in theories designed for explaining individual choice behavior the response probabilities of experimental subjects are more naturally given a personal or sometimes perhaps a psychological rather than a statistical interpretation.

[28] The terminology of certain authors differs from ours. Nagel, for instance, understands by probabilistic explanation what is here called inductive explanation; our probabilistic explanation corresponds to his concept of statistical explanation (see *Structure of Science*, pp. 22–3). Hempel also speaks of statistical explanation which can be of either deductive or inductive type; he distinguishes statistical explanation from the nomological kind of explanation (which corresponds to our probabilistic-deterministic distinction).

[29] This condition has been stated e.g. by Hempel (in *Aspects of Scientific Explanation*, p. 389) and Levi (in *Gambling with Truth*, p. 209). Obvious as it may seem to be, it is by no means philosophically unproblematic, as is shown by David Miller's 'A Paradox of Information', *The British Journal for the Philosophy of Science* **17** (1966) 59–61, and by the discussion on this paper, especially by W. Rozeboom's 'New Mysteries for Old: the Transfiguration of Miller's Paradox', *The British Journal for the Philosophy of Science* **19** (1969) 345–58.

[30] Cf. Carnap, *Logical Foundations of Probability*, pp. 495–6.

[31] De Finetti's own interpretation of his famous result concerning betting ratios on probability statements is that the assumption of (unknown) empirical probabilities is unnecessary. However, this is not the only possible interpretation, as has been argued e.g. by Hintikka ('The Philosophical Significance of de Finetti's Representation Theorem', unpublished). He sees one significance of de Finetti's result in the very fact that it shows a person who believes in the existence of objective probabilities (and surely most scientists do this) how to bet on them.

[32] This need not always be the case, however. There are good examples of what Hempel calls self-evidencing explanations, where the occurrence of the explanandum event provides the only evidential support, and where this support is nevertheless very strong (see *Aspects of Scientific Explanation*, pp. 372–3).

[33] See D. V. Lindley, 'Statistical Inference', *Journal of the Royal Statistical Society*, B, **15** (1953) 30–65. The acceptance of h means here the rejection of the alternative k.

[34] Isaac Levi's essay *Gambling with Truth* as well as many of his earlier publications on the aims of science and on the importance of decision theoretical considerations in

scientific inference and acceptance procedures are of utmost importance in this context. Unfortunately, this single reference must suffice here.

[35] For a more detailed discussion of this kind of measure of acceptability, see R. Hilpinen, *Rules of Acceptance and Inductive Logic, Acta Philosophica Fennica* **22** (1968), Ch. 9.

[36] Cf. Hempel, *Aspects of Scientific Explanation*, pp. 344–403.

ZOLTAN DOMOTOR

QUALITATIVE INFORMATION AND ENTROPY STRUCTURES

1. INTRODUCTION

1.1. *Recent Developments in Axiomatic Information Theory*

Information theory deals with the mathematical properties of communication models, which are usually defined in terms of concepts like channel, source, information, entropy, capacity, code, and which satisfy certain conditions and axioms.

Our knowledge in this field has expanded prodigiously since 1948 when C. E. Shannon gave the first sufficiently general definition of information and entropy. An indication of this expansion can be gained from the survey and extensive bibliography in Varma and Nath (1967). In particular, the last 10 years have seen a considerable interest in the abstract axiomatic treatment of the concepts of information and entropy.

Shannon's original axioms for entropy measure have been replaced by weaker conditions (see Fadeev, 1956; Khinchin, 1957; Tveberg, 1958; Kendall, 1964; and others). The weakest set of axioms known seems to be that given by Lee (1964).

Rényi (1961), on the other hand, has extended the notion of entropy to the algebra of incomplete experiments by using the concept of a generalized probability distribution.

The above characterizations of entropy all involve essentially probabilistic notions.

Ingarden and Urbanik (1962), Ingarden (1963, 1965), De Fériet and Forte (1967), and Forte and Pintacuda (1968a, b) have given axiomatic definitions of information and entropy measures without using probability measures. Similarly Kolmogorov (1965, 1967) has shown that the basic information-theoretic concepts can be formulated without recourse to probability theory.

Ingarden and Urbanik need to assume for their definition of entropy a sufficiently large pseudometric space of finite Boolean rings, in order

J. Hintikka and P. Suppes (eds.), Information and Inference, 148–194.

to be able to state the continuity of the entropy measure. On the other hand, Kolmogorov uses the concepts of recursive function and complexity.

Adler *et al.* (1965) have introduced the notion of a 'topological entropy', which is an invariant with respect to continuous mappings. They indicate also how to define the notion of entropy in various fundamental mathematical structures.

Quite recently, several information-theorists have tried to construct the information-theoretic notions by using techniques from statistical decision theory. For example, Belis and Guiasu (1967) propose a notion of a 'qualitative-quantitative information measure', defined in terms of utility. The idea is this. Given a probability space $\langle \Omega, \mathfrak{A}, P \rangle$, they introduce, besides the probability measure P on the algebra of events $\mathfrak{A}$, a utility function U, which assigns to each element of a partition $\mathscr{P}$ of Ω a non-negative real number; the entropy measure H of the partition $\mathscr{P}$ is then given by

$$H(\mathscr{P}) = - \sum_{A \in \mathscr{P}} U(A) \cdot P(A) \cdot \log_2 P(A).$$

In connection with this definition see also Suppes (1961), where it is shown that the entropy measure of a partition $\mathscr{P}$ is equal to the negative linear transformation of the expected utility of $\mathscr{P}$.

Weiss (1968) gives an axiomatic system for subjective information which is almost identical with the theories of probability and utility of Savage (1954) and Raiffa *et al.* (1964).

In a related field, that of semantic information theory (in the sense of Bar-Hillel and Carnap, 1953), there have also been advances (see especially Hintikka and Pietarinen, 1966; Hintikka, 1968).

As can be seen even from this cursory review of recent developments, there is available an immense wealth of axiomatic material dealing with purely logical and foundational aspects of information theory. It should be noted, however, that all the foundational attempts mentioned have a common aspect; they are all directed in the main towards axiomatizing the basic information-theoretic notions in the form of *functional equations*.

In this paper another approach is proposed. We shall advocate, instead of the *analytic or measure-theoretic approach*, an *algebraic approach* in terms of *relational structures*. The latter approach is more relevant to *measurement* or, generally, *epistemic aspects* of information.

In fact, the main purpose of this paper is to give axiomatic definitions of the concepts of *qualitative information* and *qualitative entropy structure*, and to study some of their basic properties. The particular sections culminate in proving certain representation theorems that elucidate the relations these notions bear to the standard concepts of information and entropy. More specifically, given a Boolean algebra $\mathfrak{A}$ together with a binary relation $\preccurlyeq$ on $\mathfrak{A}$ (interpreted as a qualitative comparison of two Boolean elements with respect to their information), sufficient conditions are given in terms of $\mathfrak{A}$ and $\preccurlyeq$ for the existence of an information measure I on $\mathfrak{A}$ such that

$$A \preccurlyeq B \Leftrightarrow I(A) \leqslant I(B) \quad \text{for all } A, B \in \mathfrak{A}.$$

For finite lattices of probabilistic experiments a similar but somewhat stronger theorem is proved. Namely, given a finite algebra of experiments $\mathbf{P}$ together with a binary relation $\preccurlyeq$ on $\mathbf{P}$ (interpreted as a qualitative comparison of two experiments with respect to their entropy), necessary and sufficient conditions are given in terms of $\mathbf{P}$ and $\preccurlyeq$ for the existence of an entropy measure H on $\mathbf{P}$ such that

$$\mathscr{P}_1 \preccurlyeq \mathscr{P}_2 \Leftrightarrow H(\mathscr{P}_1) \leqslant H(\mathscr{P}_2)$$

for all $\mathscr{P}_1, \mathscr{P}_2 \in \mathbf{P}$.

1.2. *Motivations for Basic Notions of Information Theory*

The standard notion of *information* is usually introduced in order to answer the following somewhat abstract question: How much information do we get about a point $\omega \in \Omega$ from the news that belongs to a subset A of Ω, that is that $\omega \in A$ and $A \subseteq \Omega$? It is rather natural to assume that the answer should depend on, and only on, the size of A, that is to say, on $P(A)$, where P is a standard probability measure on the Boolean algebra $\mathfrak{A}$ of subsets of Ω. In other words, the answer should be given in terms of a real-valued function I, defined on the unit interval [0, 1]. Hence, the amount of information conveyed by the statement $\omega \in A$ will be $I \circ P(A)$[1], or in a simpler notation, $I_P(A)$. It is also natural to require I to be non-negative and continuous on [0, 1]. Moreover, given two 'independent experiments' described by statements $\omega \in A$ and $\omega \in B$ ($A, B \in \mathfrak{A}$, $\omega \in \Omega$), it is reasonable to expect that the amount of information given by the experiment described by $\omega \in A$ & $\omega \in B$, that is $\omega \in A \cap B$, will be the sum of the amounts of information of the experiments taken separately.

Given a probability space $\mathbf{A} = \langle \Omega, \mathfrak{A}, P \rangle$, let $A \coprod B$ mean that the 'experiments' with outcomes $\omega \in A$ and $\omega \in B$ are probabilistically independent ($A, B \in \mathfrak{A}$); then we can collect our previous ideas in the following assumptions:

(i) The diagram

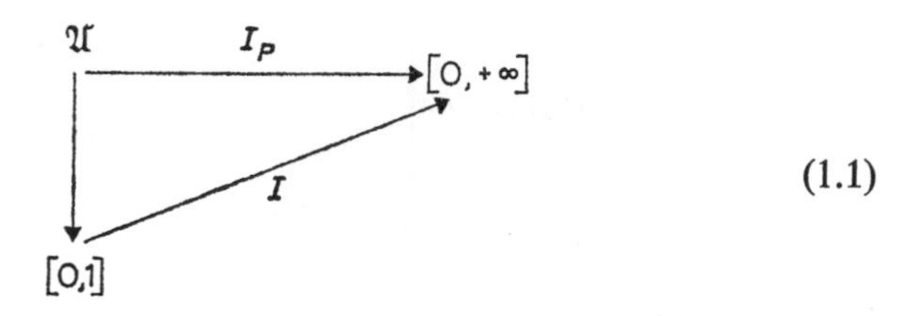

(1.1)

is commutative, that is, $I \circ P = I_P$, and I is continuous;

(ii) $A \coprod B \Rightarrow I_P(A \cap B) = I_P(A) + I_P(B)$, if $A, B \in \mathfrak{A}$.

It is a well-known fact that the only real function I_P which satisfies the conditions (1.1) is $I_P(A) = -\alpha \cdot \log P(A)$, where α is an arbitrary positive real constant. It is a matter of convention to choose a unit for measurement of the amount of information which makes $\alpha = 1$.

Now assume that we are given several experiments in the form of a system of mutually exclusive and collectively exhaustive events, $\mathscr{P} = \{A_i\}_{i=1}^n$, where

$$\bigcup_{i=1}^{m} A_i = \Omega, \quad \text{and} \quad A_i \cap A_j = \emptyset \quad \text{for} \quad i \neq j \quad \text{and} \quad i, j \leqslant m.$$

What we may well ask is the *average amount of information conveyed* by the system of experiments $\mathscr{P}$.

Since we are assuming the probabilistic frame $\mathbf{A}$, there is nothing more natural than to take the average amount of information, called the *entropy* H, to be the expected value of the amount of information:

$$H_P(\mathscr{P}) = \sum_{A \in \mathscr{P}} P(A) \cdot I_P(A), \quad \text{where} \quad I_P(A) = -\log_2 P(A). \tag{1.2}$$

The entropy measure H is usually characterized by a system of functional equations using more or less plausible ideas about the properties of H.

Let **P** be the set of all possible partitions of the basic set of elementary events Ω of the structure **A**. For simplicity the elements of **P** will be called *experiments*, for $\mathscr{P}\in\mathbf{P}$, where $A\in\mathscr{P}$ is an event, representing a possible realization of the experiment $\mathscr{P}$. Then the functional equations for H have the following form:

(i) The diagram

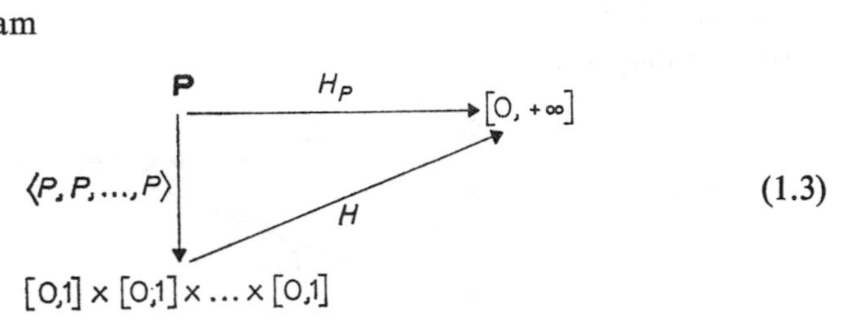

(1.3)

is commutative, that is, $H\circ\langle P, P, \ldots, P\rangle = H_{\mathbf{P}}$, and H is continuous;

(ii) $H_{\mathbf{P}}(\{A, \bar{A}\})=1$ if $P(A)=P(\bar{A})$;

(iii) $H_{\mathbf{P}}([B\,|\,A\cap B,\ \bar{A}\cap B]\,\mathscr{P})=H_{\mathbf{P}}(\mathscr{P})+P(B)\cdot H_{\mathbf{P}}(\{A, \bar{A}\})$ if $A\coprod B$; here $A, B\in\mathfrak{A}$ and $[B\,|\,A\cap B, \bar{A}\cap B]\,\mathscr{P}$ is the experiment which is the result of replacing B in the partition $\mathscr{P}$ by two disjoint events $A\cap B$, $\bar{A}\cap B$. It is assumed, of course, that $B\in\mathscr{P}$.

Using Erdös' (1946) famous number-theoretic lemma about additive arithmetic functions Fadeev (1956) showed that the only function $H_{\mathbf{P}}$ which satisfies the conditions (1.3) has the form (1.2).

What has been said so far is standard and well known. In the sequel we shall point out a different and probably new approach. Instead of constructing functional equations, proving the validity of the formula (1.2) and showing that they adequately mirror our ideas about the concepts of information and entropy, we propose here to approach the problem qualitatively.

Following De Finetti, Savage (1954) and others, we shall assume that our probabilistic frame is a qualitative probability structure $\langle\Omega, \mathfrak{A}, \preccurlyeq\rangle$, where $A\preccurlyeq B$ means that the event A is not more probable than the event B ($A, B\in\mathfrak{A}$). In the general case there is no need to associate the binary relation $\preccurlyeq$ with any subjectivist interpretation of probability.

The question arises whether we can introduce a binary relation $\preccurlyeq$ on the set of experiments **P** in such a way that this relation will express satisfactorily our intuitions and experiences about the notion of entropy. In other words, under what conditions on $\preccurlyeq$ do we have:

$\mathscr{P}_1 \preccurlyeq \mathscr{P}_2 \Leftrightarrow$ Experiment $\mathscr{P}_1$ does not have more entropy than the experiment $\mathscr{P}_2$.

In a way this question belongs to measurement theory (see Suppes and Zinnes, 1963). When we study any property of a given family of empirical objects, or a relation among these objects, one of the basic epistemological problems is to find under what conditions the given property or relation is measurable; more specifically, what are the necessary and sufficient conditions for there to exist a real valued function on the family of empirical objects whose range is a homomorphic image of the set of empirical objects in accordance with the given property or relation.

In the case of entropy this amounts to knowing the restrictions to be imposed on $\preccurlyeq$ in order that H_P of (1.2) exists and, furthermore, satisfies the following homomorphism condition:

$$\mathscr{P}_1 \preccurlyeq \mathscr{P}_2 \Leftrightarrow H_P(\mathscr{P}_1) \leqslant H_P(\mathscr{P}_2), \quad \text{if} \quad \mathscr{P}_1, \mathscr{P}_2 \in \mathbf{P}. \tag{1.4}$$

It is a trivial matter to notice that the relation $\preccurlyeq$ has to be reflexive, transitive, connected, and antisymmetric with respect to the relation $\sim$ (defined by $\mathscr{P}_1 \sim \mathscr{P}_2 \Leftrightarrow \mathscr{P}_1 \preccurlyeq \mathscr{P}_2$ & $\mathscr{P}_2 \preccurlyeq \mathscr{P}_1$, if $\mathscr{P}_1, \mathscr{P}_2 \in \mathbf{P}$). In other words, $\preccurlyeq$ has at least to be a linear ordering modulo the relation $\sim$. But these trivial assumptions are obviously insufficient to guarantee the existence of a function as complicated as H_P.

Likewise we can introduce a binary relation $\preccurlyeq$[2] on the Boolean algebra $\mathfrak{A}$, and consider the intended interpretation

$A \preccurlyeq B \Leftrightarrow$ Event A does not convey more information than event B.

Again, we shall try to formulate the conditions on $\preccurlyeq$, which allow us to find an information function I_P satisfying both (1.1), and the following homomorphism condition:

$$A \preccurlyeq B \Leftrightarrow I_P(A) \leqslant I_P(B), \quad \text{if} \quad A, B \in \mathfrak{A}. \tag{1.5}$$

Hence our problem is to discover some conditions which, though expressible in terms of $\preccurlyeq$ only, allow us to find a function $I_P(H_P)$ satisfying (1.1), (1.5) ((1.3), (1.4)).

This approach is interesting not only theoretically but also from the point of view of applications. In social, behavioral, economic, and biological sciences there is often no plausible way of assigning probabilities to events. But the subject or system in question may be pretty well able to order the events according to their probabilities, informations, or entropies, in a certain qualitative sense.

Of course, it is an empirical problem whether the qualitative probability, information, or entropy determined by the given subject or system actually satisfies the required axioms. But in any case, the qualitative approach gives the measurability conditions for the analyzed probabilistic or information-theoretic property. To be more specific, the structure $\langle \mathfrak{A}, \preccurlyeq \rangle$ is turned into an *empirical model*, if, in addition, an *empirical interpretation* $\| \ \|$ of the elements of $\mathfrak{A}$ and the atomic formulas $A \preccurlyeq B (A, B \in \mathfrak{A})$ is given. In such a case, the theory about $\langle \mathfrak{A}, \preccurlyeq \rangle$ becomes empirically testable with respect to the interpretation $\| \ \|$.

1.3. *Methodological Remarks*

One of the more fruitful ways of analyzing the mathematical structure of any concept is by using what we here call the *representation method.* This method consists of determining the entire family of homomorphisms or isomorphisms from the analyzed structure into a suitable well-known concrete mathematical structure. The work is usually done in two steps: first, the existence of at least one homomorphism is proved; secondly, one finds a set or group of transformations up to which the given homomorphism is exactly specified. The unknown and analyzed structure is then *represented* by a better known and more familiar structure so that, eventually, the unknown problem can be reduced to one perhaps already solved.

Another advantage of this method is that it handles problems of empirical 'meaning' and content in an extensional way. For it is a rather trivial fact that any mathematical approach to such a problem will give the answer at most up to isomorphism, which implies that many 'meaning' problems are extramathematical questions. For example, interpretation of the concept of probability is beyond the scope of the Kolmogorov axioms.

Yet, we would like to indicate by an example (used in the sequel) how by using the idea of representation of one structure by another one can handle the 'meaning' problem *inside* mathematics.

The next three sections will deal with certain mathematical structures. The problems these structures pose are too difficult to answer immediately, and we shall therefore in each case translate the problem into geometric language by means of the representation of *relations* by *cones* in a vector space. From this geometric language we translate the problem again into functional language by means of the representation of *cones* by *positive functionals*. Here the problem is solved, and we translate the result back into the original language of relations.

It should be noted that the translation is not always reversible. The representing structure may keep only one aspect of the original structure, but this has the advantage that the problem may be stripped of inessential features, and replaced by a familiar type of problem, hopefully easier to solve. Of course, we cannot guarantee in advance that essential features will be not lost.

Take as a concrete example the relational structure of the qualitative information $\langle\mathfrak{A}, \preccurlyeq\rangle$ which will be discussed extensively in Section 2. Any empirical content assigned to the information structure $\langle\mathfrak{A}, \preccurlyeq\rangle$ is carried through the chain of homomorphisms: relational entity $\rightsquigarrow$ geometric entity $\rightsquigarrow$ functional entity, to the information measure I on $\mathfrak{A}$. The measure I may thus acquire empirical content on the basis of the structure $\langle\mathfrak{A}, \preccurlyeq\rangle$, which we assume already to have empirical content via other structures or directly, by stipulation.

In general, the empirical meaning or content of an abstract, or so-called *theoretical* structure (model) is given through a more or less complicated tree or lattice of structures together with their mutual homomorphisms (satisfying certain conditions), where some of them, the initial, concrete, or so-called *observational* ones, are endowed with empirical meanings by postulates.

Note that the homomorphism is here always a *special function*. For example, in the case of information, I satisfies *not only* the *homomorphism condition* (which is relatively simple), but also the *axioms* for the information measure. Thus the axioms for the given structure are *essentially involved* in the existence of the homomorphism. In this respect, the representation method goes far beyond the ordinary homomorphism technique between similar structures, or the theory of elementarily equivalent models.

The 'meaning' of a given concept can be expressed extensionally by the lattice of possible representation structures that are connected by homomorphisms (with additional properties) and that represent always one particular aspect of the concept.

We do not intend to go into this intricate philosophical subject here. The only point of this discussion is to emphasize the methodological importance of our approach to concepts like qualitative probability, information and entropy.

2. QUALITATIVE PROBABILITY STRUCTURES

2.1. *Basic Facts About Qualitative Probability Structures*

In this section we briefly review the most important facts we shall need.

In 1937 De Finetti raised the following question for a finite Boolean algebra $\mathfrak{A}$ with unit element Ω enriched by a binary relation $\preccurlyeq$ with the intended interpretation: $A \preccurlyeq B \Leftrightarrow$ event A is not more probable (in some, perhaps subjective, sense) than event B. Is it possible to find a probability measure P on $\mathfrak{A}$ such that

$$A \preccurlyeq B \Leftrightarrow P(A) \leqslant P(B) \quad \text{for all } A, B \in \mathfrak{A}?$$

The necessary and sufficient conditions for the existence of P which we must impose on $\langle \mathfrak{A}, \preccurlyeq \rangle$ were first given by Kraft *et al.* (1959). Scott (1964a) generalized this result to arbitrary Boolean algebras: in the finite case his conditions reduce to those of Kraft *et al.*

The conditions for the general case are quite complicated and somewhat unintuitive, whereas the conditions for atomless or finite Boolean algebras are rather simple.

A similar problem for conditional events, and its generalizations to arbitrary Boolean algebras and to the case of algebraic error, were solved in Domotor (1969).

The necessary and sufficient conditions for the existence of a probability measure P such that $\langle \Omega, \mathfrak{A}, P \rangle$ *is a finitely additive probability space and*

$$A \preccurlyeq B \Leftrightarrow P(A) \leqslant P(B) \quad \text{for all } A, B \in \mathfrak{A}$$

are the following:

(i) $\quad \emptyset \prec \Omega$

(ii) $\emptyset \preccurlyeq A,$

(iii) $A \preccurlyeq B \vee B \preccurlyeq A,$

(iv) $$\underset{i<n}{\forall} [A_i \preccurlyeq B_i] \Rightarrow B_n \preccurlyeq A_n, \text{ if} \tag{2.1}$$

$$\underset{i \leqslant n}{\dot{\bigcup}} A_i = \underset{i \leqslant n}{\dot{\bigcup}} B_i \ \& \underset{\substack{i, j \leqslant n \\ i<j}}{\dot{\bigcup}} A_i \cap A_j = \underset{\substack{i, j, \leqslant n \\ i<j}}{\dot{\bigcup}} B_i \cap B_j \ \& \ldots \&$$

$$\& A_1 \cap A_2 \cap \ldots \cap A_n = B_1 \cap B_2 \cap \ldots \cap B_n,$$

where $A, B, A_i, B_i \in \mathfrak{A}$ *for* $i = 1, 2, \ldots, n$ *and* $\dot{\bigcup}_{i \leqslant n} A_i$ *denotes the symmetric difference of the n sets* $A_1, A_2, \ldots, A_n$. (For two sets A_1 and A_2, $A_1 \dot{\cup} A_2$ is of course $(A_1 \cap \bar{A}_2) \cup (\bar{A}_1 \cap A_2)$.)

Axiom (iv) can be written much more simply in terms of characteristic functions as

$$\underset{i \leqslant n}{\forall} [A_i \preccurlyeq B_i] \Rightarrow B_n \preccurlyeq A_n, \quad \text{if} \quad \sum_{i \leqslant n} \hat{A}_i = \sum_{i \leqslant n} \hat{B}_i,$$

where $\hat{A}_i(\omega) = 1$ if $\omega \in A_i$ and $\hat{A}_i(\omega) = 0$ if $\neg \omega \in A_i (1 \leqslant i \leqslant n)$. It is taken for granted that $A \prec B$ means $\neg B \preccurlyeq A$ and we put $A \sim B$ for $A \preccurlyeq B \,\&\, B \preccurlyeq A$.

The proof of this theorem is based on the well-known Separation Theorem and applies the technique explained in Section 1.3.

In Section 1.2 we saw that the relation of probabilistically independent events $(A \coprod B)$ played an essential role in the definition of information and entropy (see formulas (1.1) and (1.3)).

The independence relation $\coprod$ is defined entirely in terms of the probability measure P:

$$A \coprod B \Leftrightarrow P(A \cap B) = P(A) \cdot P(B) \quad (A, B \in \mathfrak{A}). \tag{2.2}$$

One wonders if it is possible to give a definition of the binary relation $\coprod$ on $\mathfrak{A}$ in terms of the relation $\preccurlyeq$. The positive answer (to be given later) will be useful in the qualitative definition of information and entropy.

A satisfactory qualitative independence relation can be relevant also in applied probability theory, where one does not care too much about the underlying probability structure $\mathbf{A} = \langle \Omega, \mathfrak{A}, P \rangle$, and emphasizes rather the analytic properties of random variables. Under these circumstances the independent random variables could be handled using the simple properties of $\coprod$ without explicit reference to the probability measure P satisfying the condition (2.2).

This problem apparently is not easy since we are working in a finite structure $\langle \Omega, \mathfrak{A}, \preccurlyeq \rangle$, and yet we have to guarantee the existence of the arithmetic multiplication in (2.2).

One of the ideas which comes naturally to hand is the assumption of such a binary relation $\preccurlyeq$ (we keep the same notation) on the set of Cartesian product of events, which would allow us to prove the following condition:

$$A \times B \preccurlyeq C \times D \Leftrightarrow P(A) \cdot P(B) \leqslant P(C) \cdot P(D), \quad \text{for all } A, B, C, D \in \mathfrak{A}. \tag{2.3}$$

Then we would put simply

$$A \coprod B \Leftrightarrow (A \cap B) \times \Omega \sim A \times B. \tag{2.4}$$

This might be a possible way of solving the problem; but what must we require of the relation $\preccurlyeq$ among the Cartesian products of events? The next section deals with this problem.

2.2. *Quadratic Qualitative Probability Structures*

Luce and Tukey (1964) gave a formal presentation of what they called *conjoint measurement structures*. Such structures are linear. Here, by contrast, *nonlinear* (*quadratic*) measurement structures will be introduced for probability. More concretely, given a finite Boolean algebra $\mathfrak{A}$ of subsets of Ω and a binary relation $\preccurlyeq$[3] on the set of Cartesian products of elements from $\mathfrak{A}$, we shall give the necessary and sufficient conditions for the existence of a probability measure P on $\mathfrak{A}$ such that the condition in (2.3) will be satisfied.

As will be seen later, the appearance of Cartesian products $A \times B$, $C \times D$ here is not essential; we could as well consider the ordered couples $\langle A, B \rangle$, $\langle C, D \rangle$. Structures of this sort differ from Luce's conjoint measurement structures in three respects: they are *finite*, the representing function has a special property, namely, it is *additive*, and finally, the representation is *quadratic* and not linear. Since most of the laws of classical physics can be represented (using the so-called π-theorem) by equations between a given (additive) empirical quantity and the *product* of other (additive) empirical quantities (possibly with rational exponents), such a structure is of basic importance in algebraic measurement theory.

For instance, for Ohm's law we might hope to give, for the system of current sources $\{c_i\}_{i \leqslant n}$ and resistors $\{r_i\}_{i \leqslant m}$, a representation theorem in the form:

$$\langle c_i, r_i \rangle \preccurlyeq \langle c_j, r_j \rangle \Leftrightarrow I_i \cdot R_i \leqslant I_j \cdot R_j ,$$

where on the right we have well-known physical quantities, namely, current and resistance ($i \leqslant n, j \leqslant m$).

Returning to quadratic probability structures, the reader may wonder in what way the formula $A \times B \preccurlyeq C \times D$ ($A, B, C, D \in \mathfrak{A}$) in (2.3) can be interpreted.

There are several partial interpretations which will be discussed in the sequel:

(a) *Qualitative probabilistic independence relation* $\coprod$:

$$A \coprod B \Leftrightarrow AB \times \Omega \sim A \times B ,$$

where, as usual, $A, B \in \mathfrak{A}$ and $\sim$ is the standard equivalence relation induced by $\preccurlyeq$.

(b) *Qualitative conditional probability relation* $\preccurlyeq$:

$$A/B \preccurlyeq C/D \Leftrightarrow AB \times D \preccurlyeq CD \times B , \quad \text{if} \quad \emptyset \times \Omega \prec B \times D ,$$

where $A, B, C, D \in \mathfrak{A}$ and $\prec$ is the *strict* counterpart of $\preccurlyeq$. The entities A/B, C/D can be considered here as primitive.

(c) *Relevance (positive and negative dependence) relations* C_+, C_-:

$$AC_+B \Leftrightarrow A \times B \prec AB \times \Omega ;$$
$$AC_-B \Leftrightarrow AB \times \Omega \prec A \times B ,$$

where $A, B \in \mathfrak{A}$. These notions may be of some help in analyzing causality problems. It is immediately obvious that $A\ C_+\ B \Leftrightarrow A/\Omega \prec A/B$ and $A\ C_-\ B \Leftrightarrow A/B \prec A/\Omega$.

(d) *Qualitative conditional independence relation* $\coprod$:

$$A/C \coprod B/C \Leftrightarrow AC \times BC \sim ABC \times C , \quad \text{if} \quad \emptyset \prec C ,$$

where $A, B, C \in \mathfrak{A}$.

Since there are several important interpretations of the formula $A \times B \preccurlyeq C \times D$, we shall study the structure of the 'quadratic' relation $\preccurlyeq$ in considerable detail.

DEFINITION 1: *A triple* $\langle \Omega, \mathfrak{A}, \preccurlyeq \rangle$ *is said to be a finitely additive quadratic qualitative probability structure* (*FAQQP*-structure) if and only if *the following conditions are satisfied:*

(Q_0) *Ω is a nonempty finite set; $\mathfrak{A}$ is the Boolean algebra of subsets of Ω; and $\preccurlyeq$ is a binary relation on* $\{A \times B : A \in \mathfrak{A} \,\&\, B \in \mathfrak{A}\}$.

(Q_1) $\emptyset \times \Omega \prec \Omega \times \Omega$;

(Q_2) $\emptyset \times A \preccurlyeq B \times C$;

(Q_3) $A \times B \preccurlyeq B \times A$;

(Q_4) $A \times B \preccurlyeq C \times D \vee C \times D \preccurlyeq A \times B$;

(Q_5) $\underset{i<n}{\forall} (A_i \times B_i \preccurlyeq A_{\alpha_i} \times B_{\beta_i}) \Rightarrow A_{\alpha_n} \times B_{\beta_n} \preccurlyeq A_n \times B_n$;

(Q_6) $\underset{i<n}{\forall} (C_i \times D_i \preccurlyeq E_i \times F_i) \Rightarrow E_n \times F_n \preccurlyeq C_n \times D_n$;

where

$$\underset{i<n}{\forall} (\emptyset \times \Omega \prec A_i \times B_i); \quad \sum_{i\leqslant n} (C_i \times D_i)^{\wedge} = \sum_{i\leqslant n} (E_i \times F_i)^{\wedge};$$

$A, B, C, D, A_i, B_i, C_i, D_i, E_i, F_i \in \mathfrak{A} (i \leqslant n)$; α, β are permutations on $\{1, 2, ..., n\}$, and $(C \times D)^{\wedge}$ denotes the characteristic function of the set $C \times D$.

Remarks:

(i) We define

$$A \preccurlyeq B \Leftrightarrow A \times \Omega \preccurlyeq B \times \Omega;$$
$$A \times B \prec C \times D \Leftrightarrow \neg\, C \times D \preccurlyeq A \times B;$$
$$A \times B \sim C \times D \Leftrightarrow A \times B \preccurlyeq C \times D \,\&\, C \times D \preccurlyeq A \times B;$$
$$(A \times B)^{\wedge} (\omega_1, \omega_2) = 1, \quad \text{if} \quad \omega_1 \in A \,\&\, \omega_2 \in B;$$
$$\text{otherwise } (A \times B)^{\wedge} (\omega_1, \omega_2) = 0 \, (\omega_1, \omega_2 \in \Omega).$$

(ii) The formula concerning characteristic functions in axiom (Q_6) can easily be translated into a system of identities among sets; a similar transformation was made in the case of the qualitative probability axioms listed in Section 2.1. Thus the axioms for $\preccurlyeq$ contain as primitives *only* the relation $\preccurlyeq$ and the algebra $\mathfrak{A}$.

The content of the above definition is laid bare in the following easily proved theorem.

THEOREM 1: *Let $\langle \Omega, \mathfrak{A}, \preccurlyeq \rangle$ be a FAQQP*-structure. *Then the following* *formulas are valid for all $A, B, C, D, E, F \in \mathfrak{A}$:*

(1) $A \times B \sim A \times B$;

(2) $A \times B \sim B \times A$;

(3) $A \times B \preccurlyeq C \times D \;\&\; C \times D \preccurlyeq E \times F \Rightarrow A \times B \preccurlyeq E \times F$;

(4) $A \times C \preccurlyeq B \times C \Leftrightarrow A \times D \preccurlyeq B \times D$, if $\emptyset \times \Omega \prec C \times D$;

(5) $A \times B \preccurlyeq C \times D \;\&\; E \times C \preccurlyeq F \times B \Rightarrow A \times E \preccurlyeq F \times D$, if $\emptyset \times \Omega \prec B \times C$;

(6) $A \times B \preccurlyeq C \times D \Leftrightarrow B \times A \preccurlyeq D \times C$;

(7) $\Omega \times A \preccurlyeq B \times \Omega \;\&\; C \times \Omega \preccurlyeq \Omega \times D \Rightarrow A \times C \preccurlyeq B \times D$;

(8) $A \times \Omega \preccurlyeq B \times \Omega \Leftrightarrow A \times A \preccurlyeq B \times B$;

(9) $A \times B \sim C \times D \Rightarrow (A \preccurlyeq C \Leftrightarrow D \preccurlyeq B)$;

(10) $(A \times A \preccurlyeq F \times F \;\&\; A \times E \preccurlyeq D \times D \;\&\; E \times E \preccurlyeq D \times F) \Rightarrow A \times E \preccurlyeq D \times F$;

(11) $A \preccurlyeq B \Leftrightarrow A \times \bar{B} \preccurlyeq \bar{A} \times B$;

(12) $\emptyset \prec A \;\&\; \emptyset \prec B \Rightarrow \emptyset \times \Omega \prec A \times B$;

(13) $\forall_{i \leqslant n} (A_i \times B_i \preccurlyeq C_i \times D_i) \;\&\; \forall_{i \leqslant n} (C_{\gamma_i} \times D_{\delta_i} \preccurlyeq A_{\alpha_i} \times B_{\beta_i}) \Rightarrow A_{\alpha_n} \times B_{\beta_n} \preccurlyeq C_{\gamma_n} \times D_{\delta_n}$, if $\forall_{i<n} (\emptyset \times \Omega \prec C_{\gamma_i} \times D_{\delta_i})$,

where $A_i, B_i, C_i, D_i \in \mathfrak{A} (i \leqslant n)$, and $\alpha, \beta, \gamma, \delta$ are permutations on $\{1, 2, \ldots, n\}$;

(14) *If $A \preccurlyeq_0 B \Leftrightarrow A \times \Omega \preccurlyeq B \times \Omega$, then $\langle \Omega, \mathfrak{A}, \preccurlyeq_0 \rangle$*

is a finitely additive qualitative probability structure in the sense of axioms (2.1).

The reader may find immediately the corresponding appropriate intended interpretations. For example, (2) means commutativity of $\sim$, (3) means transitivity of $\preccurlyeq$, (4) means monotonicity of $\preccurlyeq$, (5) means cancellation property of $\preccurlyeq$, etc.

Theorem 1 will be useful in several ways. In particular, the properties of $\coprod$ will be derived from it.

THEOREM 2 (Representation Theorem): *Let $\langle \Omega, \mathfrak{A}, \preccurlyeq \rangle$ be a structure, where Ω is a nonempty finite set; $\mathfrak{A}$ is the Boolean algebra of subsets of Ω, and $\preccurlyeq$ is a binary relation on $\{A \times B: A \in \mathfrak{A} \;\&\; B \in \mathfrak{A}\}$.*

Then $\langle \Omega, \mathfrak{A}, \preccurlyeq \rangle$ is a FAQQP-structure if and only if *there exists a finitely additive probability measure P such that $\langle \Omega, \mathfrak{A}, P \rangle$ is a probability* *space, and for all* $A, B, C, D \in \mathfrak{A}$,

$$A \times B \preccurlyeq C \times D \Leftrightarrow P(A) \cdot P(B) \leqslant P(C) \cdot P(D).$$

PROOF: It is a routine matter to show that the axioms (Q_0)–(Q_6) in Definition 1 are necessary.

The details of the proof of sufficiency are given in Domotor (1969) and we shall present here only the main ideas.

(a) Translation of the problem from the language of relations into geometric language.

We identify the Cartesian products $A \times B$ with tensor products $\hat{A} \otimes \hat{B}$, where $\hat{A}$ is the vector

$$\hat{A} = \langle c_A(\omega_1), c_A(\omega_2), \ldots, c_A(\omega_n) \rangle,$$

with $\{\omega_i\}_{i=1}^n = \Omega$, and $c_A(\omega) = 1$, if $\omega \in A$, otherwise $c_A(\omega) = 0$.

Defining $\hat{A} + \hat{B}$, $\alpha \cdot \hat{A}$ (α is a real number) in an obvious way, we generate from Ω a vector space $\mathscr{V}(\Omega)$ with $\hat{\mathfrak{A}} \subseteq \mathscr{V}(\Omega) = \mathscr{V}$, where $\hat{\mathfrak{A}} = \{\hat{A} \colon A \in \mathfrak{A}\}$. Then of course $\hat{A} \otimes \hat{B} \in \mathscr{V} \otimes \mathscr{V} = \mathscr{W}$. Now we put

$$\hat{A} \otimes \hat{B} \preccurlyeq \hat{C} \otimes \hat{D} \Leftrightarrow A \times B \preccurlyeq C \times D;$$

$\preccurlyeq$ on $\mathscr{W}$ will obviously determine a unique convex cone $\mathscr{W}^+$. This completes the translation.

(b) Translation of the problem from geometric language into functional language.

Using (Q_4) and (Q_6), the Separation Theorem (see Scott 1964b, p. 236, Theorem 1.3) will allow us to find a linear functional $\psi \colon \mathscr{W} \to Re^4$ such that

$$\hat{A} \otimes \hat{B} \preccurlyeq \hat{C} \otimes \hat{D} \Leftrightarrow \psi(\hat{A} \otimes \hat{B}) \leqslant \psi(\hat{C} \otimes \hat{D}). \tag{2.5}$$

This can be translated into a bilinear functional $\varphi \colon \mathscr{V} \times \mathscr{V} \to Re$ also satisfying (2.5). (Q_1) allows us to normalize φ, that is, choose it so that $\varphi(\hat{\Omega}, \hat{\Omega}) = 1$; we thus avoid the trivial case, where φ is identically zero. (Q_2) forces φ to be non-negative on $\{\hat{A} \otimes \hat{B} \colon A \in \mathfrak{A} \;\&\; B \in \mathfrak{A}\}$. By ($Q_3$) φ is symmetric. Finally, (Q_5) allows us to show that φ has rank 1; that is, that $\varphi(\mathbf{v}_1, \mathbf{v}_2) = \varphi_1(\mathbf{v}_1) \cdot \varphi_2(\mathbf{v}_2)$ for $\mathbf{v}_1, \mathbf{v}_2 \in \mathscr{V}$. ($Q_3$) implies that $\varphi_1 = \varphi_2 = \varphi_0$. Hence

$$\hat{A} \otimes \hat{B} \preccurlyeq \hat{C} \otimes \hat{D} \Leftrightarrow \varphi_0(\hat{A}) \cdot \varphi_0(\hat{B}) \leqslant \varphi_0(\hat{C}) \cdot \varphi_0(\hat{D}).$$

That rank $\varphi = 1$ follows from a result of Aczél *et al.* (1960).

(c) Translation of the problem from functional language back to the language of relations.

We switch from $\hat{A} \in \mathscr{V}$ and φ_0 on $\mathscr{V}$ to $A \in \mathfrak{A}$ and P on $\mathfrak{A}$. This completes the proof.

It should perhaps be pointed out that the *FAQQP*-structures exemplify an important class of finite quadratic measurement structures not previously discussed in the literature.

3. QUALITATIVE INFORMATION STRUCTURES

3.1. *Probabilistically Independent Events*

As is well known, probabilistically independent events play an essential role in the definitions of information and entropy. The independence relation between events is defined entirely in terms of the probability measure $P: P(A \cap B) = P(A) \cdot P(B)$. One wonders whether it is possible to give a definition of a corresponding binary relation $\coprod$ on $\mathfrak{A}$ in terms of the qualitative probability relation $\preccurlyeq$ on $\mathfrak{A}$. It is trivial to see by constructing a model that this is not possible in terms of finite probability structures. But, as it has been pointed out, such a relation can be defined in terms of *FAQQP*-structures by (2.4).

In this paragraph we state a theorem about the basic properties of $\coprod$. The content of the clauses should be clear.

THEOREM 3: *If* $\langle \Omega, \mathfrak{A}, \preccurlyeq \rangle$ *is a FAQQP*-structure, *then the following formulas are valid when all variables run over* $\mathfrak{A}$*:*

(1) $\emptyset \coprod A;$

(2) $\Omega \coprod A;$

(3) $A \coprod A \Leftrightarrow (A \sim \Omega \vee A \sim \emptyset);$

(4) $A \coprod A \Rightarrow A \coprod B;$

(5) $A \coprod B \ \& \ A \perp B \Rightarrow (A \sim \emptyset \vee B \sim \emptyset);$

(6) $A \coprod B \ \& \ A \subseteq B \Rightarrow (A \sim \emptyset \vee B \sim \Omega);$

(7) $A \coprod B \ \& \ A \sim B \Rightarrow \bar{A}B \sim A\bar{B};$

(8) $A \coprod B \Leftrightarrow B \coprod A;$

(9) $A \coprod B \Leftrightarrow A \coprod \bar{B};$

(10) $A \coprod B \Leftrightarrow \bar{A} \coprod \bar{B};$

(11) $A \coprod B \Rightarrow AB \prec B$, if $A \prec \Omega \; \& \; \emptyset \prec B$;

(12) $A \coprod B \Rightarrow (\emptyset \prec A \; \& \; \emptyset \prec B \Rightarrow \emptyset \prec AB)$;

(13) $A \coprod B \; \& \; B \coprod C \Rightarrow (AB \coprod C \Leftrightarrow A \coprod BC)$;

(14) $A \coprod B \; \& \; C \coprod D \Rightarrow (A \preccurlyeq C \; \& \; B \preccurlyeq D \Rightarrow AB \preccurlyeq CD)$;

(15) $A \coprod B \; \& \; A \coprod C \Rightarrow A \coprod B \cup C$, if $B \perp C$;

(16) $A \coprod B \; \& \; A \coprod C \Rightarrow A \coprod B \cap C$, if $B \cup C = \Omega$;

(17) $A \coprod B \; \& \; A \coprod C \Rightarrow (B \preccurlyeq C \Leftrightarrow A \cup B \preccurlyeq A \cup C)$, if $A \prec \Omega$;

(18) $A \coprod B \; \& \; A \coprod C \Rightarrow (AB \preccurlyeq AC \Leftrightarrow B \preccurlyeq C)$, if $\emptyset \prec A$;

(19) $A \sim C \; \& \; AB \sim CB \Rightarrow (A \coprod B \Leftrightarrow C \coprod B)$;

(20) $\forall_{i<n} (A_i B_i \preccurlyeq A_{\alpha_i} B_{\beta_i}) \Rightarrow A_{\alpha_n} B_{\beta_n} \preccurlyeq A_n B_n$, if

$\forall_{i \leqslant n} (A_i \coprod B_i \; \& \; A_{\alpha_i} \coprod B_{\beta_i} \; \& \; \emptyset \prec A_i B_i)$, *and*

α, β *are permutations on* $\{1, 2, \ldots, n\}$.

The proof is a routine application of Theorem 1. We shall use this theorem throughout Section 3.2. It is rather disappointing that the qualitative independence relation $\coprod$, which plays a central role in probability theory, has such complicated properties.

Marczewski (1958) argued that probabilistic independence has a different nature from the notions of algebraic, logical, and set-theoretic independence. That this is not precisely true was demonstrated by Maeda (1963).

The independence relation $\coprod$ can be extended to any (finite) family of events $\{A_i\}_{i \in I} \subseteq \mathfrak{A}$ with more than two elements in such a way that the following equivalence is preserved:

$$\{A_i\} \coprod_{i \in I} \Leftrightarrow \forall_{\phi \neq I_0 \subseteq I} \left[P\left(\bigcap_{i \in I_0} A_i\right) = \prod_{i \in I_0} P(A_i)\right].$$

It is sufficient to put

(i) $\{A, B\} \coprod \Leftrightarrow A \coprod B$;

(ii) $\{A_i\} \coprod_{i \in I} \Leftrightarrow \forall_{\phi \neq I_0 \subset I} \left[\{A_i\} \coprod_{i \in I_0} \; \& \; \bigcap_{i \in I_0} A_i \coprod \bigcap_{i \in I - I_0} A_i\right]$.

It can be shown easily that

(1) $\{A_i\} \coprod_{i \in I} \Rightarrow A_i \coprod_{\substack{i, j \in I \\ i \neq j}} A_j$;

$$(2) \qquad \{A_i\} \coprod_{i \in I} \Rightarrow \{B_i\} \coprod_{i \in I}, \quad \text{if} \quad \forall_{i \in I} [B_i = A_i \vee B_i = \bar{A}_i];$$

$$(3) \qquad \{A_i\} \coprod_{i \in I} \Leftrightarrow \forall_{J \subseteq I} [\{A_i\} \coprod_{i \in J}];$$

$$(4) \qquad \{A_i\} \coprod_{i \in I} \Rightarrow \forall_{I_1, I_2 \subseteq I \,\&\, I_1 \perp I_2} [\bigcap_{i \in I_1} A_i \coprod \bigcap_{i \in I_2} A_i].$$

We shall not need these rather general properties; further details about $\coprod_{i \in I}$ are therefore omitted.

Perhaps we should point out that the I-place relation $\coprod_{i \in I}$ enables us to treat probabilistic independence in lattice-theoretic terms. In particular, the lattice-theoretic notion of independence coincides under certain reasonable conditions with the probabilistic relation $\coprod$. As mentioned before, this is contrary to what Marczewski (1958) maintained.

3.2. *Qualitative Information Structures*

In this section, unlike the earlier ones, we shall work with *infinite* Boolean algebras; as we shall see, the results will be somewhat more interesting. We are able to give a definition of information measure without *any recourse* to probabilistic notions.

The structure to be studied is a Boolean algebra $\mathfrak{A}$ enriched by two binary relations $\coprod$ and $\preccurlyeq$; the relation $\coprod$ can be interpreted as follows:

$$A \coprod B \Leftrightarrow \text{Event } A \text{ is independent of event } B (A, B \in \mathfrak{A}),$$

and the $\preccurlyeq$ is interpreted as before:

$$A \preccurlyeq B \Leftrightarrow \text{Event } A \text{ does not have more information than event } B (A, B \in \mathfrak{A}).$$

The novelty here is that we give axioms for $\coprod$, $\preccurlyeq$, and $\mathfrak{A}$ which, without recourse to probability theory, ensure the existence of an information measure in the standard sense.

A formalization of a notion of qualitative independence to match the standard probabilistic notion has been needed for a long time, but I am not aware of any serious previous attempts to solve this problem. In this section we shall try to work out such a formalization. First, perhaps, we should turn to the definition:

DEFINITION 2: *Let Ω be a nonempty set, $\mathfrak{A}$ a nonempty family of subsets* *of Ω such that it is a Boolean algebra, and* $\coprod$ *and* $\preccurlyeq$ *binary relations on* $\mathfrak{A}$.

Then the quadruple $\langle \Omega, \mathfrak{A}, \preccurlyeq, \coprod \rangle$ *is called a qualitative information structure* (*QI*-structure) if and only if *the following conditions are satisfied when all variables run over* $\mathfrak{A}$;

(I_1) $\emptyset \coprod A$;

(I_2) $A \coprod B \Rightarrow B \coprod A$;

(I_3) $A \coprod B \Rightarrow \bar{B} \coprod A$;

(I_4) $A \coprod B \;\&\; A \coprod C \Rightarrow A \coprod B \cup C$, *if* $B \perp C$;

(I_5) $\Omega \prec \emptyset$;

(I_6) $A \preccurlyeq \emptyset$;

(I_7) $A \preccurlyeq B \vee B \preccurlyeq A$;

(I_8) $A \preccurlyeq B \;\&\; B \preccurlyeq C \Rightarrow A \preccurlyeq C$;

(I_9) $A \coprod B \;\&\; A \perp B \Rightarrow (A \sim \emptyset \vee B \sim \emptyset)$;

(I_{10}) $A \preccurlyeq B \Leftrightarrow A \cup C \preccurlyeq B \cup C$, *if* $C \perp A, B$;

(I_{11}) $A \preccurlyeq B \Leftrightarrow A \cap C \preccurlyeq B \cap C$, *if* $C \coprod A, B \;\&\; C \prec \emptyset$;

(I_{12}) $A \preccurlyeq B \;\&\; C \preccurlyeq D \Rightarrow A \cup C \preccurlyeq B \cup D$, *if* $B \perp D$;

(I_{13}) $A \preccurlyeq B \;\&\; C \preccurlyeq D \Rightarrow A \cap C \preccurlyeq B \cap D$, *if* $A \coprod C \;\&\; B \coprod D$;

(I_{14}) *If* $A_i \coprod A_j$ *for* $i \neq j \;\&\; i, j \leqslant n$, *then*

$$\forall B \exists A_{n+1} \underset{i \leqslant n}{\forall} (A_i \coprod A_{n+1} \;\&\; B \sim A_{n+1}),$$

where $A \perp B$ for $A, B \in \mathfrak{A}$ means $A \cap B = \emptyset$.

Remarks:

(i) All axioms but the last one, which forces $\mathfrak{A}$ to be infinite if $\mathfrak{A}/\sim$ has more than two elements, are plausible enough. Axiom (I_{14}) could be replaced by some kind of Archimedean axiom. Moreover, the reader may find some relationship to Luce's *extensive measurement system* (see Luce, 1967, p. 782). His set **B** corresponds here to the set of independent couples $\langle A, B \rangle$ ($A, B \in \mathfrak{A}$) and his operation $\circ$ corresponds to the Boolean operation $\cap$.

(ii) The axioms may be divided into three classes: First, those that point out the properties of $\coprod$; second, the axioms for $\preccurlyeq$, and third, the interacting axioms giving the relationship between $\coprod$ and $\preccurlyeq$. There is no doubt about their consistency.

(iii) Instead of taking a Boolean algebra $\mathfrak{A}$, we could consider a complete complemented modular lattice in which the relation of disjointness would become a new primitive notion. In this case our axioms for disjointness and $\preccurlyeq$ come rather close to dimension theory of continuous geometry. In fact, the proof of the Representation Theorem for *QI*-structures will be somewhat similar to the proof of the existence of a dimension function on a lattice (see Von Neumann, 1960, pp. 42–53, and Skornyakov, 1964, pp. 88–97).

It is easy to show that Definition 2 implies Theorem 3, if we replace the qualitative probability relation in Theorem 3 by the qualitative information relation.

For purposes of representation we shall need a couple of notions that will be developed in the sequel.

Let $\langle \Omega, \mathfrak{A}, \preccurlyeq, \coprod \rangle$ be a *QI*-structure. Then $\mathfrak{A}/\sim = \{[A]_\sim : A \in \mathfrak{A}\}$, where $[A]_\sim = \{B : A \sim B\}$. For simplicity we put $[A] = [A]_\sim$.

Axioms (I_7) and (I_8) force $\sim$ to be an equivalence relation on $\mathfrak{A}$. Therefore $\mathfrak{A}/\sim$ is a well-defined quotient structure. It is natural to define for $A, B \in \mathfrak{A}$:

$$[A] \leqslant [B] \Leftrightarrow A \preccurlyeq B .$$

Using (I_8), it is easy to see that this definition of $\leqslant$ does not depend on the particular choice of representatives A, B, that is,

$$[A] \leqslant [B] \Leftrightarrow [A_1] \leqslant [B_1], \quad \text{if} \quad A \sim A_1 \quad \text{and} \quad B \sim B_1 .$$

It is immediately seen that $\leqslant$ is reflexive, antisymmetric, transitive, and connected. Moreover, $[\Omega] \leqslant [A] \leqslant [\emptyset]$.

Another natural step is the definition of the following binary operation on $\mathfrak{A}/\sim$:

$$[A] \cdot [B] = [A_1 \cap B_1] ,$$

where A_1, B_1 are such elements of $\mathfrak{A}$ that $A_1 \sim A$, $B_1 \sim B$, and $A_1 \coprod B_1$. Axioms (I_{12})–(I_{14}) assure us that the term $[A] \cdot [B]$ is defined for all couples $\langle A, B \rangle$ $(A, B \in \mathfrak{A})$ and that it does not depend on the choice of representatives.

Using the axioms for *QI*-structure, the following lemma is almost immediate.

LEMMA 1: *Let* $\langle \Omega, \mathfrak{A}, \preccurlyeq, \coprod \rangle$ *be a QI*-structure. *Then the following eight clauses are valid when all variables run over* $\mathfrak{A}$*:*

(1) $[A]\cdot[B]=[B]\cdot[A]$;
(2) $[A]\cdot([B]\cdot[C])=([A]\cdot[B])\cdot[C]$;
(3) $[\emptyset]\cdot[A]=[\emptyset]$;
(4) $[\Omega]\cdot[A]=[A]$;
(5) $[A]\leqslant[B]\Leftrightarrow[A]\cdot[C]\leqslant[B]\cdot[C], [C]<[\emptyset]$;
(6) $[A]\leqslant[B]\ \&\ [C]\leqslant[D]\Rightarrow[A]\cdot[C]\leqslant[B]\cdot[D]$;
(7) $[A]\leqslant[A]\cdot[B]$;
(8) $[A]\leqslant[B]\Rightarrow[\bar{B}]\leqslant[\bar{A}]$.

As a special case of the previous 'multiplication' operation is the following 'exponentiation' operation:

$$[A]^n=\begin{cases}[\emptyset] & \text{if } n=0;\\ [A]^{n-1}\cdot[A] & \text{if } n>0.\end{cases}$$

It is easy to check that this definition is again meaningful for all $A\in\mathfrak{A}$ and $n\geqslant 0$. Besides that, it does not depend on the choice of representatives.

The following lemma is a trivial consequence of the above definition and the axioms (I_1)–(I_{14}).

LEMMA 2: *Let* $\langle\Omega,\mathfrak{A},\preccurlyeq,\coprod\rangle$ *be a QI*-structure. *Then the following six clauses are valid when all variables run over* $\mathfrak{A}$:

(1) $[\Omega]^n=[\Omega]$;
(2) $[\emptyset]^n=[\emptyset]$;
(3) $([A]^n)^m=[A]^{n\cdot m}$;
(4) $[A]^n\cdot[A]^m=[A]^{n+m}$;
(5) $[A]^n\leqslant[A]^m\Leftrightarrow n\leqslant m$, if $[\Omega]<[A]<[\emptyset]$;
(6) $[A]^n=[B]^n\Rightarrow[A]=[B]$, if $n>0$.

LEMMA 3: *The set* $C(A)=\{m/n\colon [U]^n\leqslant[A]^m\}$ *for some fixed U such that* $\Omega\prec U\prec\emptyset$ *and* $A\in\mathfrak{A}$ *defines a Dedekind cut in the set of nonnegative rational numbers* Ra^+, *that is,*

(i) $p\in C(A)\ \&\ p\leqslant q\Rightarrow q\in C(A)$, *if* $\Omega\prec A\prec\emptyset$;
(ii) $C(\emptyset)=Ra^+$;
(iii) $C(\Omega)=\emptyset$.

PROOF: Take $p\in C(A)$ and $p\leqslant q$. Then $p=m/n$ and $q=s/t$ for some natural numbers $m, n, s, t (n \neq 0, t \neq 0)$. By the assumption we have $m \cdot t < s \cdot n$ and $[U]^n \leqslant [A]^m$. Lemma 2 implies $[U]^{n \cdot t} < [A]^{m \cdot t}$. Suppose that $\neg p \in C(A)$. Then $[A]^s < [U]^t$ and therefore also $[A]^{s \cdot n} < [U]^{t \cdot n} \leqslant [A]^{m \cdot t}$. Using Lemma 2 we get $s \cdot n \leqslant m \cdot t$, which is contrary to the supposition. Clauses (ii) and (iii) follow from the definition of $\leqslant$.

Note that if there is no U such that $\Omega \prec U \prec \emptyset$, then $\mathfrak{A}/\sim$ has only two elements. The Representation Theorem which we wish to prove is trivial for this case.

It is easy to see also that

$$A \preccurlyeq B \Leftrightarrow C(A) \subseteq C(B).$$

The real number which is defined by the Dedeking cut $C(A)$ in Ra^+ will be denoted by $|C(A)|$.

Let us now define

$$\varphi([A]) = \alpha^{-|C(A)|}, \quad \text{where} \quad 1 < \alpha < \infty.$$

Clearly, $\varphi([U]) = \alpha^{-1}$, since $|C(A)| = 1$ for $[A] = [U]$. Let us put for simplicity $\varphi[A] = \varphi([A])$, if $A \in \mathfrak{A}$. The following lemma is of basic importance for further developments of our ideas.

LEMMA 4: *Let* $\langle \Omega, \mathfrak{A}, \preccurlyeq, \coprod \rangle$ *be a* QI-structure. *Then the following formulas are valid for all* $A, B \in \mathfrak{A}$:

(i) $A \preccurlyeq B \Leftrightarrow \varphi[B] \leqslant \varphi[A]$;

(ii) $A \coprod B \Rightarrow \varphi[A \cap B] = \varphi[A] \cdot \varphi[B]$;

(iii) $\varphi[\emptyset] = 0$;

(iv) $\varphi[\Omega] = 1$.

PROOF:

(i) $A \preccurlyeq B \Leftrightarrow C(A) \subseteq C(B) \Leftrightarrow -|C(B)| \leqslant -|C(A)|$.

(ii) We put $C(A) \oplus C(B) = \{m/n + m'/n' \colon m/n \in C(A) \ \& \ m'/n' \in C(B)\}$ and show that $C(A \cap B) = C(A) \oplus C(B)$, if $A \coprod B$. Then $\varphi[A \cap B] = \varphi([A] \cdot [B]) = \alpha^{-|C(A \cap B)|} = -|(C(A) \oplus C(B))| = \varphi[A] \cdot \varphi[B]$, if $A \coprod B$.

(iii) and (iv) are obvious.

We shall now define a partial binary operation in $\mathfrak{A}/\sim$ as follows:

$$[A] + [B] = [A_1 \cup B_1],$$

where $A_1, B_1 \in \mathfrak{A}$ are such that $A_1 \sim A$, $B_1 \sim B$ and $A_1 \cap B_1 = \emptyset$.

Let $\mathscr{D}$ be the set of those pairs $\langle A, B\rangle$ ($A, B \in \mathfrak{A}$) for which the definition of the term $[A]+[B]$ is meaningful. Then the following lemma is an obvious consequence of the axioms for QI-structures.

LEMMA 5: *Let* $\langle \Omega, \mathfrak{A}, \preccurlyeq, \coprod\rangle$ *be a* QI-structure. *Then the following seven clauses are valid when all variables run over* $\mathfrak{A}$*:*

$$\text{(1)} \qquad [\emptyset] + [A] = [A];$$

$$\text{(2)} \qquad [A] + [B] = [B] + [A], \quad \text{if} \quad \langle A, B\rangle \in \mathscr{D};$$

$$\text{(3)} \qquad [A] + ([B] + [C]) = ([A] + [B]) + [C], \quad \text{if} \quad \langle A, B\rangle \in \mathscr{D} \quad \text{and} \quad \langle D, C\rangle \in \mathscr{D} \quad \text{for some} \quad D \in [A] + [B].$$

$$\text{(4)} \qquad [A] \leqslant [B] \Leftrightarrow [A] + [C] \leqslant [B] + [C], \quad \text{if} \quad \langle A, C\rangle, \langle B, C\rangle \in \mathscr{D};$$

$$\text{(5)} \qquad [A] \leqslant [B] \;\&\; [C] \leqslant [D] \Rightarrow [A] + [C] \leqslant [B] + [D], \quad \text{if} \quad \langle A, C\rangle \in \mathscr{D}, \langle B, D\rangle \in \mathscr{D};$$

$$\text{(6)} \qquad [A] + [B] \leqslant [A], \quad \text{if} \quad \langle A, B\rangle \in \mathscr{D};$$

$$\text{(7)} \qquad [A]\cdot[B \cup C] = [A]\cdot[B] + [A]\cdot[C], \quad \text{if} \quad B \cap C = \emptyset \quad \text{and} \quad A \coprod B, C.$$

Lemma 5 implies the existence of a real function $\mathscr{F}$ such that

(i) $\varphi[A \cup B] = \mathscr{F}(\varphi[A], \varphi[B])$, if $A \cap B = \emptyset$;

(ii) $\mathscr{F}$ is symmetric, associative, and monotonic.

From the properties of $\mathfrak{A}$, conditions (i) and (ii), and Theorem 1 of Aczél, (1966, p. 268), one can show that $\mathscr{F}$ must be a one-one real function, say η: $[0, 1] \to [0, 1]$, such that

$$\eta \circ \varphi [A \cup B] = \eta \circ \varphi [A] + \eta \circ \varphi [B], \quad \text{if} \quad A \cap B = \emptyset.$$

Lemma 5 enables us to put

$$\varphi[CA \cup CB] = \varphi[C]\cdot\varphi[A \cup B] = \varphi[C]\cdot\eta^{-1}(\eta\circ\varphi[A] + \eta\circ\varphi[B]),$$

if $C \coprod A, B$ and $A \cap B = \emptyset$, where $AB = A \cap B$ for $A, B \in \mathfrak{A}$.

If we put $\psi[A] = \eta_0 \varphi[A]$, then we can write also

$$\psi[CA] + \psi[CB] = \eta(\varphi[C]\cdot\eta^{-1}(\psi[A] + \psi[B]))$$

under the same conditions imposed on A, B, and C as before.

By reformulating the above identity and putting

$$\chi_\gamma(\alpha) = \eta(\varphi[C]\cdot\eta^{-1}(\alpha)) \quad \text{for} \quad \alpha\in[0, 1],$$

where $\gamma=\varphi[C]$, we get

$$\chi_\gamma(\psi[A]) + \chi_\gamma(\psi[B]) = \chi_\gamma(\psi[A] + \psi[B]), \quad \text{if} \quad C\coprod A, B \quad \text{and} \quad A\cap B=\emptyset,$$

which is, in fact, Cauchy's functional equation.

Again, the properties of $\mathfrak{A}$, conditions (i) and (ii) will allow us to conclude that this equation has a solution of the form

$$\chi_\gamma(\psi[A]) = \xi(\gamma)\cdot\psi[A], \quad \text{if} \quad C\coprod A.$$

Since $\chi_\gamma(\chi_\delta(\psi[A]))=\chi_{\gamma\cdot\delta}(\psi[A])$, if $C\coprod D$, $D\coprod A$, $CD\coprod A$, $\delta=\varphi[D]$, it must be the case that $\xi(\gamma\cdot\delta)=\xi(\gamma)\cdot\xi(\delta)$. Hence $\xi(\gamma)=\gamma^\beta$, $\beta>0$. So that after substituting everything back into the original equation for $\mathscr{F}$, we obtain

$$\varphi([A\cup B]) = \sqrt[\beta]{(\varphi[A])^\beta + (\varphi[B])^\beta}, \quad \text{if} \quad A\cap B=\emptyset.$$

Now, since,

$$(\varphi[A])^\beta = (\alpha^{-|C(A)|})^\beta = (\alpha^\beta)^{-|C(A)|},$$

we can choose $\gamma=\alpha^\beta$ and put $\varphi_0([A])=\gamma^{-|C(A)|}$. Then we can show that φ_0 has the following properties:

(a) $\varphi_0[A\cup B] = \varphi_0[A] + \varphi_0[B]$, if $A\cap B=\emptyset$;

(b) $\varphi_0[A\cap B] = \varphi_0[A]\cdot\varphi_0[B]$, if $A\coprod B$;

(c) $-\log_2\varphi_0[A] \leqslant -\log_2\varphi_0[B] \Leftrightarrow A \preccurlyeq B$;

(d) $\varphi_0[\emptyset] = 0$;

(e) $\varphi_0[\Omega] = 1$.

By this conclusion we have arrived at the desired Representation Theorem:

THEOREM 4 (Representation Theorem): *Let* $\langle\Omega, \mathfrak{A}, \preccurlyeq, \coprod\rangle$ *be a QI-*structure. *Then there exists a finitely additive probability measure P on* $\mathfrak{A}$ *such that* $\langle\Omega, \mathfrak{A}, P\rangle$ *is a probability space, and*

(1) $A\preccurlyeq B \Leftrightarrow I(A)\leqslant I(B)$;

(2) $A\coprod B \Leftrightarrow I(A\cap B) = I(A) + I(B)$;

(3) $I(A) = -\log_2 P(A)$.

PROOF: We put $P(A)=\varphi_0([A])$ for $A \in \mathfrak{A}$. Then from the previous discussion of φ_0 it is easy to see that (1)–(3) are satisfied.

Clearly all the axioms (I_1)–(I_{13}) are necessary conditions for the existence of the information measure I. Axiom (I_{14}) is clearly not necessary. We leave open the problem of formulating axioms both necessary and sufficient for the existence of the measure I.

Aware of the relatively complicated necessary and sufficient conditions for the existence of a probability measure in an infinite Boolean algebra $\mathfrak{A}$, I will not go here into further details.

$I(A)=-\log_2 P(A)$ is sometimes called *self-information* of the event A. The next (slightly more general) notion is the so-called *conditional self-information of event A*, given event B: $I(A/B)=-\log_2 P(A/B)$. A further generalization leads to the *conditional mutual information of events A and B, given event C*:

$$I(A:B/C)=\log_2 \frac{P(AB/C)}{P(A/C)\cdot P(B/C)}.$$

Naturally, we also would like to give representation theorems for these more complicated measures.

In this last case, our basic structure would be the set of complicated entities $A:B/C$ $(A, B, C \in \mathfrak{A}, \emptyset \prec C)$ and two binary relations $\coprod$ and $\preccurlyeq$ on this set of entities. In fact, it would be enough to consider the formulas $A_1:B_1/C_1 \preccurlyeq A_2:B_2/C$ and $A/C \coprod B/C$, since the remainder can be defined as follows:

$$A:B \preccurlyeq C:D \Leftrightarrow A:B/\Omega \preccurlyeq C:D/\Omega\,;$$
$$A \preccurlyeq B \Leftrightarrow A:A \preccurlyeq B:B\,;$$
$$A/B \preccurlyeq C/D \Leftrightarrow A:A/B \preccurlyeq C:C/D\,;$$
$$A \coprod B \Leftrightarrow A/\Omega \coprod B/\Omega, \quad \text{where} \quad A, B, C, D \in \mathfrak{A}.$$

Some of the properties of the *qualitative conditional mutual information relation* $\preccurlyeq$ are analogous to those of the *qualitative self-information relation*. For example,

$$(A_1:C_1/E_1 \preccurlyeq A_2:C_2/E_2 \ \&\ B_1:D_1/E_1 \preccurlyeq B_2:D_2/E_2)$$
$$\Rightarrow A_1B_1:C_1D_1/E_1 \preccurlyeq A_2B_2:C_2D_2/E_2, \quad \text{if}$$
$$A_i/E_i \coprod B_i/E_i \ \&\ C_i/E_i \coprod D_i/E_i \ \&\ A_iB_i/E_i \coprod C_iD_i/E_i, \quad i=1,2.$$

We do not intend to develop further details here, because of the rather complicated nature of these properties. Note that we have several notions interacting here: conditional events, the independence relation, and the mutual information relation. From the point of view of algebraic measurement theory the problem is to give measurability conditions for very complicated relations defined on the above-mentioned complex entities.

4. QUALITATIVE ENTROPY STRUCTURES

4.1. *Algebra of Probabilistic Experiments*

In Section 1.2 we stated that the main algebraic entity to be used in the definition of an entropy structure is the partition of the set of elementary events Ω. We decided to call partitions *experiments* and the set of all possible experiments over Ω has been denoted by **P**.

We can, alternatively, analyze qualitative entropy in terms of Boolean algebras *generated by experiments* (partitions of the sample space). Experiments are the sets of atoms of these Boolean algebras, and thus there is a one-one correspondence between them. Formally we get nothing new.

If we are given two partitions $\mathscr{P}_1, \mathscr{P}_2$, we can define the so-called *finer-than relation* ($\subseteq$) between them as follows:

$$\mathscr{P}_1 \subseteq \mathscr{P}_2 \Leftrightarrow \forall A \in \mathscr{P}_1 \exists B \in \mathscr{P}_2 (A \subseteq B).$$

An equivalent definition would be:

$$\mathscr{P}_1 \subseteq \mathscr{P}_2 \Leftrightarrow \forall_A (A \in \mathscr{P}_2 \Rightarrow A = \bigcup_{i \leqslant k} B_i) \quad \text{for some } B_i\text{'s}$$

$$\text{from } \mathscr{P}_1, \quad i \leqslant k.$$

We have in particular,

$$\cdots \subseteq \{\emptyset, \bar{A}, A\bar{B}, AB\bar{C}, ABC\} \subseteq \{\emptyset, \bar{A}, A\bar{B}, AB\} \subseteq \{\emptyset, \bar{A}, A\}.$$

Now, given a relation on a set, it is natural to ask whether it is possible to define some kind of lattice operations induced by this relation. The answer here is positive. The first operation of interest is called the *product of experiments*:

$$\mathscr{P}_1 \cdot \mathscr{P}_2 = \{A \cap B : A \in \mathscr{P}_1 \ \& \ B \in \mathscr{P}_2\} \ (\mathscr{P}_1, \mathscr{P}_2 \in \mathbf{P}) \tag{4.1a}$$

or, more generally,

$$\prod_{i\leqslant n} \mathscr{P}_i = \{\bigcap_{i\leqslant n} A_i : \underset{i\leqslant n}{\forall} (A_i \in \mathscr{P}_i)\}.$$

$\mathscr{P}_1 \cdot \mathscr{P}_2$ is the *greatest experiment* which is finer than both $\mathscr{P}_1$ and $\mathscr{P}_2$; that is,

(i) $\mathscr{P}_1 \cdot \mathscr{P}_2 \subseteq \mathscr{P}_1 \;\&\; \mathscr{P}_1 \cdot \mathscr{P}_2 \subseteq \mathscr{P}_2$,

(ii) $\mathscr{P} \subseteq \mathscr{P}_1 \;\&\; \mathscr{P} \subseteq \mathscr{P}_2 \Rightarrow \mathscr{P} \subseteq \mathscr{P}_1 \cdot \mathscr{P}_2$.

Obviously $\mathscr{P}_1 \subseteq \mathscr{P}_2 \Leftrightarrow \mathscr{P}_1 \cdot \mathscr{P}_2 = \mathscr{P}_1$.

The dual operation is called the *sum of experiments* and is defined as follows:

$$\mathscr{P}_1 + \mathscr{P}_2 = \prod_{\substack{\mathscr{P}_1 \subseteq \mathscr{P} \\ \mathscr{P}_2 \subseteq \mathscr{P}}} \{\mathscr{P}\}, \tag{4.1b}$$

where $\prod$ denotes the standard generalization of the operation $\cdot$ to sets of experiments. A more concrete definition is the following:

$$\mathscr{P}_1 + \mathscr{P}_2 = \{\bigcup_{i\leqslant n} A_i : A_1 \text{ not } \perp A_2 \text{ not } \perp \ldots \text{ not } \perp A_n$$
$$\text{is a maximal chain of overlapping events in } \mathscr{P}_1 \cup \mathscr{P}_2\},$$

where A_i not $\perp A_j \Leftrightarrow \neg A_i \perp A_j$, $i, j \leqslant n$ and $A \perp B \Leftrightarrow A \cap B = \emptyset$, as before.

$\mathscr{P}_1 + \mathscr{P}_2$ is the *smallest experiment* coarser than both $\mathscr{P}_1$ and $\mathscr{P}_2$; that is,

(i) $\mathscr{P}_1 \subseteq \mathscr{P}_1 + \mathscr{P}_2 \;\&\; \mathscr{P}_2 \subseteq \mathscr{P}_1 + \mathscr{P}_2$,

(ii) $\mathscr{P}_1 \subseteq \mathscr{P} \;\&\; \mathscr{P}_2 \subseteq \mathscr{P} \Rightarrow \mathscr{P}_1 + \mathscr{P}_2 \subseteq \mathscr{P}$.

Again it is clear that $\mathscr{P} \subseteq \mathscr{P}_2 \Leftrightarrow \mathscr{P}_1 + \mathscr{P}_2 = \mathscr{P}_2$.

The partition $\mathscr{O} = \{\emptyset, \Omega\}$ is called the *maximal experiment* and the partition $\mathscr{A} = \{\{\omega\} : \omega \in \Omega\} \cup \{\emptyset\}$ is called the *minimal experiment*. Clearly $\mathscr{A} \subseteq \mathscr{P} \subseteq \mathscr{O}$ for any $\mathscr{P} \in \mathbf{P}$. Equally straightforward are

$$\mathscr{P} \cdot \mathscr{O} = \mathscr{P} \quad \text{and} \quad \mathscr{P} + \mathscr{O} = \mathscr{O},$$
$$\mathscr{P} \cdot \mathscr{A} = \mathscr{A} \quad \text{and} \quad \mathscr{P} + \mathscr{A} = \mathscr{P}.$$

The total number of experiments e_n over a finite set Ω with n elements is given by the following recursive formula:

$$e_0 = 1 \;\&\; e_{n+1} = \sum_{i=0}^{n} \binom{n}{i} e_i.$$

The reader can easily check that the structure $\langle \mathbf{P}, \mathcal{O}, \mathcal{A}, +, \cdot, \subseteq \rangle$ satisfies the lattice axioms. Unfortunately, it is not a Boolean algebra, so there is no hope of getting any useful entropy measure on it without further assumptions. The help will come from the independent relation $\coprod$ on experiments.

The structure $\langle \Omega, \mathbf{P}, \subseteq \rangle$ in which the product and sum (4.1a), (4.1b) of experiments are defined, will be called the *algebra of experiments over the set of elementary events* Ω. The reader may be familiar with the following chain of isomorphisms:

> $\mathbf{P} \cong$ Lattice of equivalence relations on $\Omega \cong$ Lattice of complete Boolean subalgebras of $\mathfrak{A} \cong$ Lattice of subgroups of a finite group $\cong$ Lattice of subgraphs of a topological graph $\cong$ Finite geometric system of lines and pencils $\cong$ Lattice of modal operators on $\mathfrak{A}$ satisfying the modal axiom system S_5.

Any one of these structures could be used as the underlying algebraic structure of the entropy measure. For example, in graph representation, the entropy measure could be viewed also as a measure of the *relative complexity of graphs*:

$$H_{\mathscr{P}}(\mathscr{G}) = -\sum_{A \in \mathscr{P}} P(A) \cdot \log_2 P(A),$$

where

$$P(A) = \left|\frac{A}{V}\right|, \quad A \in \mathscr{P}, |\quad| = \text{cardinality},$$

and $\mathscr{P}$ is the partition of the set of vertices V of the graph $\mathscr{G}$. In the same way we can talk about the *complexity of a group*. By the *complexity of a mathematical structure* we mean a function of all the elements of a (complete) set of invariants of the given structure.

4.2. *Algebra of Incomplete Experiments*

In the case that not all outcomes of an experiment are observable, it is convenient to introduce the notion of so-called *incomplete* (partial) *experiment* which is defined as a partition of an event A $(A \in \mathfrak{A})$. By introducing this notion the family of experiments $\mathbf{P}$ can be extended to a richer family, namely, to the set of incomplete experiments: $\mathbf{E}$.

$\mathscr{P} \in \mathbf{E}$ will be called a *complete experiment* if and only if $\bigcup \mathscr{P} = \Omega$.

Now, if we define for $\mathscr{P}_1, \mathscr{P}_2 \in \mathbf{E}$:

(i) *Finer-than Relation:*

$$\mathscr{P}_1 \subseteq \mathscr{P}_2 \Leftrightarrow \bigcup \mathscr{P}_1 = \bigcup \mathscr{P}_2 \;\&\; \forall A \in \mathscr{P}_1 \exists B \in \mathscr{P}_2 [A \subseteq B];$$

(ii) *Product of Experiments:*

$$\mathscr{P}_1 \cdot \mathscr{P}_2 = \{A \cap B : A \in \mathscr{P}_1 \;\&\; B \in \mathscr{P}_2\};$$

(iii) *Sum of Experiments:*

$$\mathscr{P}_1 + \mathscr{P}_2 = \prod_{\substack{\mathscr{P}_1 \leqslant \mathscr{P} \\ \mathscr{P}_2 \leqslant \mathscr{P}}} \{\mathscr{P}\} \quad \text{(see section 4.1)};$$

(iv) *Direct Sum of Experiments:*

$$\mathscr{P}_1 \oplus \mathscr{P}_2 = \mathscr{P}_1 \cup \mathscr{P}_2, \quad \text{if} \quad \bigcup \mathscr{P}_1 \cap \bigcup \mathscr{P}_2 = \emptyset,$$

then $\mathbf{E}$ becomes a lattice with respect to $\subseteq$, $\cdot$, $+$, and moreover

$$\mathscr{P} \cdot (\mathscr{Q}_1 \oplus \mathscr{Q}_2) = \mathscr{P} \cdot \mathscr{Q}_1 + \mathscr{P} \cdot \mathscr{Q}_2, \quad \text{for} \quad \mathscr{P}, \mathscr{Q}_1, \mathscr{Q}_2 \in \mathbf{E}.$$

If $n = |\mathscr{P}|$ and $\mathbf{E}_n = \{\mathscr{P} : |\mathscr{P}| = n\}$, then clearly $\mathbf{E} = \bigcup_{n \geqslant 0} \mathbf{E}_n$, where $\mathbf{E}_n$ is called the *graded lattice of experiments.*

Suppose that we are given a probability space $\langle \Omega, \mathfrak{A}, P \rangle$. Then the entropy of an incomplete experiment $\mathscr{P} \in \mathbf{E}$ is given by the following definition:

$$H(\mathscr{P}) = \frac{-\sum_{A \in \mathscr{P}} P(A) \cdot \log_2 P(A)}{W(\mathscr{P})}, \tag{4.2}$$

where $W(\mathscr{P}) = \sum_{A \in \mathscr{P}} P(A)$ is called the *weight of the experiment* $\mathscr{P}$. Obviously $H|_{\mathbf{P}}$ specializes to the standard Shannon-Rényi entropy.

It was Rényi (1961) who gave the first axiomatization of H on $\mathbf{E}$ in the form of functional equations for H.

The structure $\langle \Omega, \mathbf{E}, \subseteq \rangle$ in which the product, sum and direct sum of experiments are defined as pointed out above, will be called the *algebra of incomplete experiments over* Ω.

Note that the graded algebra of experiments $\mathbf{E}_n$ induces a *partial entropy measure* $H_n = H|_{\mathbf{E}_n}$ which is a function of n variables. We shall have more to say about this function in Section 4.4.

Naturally, we can look for the qualitative entropy relation $\preccurlyeq$ on $\mathbf{E}$, corresponding to H in the sense of the formula (1.4). Again, it is a trivial matter to notice that $\preccurlyeq$ has to be at least a linear ordering modulo $\sim$, where, as expected, $\mathscr{P}_1 \sim \mathscr{P}_2 \Leftrightarrow \mathscr{P}_1 \preccurlyeq \mathscr{P}_2 \;\&\; \mathscr{P}_2 \preccurlyeq \mathscr{P}_1$, if $\mathscr{P}_1, \mathscr{P}_2 \in \mathbf{E}$. It is rather difficult to state the axioms for $\preccurlyeq$ on $\mathbf{E}$, which will describe the *local* property of H, that is, the relationship to the underlying probability measure.

4.3. *Independent Experiments*

DEFINITION 3: *Let* $\mathbf{Q} = \langle \Omega, \mathfrak{A}, \preccurlyeq \rangle$ *be a* *FAQQP*-structure *and let* $\langle \Omega, \mathbf{P}, \subseteq \rangle$ *be the algebra of experiments over* Ω. *Then we shall say that two experiments are independent,* $\mathscr{P}_1 \coprod \mathscr{P}_2$, if and only if

$$A \in \mathscr{P}_1 \;\&\; B \in \mathscr{P}_2 \Rightarrow A \coprod B, \quad \textit{for all } A, B \in \mathfrak{A}.$$

Some of the basic properties of independent experiments are stated in the following theorem.

THEOREM 5: If $\langle \Omega, \mathfrak{A}, \preccurlyeq \rangle$ is a *FAQQP*-structure modulo $\sim$ and $\langle \Omega, \mathbf{P}, \subseteq \rangle$ is the algebra of experiments over Ω, then the following formulas are valid for all $\mathscr{P}, \mathscr{P}_1, \mathscr{P}_2 \in \mathbf{P}$:

(1) $\mathscr{O} \coprod \mathscr{P}$;

(2) $\mathscr{P} \coprod \mathscr{P} \Rightarrow \mathscr{P} = \mathscr{O}$;

(3) $\mathscr{P}_1 \coprod \mathscr{P}_2 \Leftrightarrow \mathscr{P}_2 \coprod \mathscr{P}_1$;

(4) $\mathscr{P}_1 \coprod \mathscr{P}_2 \;\&\; \mathscr{P}_2 \subseteq \mathscr{P}_3 \Rightarrow \mathscr{P}_1 \coprod \mathscr{P}_3$;

(5) $\mathscr{P} \coprod \mathscr{P} \Rightarrow \mathscr{P} \coprod \mathscr{P}_1$;

(6) $\mathscr{P}_1 \coprod \mathscr{P}_2 \;\&\; \mathscr{P}_2 \coprod \mathscr{P}_3 \Rightarrow (\mathscr{P}_1 \cdot \mathscr{P}_2 \coprod \mathscr{P}_3 \Leftrightarrow \mathscr{P}_1 \coprod \mathscr{P}_2 \cdot \mathscr{P}_3)$;

(7) $\mathscr{P}_1 \coprod \mathscr{P} \;\&\; \mathscr{P}_2 \coprod \mathscr{P} \Rightarrow \mathscr{P}_1 \cdot \mathscr{P}_2 \coprod \mathscr{P}$, if $A \cup B = \Omega,\ A \in \mathscr{P}_1,\ B \in \mathscr{P}_2$;

(8) $\mathscr{P}_1 \coprod \mathscr{P}_2 \cdot \mathscr{P}_3 \;\&\; \mathscr{P}_1 \cdot \mathscr{P}_2 \coprod \mathscr{P}_3 \Rightarrow (\mathscr{P}_1 \coprod \mathscr{P}_2 \cdot \mathscr{P}_2 \coprod \mathscr{P}_3)$;

(9) $\mathscr{P} \coprod \mathscr{P}_1 \;\&\; \mathscr{P} \coprod \mathscr{P}_2 \Rightarrow (\mathscr{P} \cdot \mathscr{P}_1 = \mathscr{P} \cdot \mathscr{P}_2 \Leftrightarrow \mathscr{P}_1 = \mathscr{P}_2)$;

(10) $\mathscr{P}_1 \coprod \mathscr{P}_2 \;\&\; \mathscr{P}_1 \subseteq \mathscr{P}_2 \Rightarrow \mathscr{P}_2 = \mathscr{O}$.

The proof is a simple application of Theorem 3. The assumption that $\langle \Omega, \mathfrak{A}, \preccurlyeq \rangle$ is a *FAQQP*-structure is inessential. We could as well assume any *FQCP*-structure or even any other structure in which the relation $\coprod$ is defined for events (see Domotor, 1969).

The reader will notice that the relation $\coprod$ on $\mathbf{P}$ is not unlike the disjointness relation $\perp$ on $\mathfrak{A}$. In particular, $\bar{A} = \bigcap \{B : A \cup B = \Omega \;\&\; A \perp B\}$.

If we define similarly:

$$\bar{\mathscr{P}} = \prod_{\substack{\mathscr{P}\cdot\mathscr{Q}=\mathscr{A} \\ \mathscr{P} \amalg \mathscr{Q}}} \{\mathscr{Q}\} \quad \text{and} \quad \mathscr{P}_1 \wedge \mathscr{P}_2 = (\bar{\mathscr{P}}_1 \cdot \bar{\mathscr{P}}_2)\, \bar{\mathscr{P}},$$

then we get a Boolean algebra of those experiments, for which $\bar{\mathscr{P}}$ exists. If $\bar{\mathscr{P}}$ exists, then it is uniquely determined, as we can easily check by using Theorem 5 (9). Analogously, $\mathscr{P}_1 \wedge \mathscr{P}_2$ is uniquely determined, provided that it exists.

It is unfortunate that the independence relation $\amalg$ on **P** generates a Boolean algebra which is only a proper subset of the lattice **P**. We would hardly want to rule out those experiments which have no complements according to definition given above; for the entropy measure H_P is defined on the whole set **P**. On the other hand, it is highly desirable to have on **P** a richer structure than a lattice.

The reader will notice in Theorem 5 that the independence relation $\amalg$ on **P** satisfies the axioms of a semi-orthogonal relation on a lattice, as given by Maeda (1960). Theorem 5 also holds for incomplete experiments, provided that we put further assumptions in some of the clauses. We shall also have properties like

$$\mathscr{P} \amalg \mathscr{Q}_1 \;\&\; \mathscr{P} \amalg \mathscr{Q}_2 \Rightarrow \mathscr{P} \amalg \mathscr{Q}_1 \oplus \mathscr{Q}_2 .$$

4.4. *Qualitative Entropy Structures*

As already mentioned in Section 1.2, we shall develop a qualitative theory of entropy based on *qualitative probability theory*. The only primitive notions used will be the qualitative entropy relation $\preccurlyeq$ and the independence relation $\amalg$, both relations over **P**, the set of experiments.

In this section we shall use the following notation: If $A \in \mathfrak{A}$, the experiment $\{A, \bar{A}\}$ is called a *Bernoulli experiment*; the variables $\mathscr{B}, \mathscr{B}_1, \mathscr{B}_2, \ldots$ will run over Bernoulli experiments. Familiar enough is the fact that each experiment $\mathscr{P} \in \mathbf{P}$ can be written as a product $\prod_{i \leqslant n} \mathscr{B}_i$, where the family $\{\mathscr{B}_i\}_{i \leqslant n}$ is so chosen that no subset of it is sufficient for the job. This representation, unfortunately, is not unique.

Experiment $\mathscr{P}$ is called (*locally*) *equiprobable* if and only if $\forall A, B \in \mathfrak{A}$ $(A, B \in \mathscr{P} \Rightarrow A \sim B)$. The variables for equiprobable experiments will be $\mathscr{E}, \mathscr{E}_1, \mathscr{E}_2, \ldots.$

We call two experiments equivalent *modulo* $\sim$, in symbols, $\mathscr{P}_1 \equiv \mathscr{P}_2$, if and only if $\mathscr{P}_2 = [B_1 | A_1]([B_2 | A_2](\ldots([B_n | A_n]\mathscr{P}_1)\ldots))$, where

$$\forall_{i \leqslant n}(A_i \sim B_i \ \& \ B_i \in \mathscr{P}_1)$$

and the right-hand side of the equation is an experiment. The relation $\equiv$ on $\mathbf{P}$ is clearly an equivalence relation.

We define as before:

$$\mathscr{P}_1 \sim \mathscr{P}_2 \Leftrightarrow (\mathscr{P}_1 \preccurlyeq \mathscr{P}_2 \ \& \ \mathscr{P}_2 \preccurlyeq \mathscr{P}_1),$$
$$\mathscr{P}_1 \preccurlyeq \mathscr{P}_2 \Leftrightarrow \neg\, \mathscr{P}_2 \preccurlyeq \mathscr{P}_1$$
$$\mathscr{O} = \{\emptyset, \Omega\},$$
$$\mathscr{A} = \{\{\omega\} : \omega \in \Omega\}.$$

Let $\mathscr{P} = \mathscr{P}_1 \odot \mathscr{P}_2 \Leftrightarrow (\mathscr{P} = \mathscr{P}_1 \cdot \mathscr{P}_2 \ \& \ \mathscr{P}_1 \coprod \mathscr{P}_2 \ \& \ \neg \mathscr{P}_1, \mathscr{P}_2 \equiv \mathscr{O})$; let $\mathbf{B} = \{\mathscr{P} \in \mathbf{P} : \neg \exists \mathscr{P}_1, \mathscr{P}_2 [\mathscr{P} = \mathscr{P}_1 \odot \mathscr{P}_2]\}$; and let $\mathscr{D}$ enumerate $\mathbf{B}$, so that $\mathbf{B} = \{\mathscr{D}_i\}_{i \leqslant k}$. Then we define: $\hat{\mathscr{P}} = \langle d_1, d_2, \ldots, d_k \rangle$, where, if $\mathscr{P} = \mathscr{D}_i$ $(i \subset \{1, 2, \ldots, k\})$, then $d_i = 1$, and $\forall_{j \neq i} \ \& \ 1 \leqslant j \leqslant k (d_j = 0)$; otherwise $\hat{\mathscr{P}} = \hat{\mathscr{Q}}_1 + \hat{\mathscr{Q}}_2$ for some $\mathscr{Q}_1, \mathscr{Q}_2 \in \mathbf{P}$. Let $\hat{\mathscr{O}} = \langle 0, 0, \ldots, 0 \rangle$ be the zero vector. In other words, $\hat{\mathscr{P}} \in \mathscr{V}(\mathbf{B})$, where $\mathbf{B}$ is the basis of the k-dimensional vector space $\mathscr{V}(\mathbf{B})$. Now we are ready for the following definition:

DEFINITION 4: *Let* $\mathbf{Q} = \langle \Omega, \mathfrak{A}, \preccurlyeq \rangle$ *be a FAQQP*-structure *or a FQCP*-structure.[5] *Then the quadruple* $\langle \Omega, \mathbf{P}, \preccurlyeq, \coprod \rangle$ *is said to be a finite qualitative quasi-entropy structure (FQQE*-structure*) over* $\mathbf{Q}$ if and only if *the following conditions are satisfied for all variables running over* $\mathbf{P}$*:*

(E$_0$) $\mathbf{P}$ *is the algebra of finite experiments over* Ω*;* $\coprod$ *denotes the probabilistic independence relation on* $\mathbf{P}$*, and* $\preccurlyeq$ *is a binary relation on* $\mathbf{P}$*;*

(E$_1$) $\mathscr{P}_1 \subseteq \mathscr{P}_2 \Rightarrow \mathscr{P}_2 \preccurlyeq \mathscr{P}_1$;

(E$_2$) $\mathscr{O} \prec \mathscr{B}$, *if* $\mathscr{B} \sim \mathscr{E}$;

(E$_3$) $\mathscr{P}_1 \equiv \mathscr{P}_2 \Rightarrow \mathscr{P}_1 \sim \mathscr{P}_2$;

(E$_4$) $\mathscr{P}_1 \preccurlyeq \mathscr{P}_2 \vee \mathscr{P}_2 \preccurlyeq \mathscr{P}_1$;

(E$_5$) $\underset{i<n}{\forall} (\mathscr{P}_i \preccurlyeq \mathscr{A}_i) \Rightarrow \mathscr{P}_n \preccurlyeq \mathscr{A}_n$, *if* $\sum_{i \leqslant n} \hat{\mathscr{P}}_i = \sum_{i \leqslant n} \hat{\mathscr{Q}}_i$.

Remarks:

(i) In axiom (E$_5$), the formula concerning characteristic functions can easily be translated into a system of identities among experiments.

(ii) There is no doubt that the axioms (E_0)–(E_5) are consistent and independent. The crucial axioms are (E_4) and (E_5). Axioms (E_1) and (E_2) give the so-called normalization conditions, whereas (E_3) forces us to consider equiprobable classes of events, rather than events themselves.

(iii) The definition of an *infinite qualitative quasi-entropy structure* for purposes of representation by an entropy measure on **P** does not cause any fundamental difficulties. The *FAQQP*-structure or the *FQCP*-structure must of course be replaced by an infinite one; otherwise, we proceed as in the finite case. In fact, the axioms are prodigiously complicated and far less intuitive than those given above. The case will be omitted here.

(iv) In axiom (E_2), we must assume the existence of an equiprobable experiment $\mathscr{E}$, for we need this axiom to show that the entropy measure H is strictly positive for at least one element from **P**. An alternative axiom might be $\mathscr{O} \preccurlyeq \mathscr{A}$, but this would rule out some elementary algebras **P**.

(v) Note that the (*global*) *entropy relation* $\preccurlyeq$ depends on two factors: on the underlying *algebra of experiments* and the *independence relation* $\coprod$ defined on this algebra (this relation is hidden in axiom (E_5)). We do not give here the link between the *FQQE*-structure (macro-structure) and the *FAQQP*-structure (micro-structure).

The following easy theorem displays the content of the above definition.

THEOREM 6: *Let* $\langle \Omega, \mathbf{P}, \preccurlyeq, \coprod \rangle$ *be a FQQE*-structure *or a FQCP*-structure. *Then for all variables running over* **P** *and* $A, B, C \in \mathfrak{A}$:

(1) $\mathscr{P} \preccurlyeq \mathscr{P}$;

(2) $\{\emptyset\} \cup \mathscr{P} \sim \mathscr{P}$;

(3) $\mathscr{P}_1 \preccurlyeq \mathscr{P}_2 \ \& \ \mathscr{P}_2 \preccurlyeq \mathscr{P}_3 \Rightarrow \mathscr{P}_1 \preccurlyeq \mathscr{P}_3$;

(4) $\mathscr{P}_1 \preccurlyeq \mathscr{P}_2 \ \& \ \mathscr{P}_2 \preccurlyeq \mathscr{P}_1 \Rightarrow \mathscr{P}_1 \sim \mathscr{P}_2$;

(5) $\sim$ *is an equivalence relation*;

(6) $\mathscr{O} \preccurlyeq \mathscr{P}$;

(7) $\mathscr{O} \preccurlyeq \{A, \bar{A}\} \preccurlyeq \{A, \bar{A}B, \overline{AB}\} \preccurlyeq \{A, \bar{A}B, \overline{AB}C, \overline{ABC}\} \preccurlyeq \cdots \preccurlyeq \mathscr{A}$;

(8) $\mathscr{P}_1 \preccurlyeq \mathscr{P}_2 \Leftrightarrow \mathscr{P}_1 \cdot \mathscr{P} \preccurlyeq \mathscr{P}_2 \cdot \mathscr{P}$, *if* $\mathscr{P} \coprod \mathscr{P}_1, \mathscr{P}_2$;

(9) $\mathscr{P}_1 \preccurlyeq \mathscr{Q}_1 \ \& \ \mathscr{P}_2 \preccurlyeq \mathscr{Q}_2 \Rightarrow \mathscr{P}_1 \cdot \mathscr{P}_2 \preccurlyeq \mathscr{Q}_1 \cdot \mathscr{Q}_2$, *if* $\mathscr{P}_1 \coprod \mathscr{P}_2 \ \& \ \mathscr{Q}_1 \coprod \mathscr{Q}_2$;

(10) $\mathscr{P}_1 \sim \mathscr{P}_2 \ \& \ \mathscr{P}_2 \cdot \mathscr{P}_3 \preccurlyeq \mathscr{P}_4 \Rightarrow \mathscr{P}_1 \cdot \mathscr{P}_3 \preccurlyeq \mathscr{P}_4$, *if* $\mathscr{P}_3 \coprod \mathscr{P}_1, \mathscr{P}_2$;

(11) $\underset{i<n}{\forall} (\mathscr{P}_i \preccurlyeq \mathscr{Q}_i) \Rightarrow \mathscr{Q}_n \preccurlyeq \mathscr{P}_n$, *if*

$$\prod_{i\leqslant n} \mathscr{P}_i \sim \prod_{i\leqslant n} \mathscr{Q}_i \;\&\; \mathscr{P}_i \coprod_{i\leqslant n} \;\&\; \mathscr{Q}_i \coprod_{i\leqslant n};$$

(12) $\underset{i\leqslant n}{\forall} (A_i \sim B_i) \Rightarrow \{A_i\}_{i\leqslant n} \sim \{B_i\}_{i\leqslant n}$, *if* $\{A_i\}_{i\leqslant n}, \{B_i\}_{i\leqslant n} \in \mathbf{P}$.

The empirical content of Theorem 6 should be clear.

THEOREM 7 (Representation Theorem): *Let* $\langle \Omega, \mathbf{P}, \preccurlyeq, \coprod \rangle$ *be a structure, where* Ω *is a nonempty finite set;* $\mathbf{P}$ *is the set of partitions of* Ω*;* $\coprod$ *is the independence relation on* $\mathbf{P}$ *in the sense of Definition 3; and* $\preccurlyeq$ *is a binary relation on* $\mathbf{P}$.

Then $\langle \Omega, \mathbf{P}, \preccurlyeq, \coprod \rangle$ *is a FQQE*-structure if and only if *there exists a quasi-entropy function* $H{:}\,\mathbf{P}\to Re$ *satisfying the following conditions for all* $\mathscr{P}_1, \mathscr{P}_2 \in \mathbf{P}$*:*

(i) $\mathscr{P}_1 \preccurlyeq \mathscr{P}_2 \Leftrightarrow H(\mathscr{P}_1) \leqslant H(\mathscr{P}_2)$;

(ii) $\mathscr{P}_1 \coprod \mathscr{P}_2 \Rightarrow H(\mathscr{P}_1 \cdot \mathscr{P}_2) = H(\mathscr{P}_1) + H(\mathscr{P}_2)$;

(iii) $\mathscr{P}_1 \subseteq \mathscr{P}_2 \Rightarrow H(\mathscr{P}_2) \leqslant H(\mathscr{P}_1)$;

(iv) $H(\mathscr{O}) = 0$;

(v) $H(\mathscr{B}) = 1$, if $\mathscr{B} \sim \mathscr{E}$.

PROOF: There is no question of the conditions' not being necessary, and we prove here only their sufficiency.

Let $\langle \Omega, \mathbf{P}, \preccurlyeq, \coprod \rangle$ be a *FQQE*-structure over $\mathbf{Q}$. Let $\mathscr{V}(\mathbf{B})$ be the k-dimensional vector space, described just before Definition 4. We can obviously make $\mathbf{P}$ into a finite subset of $\mathscr{V}(\mathbf{B})$ by assigning to each $\mathscr{P} \in \mathbf{P}$ a vector $\hat{\mathscr{P}}$, where $(\mathscr{P}_1 \odot \mathscr{P}_2)^\wedge = \hat{\mathscr{P}}_1 + \hat{\mathscr{P}}_2$. In a similar way $\preccurlyeq$ can be represented on $\mathscr{V}(\mathbf{B})$. Having done this, we are ready to use Scott's Theorem (1964b, Theorem 1.3, p. 236) taking advantage of (E_3)–(E_5) to switch to quotient structures. Scott's result tells us that there is a linear functional $\psi{:}\,\mathscr{V}(\mathbf{B})\to Re$, and thus another functional $\varphi{:}\,\mathbf{P}\to Re$, such that the conditions (i), (ii), and (iv) of Theorem 7 arc satisficd by φ. (E_1) forces φ to be non-negative on $\mathbf{P}$, and also to satisfy (iii).

Finally, (E_2) gives $\varphi(\{A, \bar{A}\})>0$, if $E\sim\bar{E}$. Hence, by putting

$$H(\mathscr{P}) = \frac{\varphi(\mathscr{P})}{\varphi(\{A, \bar{A}\})},$$

we get the desired quasi-entropy function. Q.E.D.

Condition (iii) in Theorem 7 expresses the most important property of the entropy measure, namely, its additivity. Unfortunately, this property is much weaker than (iii) in (1.3), Section 1.2. It is trivial to show that many functions besides those of (1.2) satisfy the above conditions. This lack of specificity explains the 'quasi-' prefix.

It is well known that the conditions (i)–(v) in Theorem 7 together with the condition

$$H(\mathscr{P}) = \sum_{A \in \mathscr{P}} f \circ P(A),\ f: [0, 1] \to Re,\ f(\tfrac{1}{2}) = \tfrac{1}{2}, \tag{4.3}$$

for some f continuous, are enough to specify an entropy measure H_P in the form (1.2).

In order to guarantee the existence of a continuous function f, satisfying (4.3), we further have to restrict $\preccurlyeq$, and to add more 'interacting' conditions between $\preccurlyeq$ on **P** and $\preccurlyeq$ on $\mathfrak{A}$.

The following necessary conditions are obvious candidates:

(1) $A \preccurlyeq B \Leftrightarrow \{A, \bar{A}\} \preccurlyeq \{B, \bar{B}\}$, if $A, B \preccurlyeq E \sim \bar{E}$;

(2) $A \preccurlyeq B \Leftrightarrow \{B, \bar{B}\} \preccurlyeq \{A, \bar{A}\}$, if $\bar{E} \sim E \preccurlyeq A, B$;

(3) $A \preccurlyeq C \Leftrightarrow [B \mid AB, \bar{A}B]\, \mathscr{P} \preccurlyeq [B \mid CB, \bar{C}B]\, \mathscr{P})$, if $\emptyset \prec B \in \mathscr{P}, B, C \preccurlyeq E \sim \bar{E}, \emptyset \prec B, C \prec \Omega, A \coprod B, C$;

(4) $B \preccurlyeq C \Leftrightarrow [B \mid AB, \bar{A}B]\, \mathscr{P} \preccurlyeq [C \mid AC, \bar{A}C]\, \mathscr{P}$, if $B, C \in \mathscr{P}, A \coprod B, C, \emptyset \prec A \prec \Omega$;

(5) $\mathscr{P} \preccurlyeq \mathscr{E}$, if $|\mathscr{P}| = |\mathscr{E}|$.

It would be incredible if these conditions were sufficient. At least three axioms or axiom schemas similar to (E_5) are needed to guarantee the existence of the sum, multiplication, and logarithm functions in (1.2). Over and above that we need the qualitative probability axioms, which we can assume to be given, of course.

Given these axioms, our representation theorem would also guarantee the existence of a probability measure P such that in addition to (i)–(v) in Theorem 7 we would have:

(vi) $H(\mathscr{P}) + H(\{A, \bar{A}\}) \cdot P(B) = H([B \mid AB, \bar{A}B]\, \mathscr{P})$, if $B \in \mathscr{P}$, and $A \coprod B, \emptyset \prec A \prec \mathscr{R}$;

(vii) $A \preccurlyeq B \Leftrightarrow P(A) \leqslant P(B)$;

(viii) $A \coprod B \Leftrightarrow P(A \cap B) = P(A) \cdot P(B)$;

(ix) $\mathscr{P}_1 \coprod \mathscr{P}_2 \Leftrightarrow \forall A, B (A \in \mathscr{P}_1 \,\&\, B \in \mathscr{P}_2 \Rightarrow A \coprod B)$.

It seems to be an open problem to specify the relationship between the *macro-* and *micro-structures* under the given very restrictive *finite conditions.*

FQQE-structures characterize the macro-properties of the entropy from a qualitative point of view. The reader may have noticed the following striking formal similarity between the conditional entropy measure and the (absolute) probability measure:

(1) $\mathscr{P}_1 \coprod \mathscr{P}_2 \Rightarrow H(\mathscr{P}_1/\mathscr{P}_2) = H(\mathscr{P}_1)$,

$$A \perp B \Rightarrow P(A - B) = P(A),$$

(2) $H(\mathscr{P}_1 \cdot \mathscr{P}_2) = H(\mathscr{P}_2) + H(\mathscr{P}_1/\mathscr{P}_2)$,

$P(A \cup B) = P(B) + P(A - B)$,

(3) $0 \leqslant H(\mathscr{P}_1) \leqslant H(\mathscr{P}_1 \cdot \mathscr{P}_2) \leqslant H(\mathscr{P}_1) + H(\mathscr{P}_2)$,

$0 \leqslant P(A) \leqslant P(A \cup B) \leqslant P(A) + P(B)$.

This rather primitive one-one correspondence between

$$\mathscr{P}_1 \coprod \mathscr{P}_2,\ \mathscr{P}_1 \cdot \mathscr{P}_2,\ \mathscr{P}_1/\mathscr{P}_2 \quad \text{and} \quad A \perp B,\ A \cup B,\ A - B$$

contains certainly some heuristic anticipation of a deeper relationship between the macro- and micro-structures: $\langle \Omega, \mathbf{P}, H \rangle$ and $\langle \Omega, \mathfrak{A}, P \rangle$. One can see also why the lattice operation $+$ in $\mathbf{P}$ has so little use in entropy theory. The more interesting operation on $\mathbf{P}$ would be the composition of two experiments, $\mathscr{P}_1 \wedge \mathscr{P}_2$, defined in Section 4.3. The only problem is that $\langle \mathbf{P}, \mathcal{O}, \mathscr{A}, \cdot, \wedge \rangle$ cannot be embedded into a Boolean algebra.

We shall now turn to the problem of conditional entropy. Another interesting similarity between the conditional entropy and (conditional) probability is the following:

(1) $H(\mathscr{P}_1/\mathscr{P}_2) = H(\mathscr{P}_1 \cdot \mathscr{P}_2) - H(\mathscr{P}_2)$,

$P(A/B) = P(AB)/P(B),\ P(B) > 0$,

(2) $\mathscr{P}_1 \coprod \mathscr{P}_2 \Leftrightarrow H(\mathscr{P}_1/\mathscr{P}_2) = H(\mathscr{P}_1/\mathcal{O})$,

$A \coprod B \Leftrightarrow P(A/B) = P(A/\Omega), \quad \text{if} \quad P(B) > 0$,

(3) $H(\mathscr{P}_1/\mathscr{P}_2 \cdot \mathscr{P}_3) = H(\mathscr{P}_1 \cdot \mathscr{P}_2/\mathscr{P}_3) - H(\mathscr{P}_2/\mathscr{P}_3)$,

$P(A/BC) = P(AB/C)/P(B/C), \quad \text{if} \quad P(BC) \cdot P(C) > 0$.

We shall consider these similarities as a heuristic guide to further developments of entropy structures. One can consider the entity $\mathscr{P}_1/\mathscr{P}_2$ to be a partition (experiment) in $\mathfrak{A}/\mathfrak{A}[\mathscr{P}_2]$.[6] Then $\mathscr{P}_1/\mathscr{P}_2$ is the set of experiments indistinguishable from $\mathscr{P}_1$, given $\mathscr{P}_2$.

As in the case of probability structures (see Section 2.2, Definition 1) we shall study a kind of composition of entropy structures. In particular, given the algebra of experiments $\langle \Omega, \mathbf{P}, \subseteq \rangle$, we shall study a binary relation $\preccurlyeq$ on $\mathbf{P} \times \mathbf{P}$ and a special representation function $\psi : \mathbf{P} \to Re$, which, among other things, satisfies

$$\langle \mathscr{P}_1, \mathscr{P}_2 \rangle \preccurlyeq \langle \mathscr{Q}_1, \mathscr{Q}_2 \rangle \Leftrightarrow \psi(\mathscr{P}_1) + \psi(\mathscr{P}_2) \leqslant \psi(\mathscr{Q}_1) + \psi(\mathscr{Q}_2) \quad \text{for all} \quad \mathscr{P}_1, \mathscr{P}_2, \mathscr{Q}_1, \mathscr{Q}_2 \in \mathbf{P}.$$

There are several important partial interpretations of this relation: First of all, the *qualitative conditional quasi-entropy relation* hopefully can be defined as

$$\mathscr{P}_1/\mathscr{P}_2 \preccurlyeq \mathscr{Q}_1/\mathscr{Q}_2 \Leftrightarrow \langle \mathscr{P}_1 \cdot \mathscr{P}_2, \mathscr{Q}_2 \rangle \preccurlyeq \langle \mathscr{Q}_1 \cdot \mathscr{Q}_2, \mathscr{P}_2 \rangle.$$

Naturally, we can put

$$\mathscr{P}_1 \preccurlyeq \mathscr{P}_2 \Leftrightarrow \langle \mathscr{P}_1, \mathscr{O} \rangle \preccurlyeq \langle \mathscr{P}_2, \mathscr{O} \rangle$$

and then the *probabilistic independence relation* $\coprod$ on experiments is given by

$$\mathscr{P}_1 \coprod \mathscr{P}_2 \Leftrightarrow \langle \mathscr{P}_1, \mathscr{P}_2 \rangle \preccurlyeq \langle \mathscr{P}_1 \cdot \mathscr{P}_2, \mathscr{O} \rangle.$$

It is clear that we could also talk about *positive* and *negative dependence notions* similar to those introduced for probabilities. The structure $\langle \mathbf{P} \times \mathbf{P}, \preccurlyeq \rangle$ also has independent importance in algebraic measurement theory, where the atomic formula $\langle \mathscr{P}_1, \mathscr{P}_2 \rangle \preccurlyeq \langle \mathscr{Q}_1, \mathscr{Q}_2 \rangle$ may be interpreted as a comparison of two *empirical compositions* of certain physical entities, which is representable by an inequality between the sum of magnitudes of a linear physical quantity. In this paper we shall be interested only in the *entropy*-interpretation.

DEFINITION 5: *Let* $\mathbf{Q} = \langle \Omega, \mathfrak{A}, \preccurlyeq \rangle$ *be a FAQQP*-structure. *Then the quadruple* $\langle \Omega, \mathbf{P}, \preccurlyeq, \coprod \rangle$ *is said to be a finite qualitative quasi-entropy difference structure (FQQED*-structure*) over* $\mathbf{Q}$ *if and only if the following conditions are satisfied for all variables running over* $\mathbf{P}$:

(D_0) $\mathbf{P}$ *is the algebra of finite experiments over* Ω; $\coprod$ *is the probabilistic independence relation on* $\mathbf{P}$ *and* $\preccurlyeq$ *is a relation on* $\mathbf{P} \times \mathbf{P}$;

(D_1) $\mathscr{P}_1 \subseteq \mathscr{P}_2 \Rightarrow \langle \mathscr{P}_2, \mathscr{P} \rangle \preccurlyeq \langle \mathscr{P}_1, \mathscr{P} \rangle$;

(D_2) $\langle \mathscr{O}, \mathscr{P} \rangle \prec \langle \mathscr{B}, \mathscr{P} \rangle$, *if* $\mathscr{B} \sim \mathscr{E}$;

(D_3) $\langle \mathscr{P}_1, \mathscr{P}_2 \rangle \preccurlyeq \langle \mathscr{Q}_1, \mathscr{Q}_2 \rangle \vee \langle \mathscr{Q}_1, \mathscr{Q}_2 \rangle \preccurlyeq \langle \mathscr{P}_1, \mathscr{P}_2 \rangle$;

(D_4) $\langle \mathscr{P}_1, \mathscr{P}_2 \rangle \preccurlyeq \langle \mathscr{Q}_1, \mathscr{Q}_2 \rangle \Rightarrow \langle \mathscr{Q}_2, \mathscr{Q}_1 \rangle \preccurlyeq \langle \mathscr{P}_2, \mathscr{P}_1 \rangle$;

(D_5) $\underset{i<n}{\forall} (\langle \mathscr{P}_i, \mathscr{Q}_i \rangle \preccurlyeq \langle \mathscr{R}_i, \mathscr{S}_i \rangle) \Rightarrow \langle \mathscr{S}_n, \mathscr{R}_n \rangle \preccurlyeq \langle \mathscr{P}_n, \mathscr{Q}_n \rangle$,

if $\sum_{i \leqslant n} \hat{\mathscr{P}}_i = \sum_{i \leqslant n} \hat{\mathscr{R}}_i \;\&\; \sum_{i \leqslant n} \hat{\mathscr{Q}}_i = \sum_{i \leqslant n} \hat{\mathscr{S}}_i$,

where $\hat{\mathscr{P}}_i, \hat{\mathscr{R}}_i, \hat{\mathscr{Q}}_i, \hat{\mathscr{S}}_i$ *for* $i = 1, 2, \ldots, n$

have the same meaning as in Definition 4.

The remarks to Definition 4 are relevant also to Definition 5. The content of the definition should be clear; therefore we proceed to Theorem 8.

THEOREM 8 (Representation Theorem): *Let* $\langle \Omega, \mathbf{P}, \preccurlyeq, \coprod \rangle$ *be a structure, where* Ω *is a nonempty finite set;* $\mathbf{P}$ *is the set of partitions of* Ω*;* $\coprod$ *is the independence relation on in the sense of the Definition* 5*; and* $\preccurlyeq$ *is a relation on* $\mathbf{P} \times \mathbf{P}$.

Then $\langle \Omega, \mathbf{P}, \preccurlyeq, \coprod \rangle$ *is a FQQED*-structure if and only if *there exists a quasi-entropy function* $H: \mathbf{P} \to Re$ *satisfying the following conditions for all* $\mathscr{P}_1, \mathscr{P}_2, \mathscr{Q}_1, \mathscr{Q}_2 \in \mathbf{P}$*:*

(i) $$\langle \mathscr{P}_1, \mathscr{P}_2 \rangle \preccurlyeq \langle \mathscr{Q}_1, \mathscr{Q}_2 \rangle \Leftrightarrow H(\mathscr{P}_1) - H(\mathscr{P}_2) \leqslant H(\mathscr{Q}_1) - H(\mathscr{Q}_2);$$

(ii) *H satisfies conditions* (ii)–(v) *of Theorem* 7.

PROOF: The necessity is obvious. For sufficiency, let $\langle \Omega, \mathbf{P}, \preccurlyeq, \coprod \rangle$ be a *FQQED*-structure over $\mathbf{Q}$. Let $\mathscr{V}(\mathbf{B})$ be the k-dimensional vector space, described in the proof of Theorem 7. We can transform $\mathbf{P} \times \mathbf{P}$ into a finite subset of the (external) direct sum $\mathscr{V}(\mathbf{B}) \oplus \mathscr{V}(\mathbf{B})$ by assigning to each pair $\langle \mathscr{P}, \mathscr{Q} \rangle$ a vector $\hat{\mathscr{P}} \oplus \hat{\mathscr{Q}} \in \mathscr{V}(\mathbf{B}) \oplus \mathscr{V}(\mathbf{B})$. We then proceed almost exactly as does Scott (1964b, Theorem 3.2, p. 245), so that the axioms (D_3), (D_4), (D_5) are justified. As in Theorem 7, the normalization conditions (D_1), (D_2) will allow us to construct a function H (which exists on the basis of (D_3)–(D_5)) with the desired properties (i) and (ii) in Theorem 8. Q.E.D.

Now if we put

$$\mathscr{P}_1 / \mathscr{P}_2 \preccurlyeq \mathscr{Q}_1 / \mathscr{Q}_2 \Leftrightarrow \langle \mathscr{P}_1 \cdot \mathscr{P}_2, \mathscr{P}_2 \rangle \preccurlyeq \langle \mathscr{Q}_1 \cdot \mathscr{Q}_2, \mathscr{Q}_2 \rangle \tag{4.4}$$

we can easily prove the following theorem with the help of Definition 5:

THEOREM 9: *Let* $\langle \Omega, \mathbf{P}, \preccurlyeq, \coprod \rangle$ *be a FQQED*-structure over *a FQCP*-structure. *Then the following formulas hold, when all variables ran over* $\mathbf{P}$*:*

(1) $\sim$ *is an equivalence relation;*

(2) $\mathcal{O}/\mathcal{P} \preccurlyeq \mathcal{P}_1/\mathcal{P}_2$;

(3) $\mathcal{O}/\mathcal{B} \prec \mathcal{B}/\mathcal{B}$, if $\mathcal{B} \sim \mathcal{E}$;

(4) $\mathcal{P}_1/\mathcal{P}_2 \sim \mathcal{P}_1 \cdot \mathcal{P}_2/\mathcal{P}_2$;

(5) $\mathcal{P}/\mathcal{P} \sim \mathcal{Q}/\mathcal{Q}$;

(6) $\mathcal{O}/\mathcal{P}_1 \preccurlyeq \mathcal{P}_2/\mathcal{P}_2$;

(7) $\mathcal{P}_1 \preccurlyeq \mathcal{P}_2 \Leftrightarrow \mathcal{P}_1/\mathcal{P}_2 \preccurlyeq \mathcal{P}_2/\mathcal{P}_1$;

(8) $\mathcal{P}_1/\mathcal{P}_2 \cdot \mathcal{P}_3 \preccurlyeq \mathcal{P}_1 \cdot \mathcal{P}_2/\mathcal{P}_3$;

(9) $\mathcal{P}_1/\mathcal{P}_2 \cdot \mathcal{P}_3 \preccurlyeq \mathcal{P}_1/\mathcal{P}_2$;

(10) $\mathcal{P}_1/\mathcal{P}_2 \preccurlyeq \mathcal{P}_1$;

(11) $\mathcal{P}_1 \subseteq \mathcal{P}_2 \Rightarrow \mathcal{P}_2/\mathcal{P} \preccurlyeq \mathcal{P}_1/\mathcal{P}$;

(12) $\mathcal{P} \subseteq \mathcal{P}_1 \Rightarrow \mathcal{P}_1 \cdot \mathcal{P}_2/\mathcal{P} \sim \mathcal{P}_2/\mathcal{P}$;

(13) $\mathcal{P}_1 \subseteq \mathcal{P}_2 \Rightarrow \mathcal{P}_2/\mathcal{P}_1 \sim \mathcal{O}$;

(14) $\mathcal{P}_1/\mathcal{P}_2 \sim \mathcal{P}_1 \Leftrightarrow \mathcal{P}_1 \coprod \mathcal{P}_2$;

(15) $\mathcal{P}_1/\mathcal{P} \preccurlyeq \mathcal{P}_2/\mathcal{P} \Leftrightarrow \mathcal{P}_1 \cdot \mathcal{Q}/\mathcal{P} \preccurlyeq \mathcal{P}_2 \cdot \mathcal{Q}/\mathcal{P}$, if $\mathcal{P}_1, \mathcal{P}_2 \coprod \mathcal{Q}$;

(16) $\mathcal{P}_1 \cdot \mathcal{P}_2 \sim \mathcal{Q}_1 \cdot \mathcal{Q}_2 \Rightarrow (\mathcal{P}_2 \preccurlyeq \mathcal{Q}_2 \Leftrightarrow \mathcal{Q}_1/\mathcal{Q}_2 \preccurlyeq \mathcal{P}_1/\mathcal{P}_2)$;

(17) $(\mathcal{P}_1/\mathcal{P}_2 \preccurlyeq \mathcal{Q}_1/\mathcal{Q}_2 \ \& \ \mathcal{P}_2 \preccurlyeq \mathcal{Q}_2) \Rightarrow \mathcal{P}_1 \cdot \mathcal{P}_2 \preccurlyeq \mathcal{Q}_1 \cdot \mathcal{Q}_2$;

(18) $(\mathcal{P}_1 \cdot \mathcal{P}_2 \preccurlyeq \mathcal{Q}_1 \cdot \mathcal{Q}_2 \ \& \ \mathcal{Q}_2 \preccurlyeq \mathcal{P}_2) \Rightarrow \mathcal{P}_1/\mathcal{P}_2 \preccurlyeq \mathcal{Q}_1/\mathcal{Q}_2$;

(19) $(\mathcal{P}_1/\mathcal{P}_2 \cdot \mathcal{P}_3 \preccurlyeq \mathcal{Q}_1/\mathcal{Q}_2 \cdot \mathcal{Q}_3 \ \& \ \mathcal{P}_2/\mathcal{P}_3 \preccurlyeq \mathcal{Q}_2/\mathcal{Q}_3)$
$\Rightarrow \mathcal{P}_1 \cdot \mathcal{P}_2/\mathcal{P}_3 \preccurlyeq \mathcal{Q}_1 \cdot \mathcal{Q}_2/\mathcal{Q}_3$;

(20) $\mathcal{P}_1 \cdot \mathcal{P}_2/\mathcal{P}_3 \sim \mathcal{Q}_1 \cdot \mathcal{Q}_2/\mathcal{Q}_3$
$\Rightarrow (\mathcal{P}_1/\mathcal{P}_2 \cdot \mathcal{P}_3 \preccurlyeq \mathcal{Q}_1/\mathcal{Q}_2 \cdot \mathcal{Q}_3 \Rightarrow \mathcal{Q}_2/\mathcal{Q}_3 \preccurlyeq \mathcal{P}_2/\mathcal{P}_3)$;

(21) $(\mathcal{P}_1/\mathcal{P}_2 \cdot \mathcal{P}_3 \preccurlyeq \mathcal{Q}_2/\mathcal{Q}_3 \ \& \ \mathcal{P}_2/\mathcal{P}_3 \preccurlyeq \mathcal{Q}_1/\mathcal{Q}_2 \cdot \mathcal{Q}_3)$
$\Rightarrow \mathcal{P}_1 \cdot \mathcal{P}_2/\mathcal{P}_3 \preccurlyeq \mathcal{Q}_1 \cdot \mathcal{Q}_2/\mathcal{Q}_3$;

(22) $(\mathcal{P}_1/\mathcal{P}_2 \preccurlyeq \mathcal{Q}_1/\mathcal{Q}_2 \ \& \ \mathcal{Q}_1/\mathcal{Q}_2 \preccurlyeq \mathcal{R}_1/\mathcal{R}_2) \Rightarrow \mathcal{P}_1/\mathcal{P}_2 \preccurlyeq \mathcal{R}_1/\mathcal{R}_2$;

(23) $\prod_{i=1}^{n} \mathcal{P}_i \sim \prod_{i=1}^{n} \mathcal{Q}_i \ \& \ \forall_{1 \leqslant i \leqslant n} \left(\mathcal{P}_i / \prod_{j=0}^{i-1} \mathcal{P}_j \preccurlyeq \mathcal{Q}_i / \prod_{j=0}^{i-1} \mathcal{Q}_j\right)$
$\Rightarrow \mathcal{Q}_n / \prod_{j=0}^{n-1} \mathcal{Q}_j \preccurlyeq \mathcal{P}_n / \prod_{j=0}^{n-1} \mathcal{P}_j$ where $\mathcal{P}_0 \sim \mathcal{O}$.

No more than with Definition 4 can we hope to show that

$$H(\mathscr{P}_1/\mathscr{P}_2) = -\sum_{A \in \mathscr{P}_1,\, B \in \mathscr{P}_2} P(AB) \cdot \log_2 P(A/B) \tag{4.5}$$

without giving some further axioms to link $\preccurlyeq$ on $\mathbf{P}$ with the probability relation $\preccurlyeq$ on $\mathfrak{A}$.

Khinchin (1957) showed that the conditions

(a) $H(\mathscr{P}_1 \cdot \mathscr{P}_2) - H(\mathscr{P}_2) = H(\mathscr{P}_1/\mathscr{P}_2)$;

(b) $H(\mathscr{P}) \leqslant H(\mathscr{E})$, if $\mathscr{P} = \mathscr{E}$;

(c) $H(\mathscr{P} \cup \{\emptyset\}) = H(\mathscr{P})$;

imply the identity

$$H(\mathscr{P}) = -\sum_{A \in \mathscr{P}} P(A) \cdot \log_2 P(A),$$

and therefore, also the identity (4.5). In our case (a) is true by definition, and (b) and (c) become valid by adding the following two axioms:

(D_6) $\langle \mathscr{P}, \mathscr{O} \rangle \preccurlyeq \langle \mathscr{E}, \mathscr{O} \rangle$, if $|\mathscr{P}| = |\mathscr{E}|$;

(D_7) $\langle \mathscr{P} \cup \{\emptyset\}, \mathscr{O} \rangle \sim \langle \mathscr{P}, \mathscr{O} \rangle$.

Naturally $\mathscr{E}$ must exist, otherwise the axiom (D_6) would be vacuously true. Given that, (D_0)–(D_7) imply the conditions (a), (b), (c) for *finite qualitative conditional entropy relations.*

Forte and Pintacuda (1968a) showed that if $H: \mathbf{E} \to Re^+$ satisfies the conditions:

(i) *Monotony:*

$$\mathscr{P}_1 \subseteq \mathscr{P}_2 \Rightarrow H(\mathscr{P}_2) \leqslant H(\mathscr{P}_1);$$

(ii) *Additivity:*

$$\mathscr{P}_1 \coprod \mathscr{P}_2 \Rightarrow H(\mathscr{P}_1 \cdot \mathscr{P}_2) = H(\mathscr{P}_1) + H(\mathscr{P}_2);$$

(iii) *Local Property:*

$$H(\mathscr{P}_1 \oplus \mathscr{P}_2) + H(\mathscr{Q}_1 \oplus \mathscr{Q}_2) = H(\mathscr{P}_1 \oplus \mathscr{Q}_2) + H(\mathscr{Q}_1 \oplus \mathscr{P}_2),$$

if the terms involved in the equation are meaningful, and if $H_n = H \upharpoonright \mathbf{E}_n$, then $H_n(\mathscr{P}) = H_1(\bigcup \mathscr{P}) + G(\mathscr{P})$ for some G and $\mathscr{P} \in \mathbf{E}_n$, where

$$G(\mathscr{P}) = \sum_{i=0}^{n-2} F\left(\bigcup_{k=0}^{i} A_k,\ A_{i+1},\ \bigcup_{j=i+2}^{n} A_j\right)$$

for some F. G satisfies again the conditions (i)–(iii). By imposing further conditions on F one can show that H must have the form (4.2) (see Forte and Pintacuda 1968b).

If we check the definitions and the representation theorems for quasi-entropy structures, it is not hard to see that the generalization of results from complete experiments $\mathbf{P}$ to incomplete experiments $\mathbf{E}$ is a matter of slight changes. We shall leave this out, because conceptually one gets nothing new.

The notion of entropy had long been used in thermodynamics without anyone noticing its applications in communication theory. Shannon discovered its importance for the theory of information. One wonders if there is any other notion or quantity which is used in classical thermodynamics and has a meaningful interpretation in information theory. Onicescu (1966) gave a positive answer to this question by showing that *energy* is also such a quantity.

Let $\langle \Omega, \mathfrak{A}, P \rangle$ be a probability space and P the algebra of experiments over Ω. Then by (*informational*) *energy* E we shall understand the function $E : \mathbf{P} \to Re$ defined as follows:

$$E(\mathscr{P}) = \sum_{A \in \mathscr{P}} (P(A))^2 \quad (\mathscr{P} \in \mathbf{P}).$$

The *conditional* (*informational*) *energy* is defined as follows:

$$E(\mathscr{P}_1 / \mathscr{P}_2) = \sum_{\substack{A \in \mathscr{P}_1 \\ B \in \mathscr{P}_2}} P(B) \cdot (P(A/B))^2 \quad (\mathscr{P}_1, \mathscr{P}_2 \in \mathbf{P}).$$

Unlike the entropy, (informational) energy measures the degree of organization of the experiment. It grows from value $1/n$ which corresponds to the maximally disorganized (homogenized) experiment ($|\mathscr{P}| = n$) to value 1 which corresponds to a fully organized (concentrated) experiment (all probabilities but one are zero).

The following theorem makes this interpretation more clear.

THEOREM 10: Let $\langle \Omega, \mathfrak{A}, P \rangle$ *be a probability space and* $\mathbf{P}$ *the algebra of experiments over* Ω. *Then the following clauses are valid when all variables run over* $\mathbf{P}$:

(1) $E(\mathscr{O}) = 1$;

(2) $0 \leqslant E(\mathscr{P}) \leqslant 1$;

(3) $E(\mathscr{P}) = E(\mathscr{P}^*)$, *if* $\mathscr{P}^* = \mathscr{P} \cup \{\emptyset\}$;

(4) $\mathscr{P}_1 \subseteq \mathscr{P}_2 \Rightarrow E(\mathscr{P}_1) \leqslant E(\mathscr{P}_2)$ (*monotonicity*);

(5) $\mathscr{P}_1 \coprod \mathscr{P}_2 \Rightarrow E(\mathscr{P}_1 \cdot \mathscr{P}_2) = E(\mathscr{P}_1) \cdot E(\mathscr{P}_2)$ (*multiplicativity*);

(6) $E(\mathscr{E}) = 1/n, \quad \textit{if} \quad |\mathscr{E}| = n;$

(7) $E(\mathscr{P}) < n \cdot E(\mathscr{P} \cdot \mathscr{Q}), \quad \textit{if} \quad n = |\mathscr{Q}| > 1;$

(8) $E(\mathscr{P}_1 \cdot \mathscr{P}_2) \leqslant E(\mathscr{P}_1 / \mathscr{P}_2).$

Let $\langle \Omega, \mathfrak{A}, P \rangle$ be a probability space and $\mathbf{P}$ the algebra of experiments over Ω. Then the following conditions are necessary and sufficient for the mapping $E_P : \mathbf{P} \rightarrow Re^+$ to be an (informational) *energy* measure:

(i) The diagram

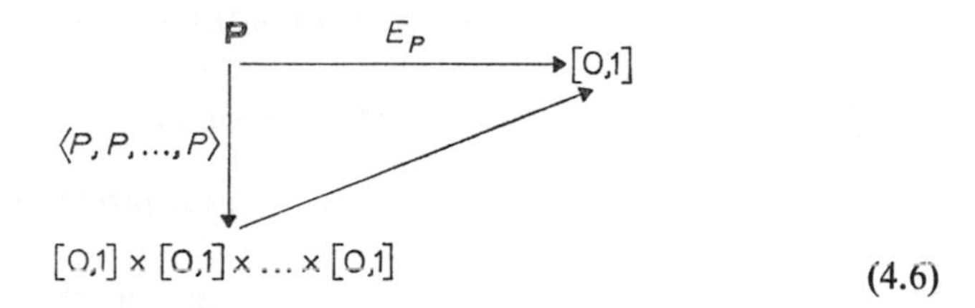

(4.6)

is commutative, that is, $E \circ \langle P, P, \ldots, P \rangle = E_P$ and E is continuous;

(ii) $E_P(\{A, \bar{A}\}) = \frac{1}{2}$, if $P(A) = P(\bar{A})$;

(iii) $E_P([A/AB, A\bar{B}]\mathscr{P}) = E_P(\mathscr{P}) - P^2(A) \cdot (1 - E_P(\{B, \bar{B}\}))$, if $A \coprod B$, $\mathscr{P} \in \mathbf{P}$.

The notion of energy is easily generalized for the case of incomplete experiments. We define

$$E(\mathscr{P}) = \frac{\sum_{A \in \mathscr{P}} P^2(A)}{W(\mathscr{P})},$$

where

$$W(\mathscr{P}) = \sum_{A \in \mathscr{P}} P(A) \quad \text{and} \quad \mathscr{P} \in \mathbf{E}.$$

Theorem 10 remains unchanged. In particular, we shall have also

$$E(\mathscr{P}_1 \oplus \mathscr{P}_2) = \\ = \frac{W(\mathscr{P}_1) \cdot E(\mathscr{P}_1) + W(\mathscr{P}_2) \cdot E(\mathscr{P}_2)}{W(\mathscr{P}_1) + W(\mathscr{P}_2)} \quad \text{if} \quad \mathscr{P}_1 \oplus \mathscr{P}_2 \in \mathbf{E}.$$

The conditions (4.6) can be generalized for the case of energy of incomplete experiments. The ideas run along lines of Rényi (1961).

Using the technique from Section 2.2 in connection with quadratic qualitative probability structures, one can prove Representation Theorems for *qualitative* (*informational*) *quasi-energy structures*. Analogously, like in the case of the entropy measure, the multiplicativity condition is much weaker than condition (iii) in (4.6). Therefore the representation theorem for qualitative energy structures will be quite complicated. We have to add very strong axioms which will link the micro-structure $\langle \Omega, \mathfrak{A}, \preccurlyeq \rangle$ with the macro-structure $\langle \Omega, \mathbf{P}, \preccurlyeq \rangle$. The relation-theoretic description of Equation (4.6) (iii) will do the work.

5. SUMMARY AND CONCLUSIONS

5.1. *Concluding Remarks*

Throughout this paper we have been studying the properties of two binary relations – one on a Boolean algebra $\mathfrak{A}$ and one on a finite lattice of partitions $\mathbf{P}$, both generated by the set of sample points Ω. The structure $\langle \Omega, \mathfrak{A}, \preccurlyeq, \coprod \rangle$ has been called an *information structure*, provided that the properties of $\preccurlyeq$ and $\coprod$ allowed us to show the existence of an *information* measure I satisfying

$$A \preccurlyeq B \Leftrightarrow I(A) \leqslant I(B) \quad \text{for all } A, B \in \mathfrak{A}.$$

Similarly, we called the structure $\langle \Omega, \mathbf{P}, \preccurlyeq, \coprod \rangle$ an *entropy structure*, if an *entropy* measure H could be found for which

$$\mathscr{P}_1 \preccurlyeq \mathscr{P}_2 \Leftrightarrow H(\mathscr{P}_1) \leqslant H(\mathscr{P}_2) \quad \text{for all} \quad \mathscr{P}_1, \mathscr{P}_2 \in \mathbf{P}.$$

The results are quite general and simple, mostly because we decided to work with finite structures. An extension to infinite structures does not cause any problems, as we have pointed out. Our basic concern was to show that both entropy and information measures can be studied in the spirit of *Representational* or *Algebraic Measurement Theory*, as given, e.g., in Suppes and Zinnes (1963).

Another important problem attacked was the *independence relation*. In empirical theories the notion of independence is mostly related to a particular probability measure. I emphasized that this notion should be handled independently as an additional primitive relation on the algebra of events or experiments; even though it may, eventually, have little in common with the more familiar probabilistic notions. I have a similar opinion about the various conditional entities. Their properties are, indeed, quite complicated; I therefore left out their detailed analysis.

As noted in Section 1, several people have tried to develop *semantic information theory*. In my view, it can be reduced to the standard information theory, because the set of propositions on which semantic information measures are defined forms, under certain rather weak conditions, a Boolean algebra. We do not think that there is much more to learn about information measures on propositions, before a satisfactory theory of probability on first-order languages is developed. Probabilities of quantified formulas may then give something new. Beyond this lies the prospect of studying entropies in first-order theories and, perhaps, of answering some of the methodological questions posed by empirical theories. But any such advances will have to be preceded by elucidation of the structure of the independence relation on the set of quantified formulas, the structure of the set of conditional formulas, and so on. It may be that a purely qualitative approach will be more fruitful to begin with.

5.2. *Open Problems*

Among the basic questions of the *Representation Theory of measures*, which we have been advocating here, is the problem of the *uniqueness* of the measure. That is to say, we would like to know the structure of the class of measures μ satisfying the representation condition:

$$A \preccurlyeq B \Leftrightarrow \mu(A) \leqslant \mu(B), \quad \text{for all} \quad A, B \in \mathfrak{A},$$

where $\preccurlyeq$ is a relation and μ is a measure on $\mathfrak{A}$.

Unfortunately, no structure studied here has the simple answer. For example, in the case of finite qualitative probability structures, specified by conditions (2.1), we can check easily that the set of probability measures corresponding to the qualitative probability relation $\preccurlyeq$ forms a convex polyhedron (from $A \preccurlyeq B \Leftrightarrow P_i(A) \leqslant P_i(B)$, $\alpha_i \geqslant 0$, $\sum_{i \leqslant n} \alpha_i = 1$, $i = 1, 2, \ldots, n$, we can infer $A \preccurlyeq B \Leftrightarrow \sum_{i \leqslant n} \alpha_i \cdot P_i(A) \leqslant \sum_{i \leqslant n} \alpha_i \cdot P_i(B)$). But this observation is very general. By further analysis we shall find that the cardinality of the set generating the convex polyhedron is in a simple relationship to the cardinality of the set of atoms of the Boolean algebra of the probability space. This starts to throw some light on the uniqueness problem.

It is known that in atomless Boolean algebras the representing measure is unique. On the other hand, atomless Boolean algebras are not the most important ones.

The problem of uniqueness in general nonlinear measurement structures certainly deserves some further study.

Another problem is to find those conditions that must be imposed on $\langle \Omega, \mathfrak{A}, \preccurlyeq, \coprod \rangle$ or $\langle \Omega, \mathbf{P}, \preccurlyeq, \coprod \rangle$ for the probability occurring in the information or entropy measure to have a *specific distribution* (e.g. Bernoulli, Binomial, Gaussian). In this case we might hope that the measures would be unique up to some reasonable group of transformations; moreover, the qualitative way of proving theorems may be more straightforward.

Yet another question is to determine the conditions to be imposed on the structures $\langle \Omega, \mathfrak{A}, \preccurlyeq, \coprod \rangle$ and $\langle \Omega, \mathbf{P}, \preccurlyeq, \coprod \rangle$ to ensure that the representation by information I and entropy H have the more specific form:

$$A \prec B \Leftrightarrow \varepsilon + I(A) \leqslant I(B), 0 < \varepsilon < \infty, A, B \in \mathfrak{A};$$
$$\mathscr{P}_1 \prec \mathscr{P}_2 \Leftrightarrow \varepsilon + H(\mathscr{P}_1) \leqslant H(\mathscr{P}_2), 0 < \varepsilon < \infty, \mathscr{P}_1, \mathscr{P}_2 \in \mathbf{P}.$$

This question is motivated by the problem that arises in algebraic measurement-error theory. A further generalization of the problem occurs when the error, rather than being constant, is taken as a function of the event $A (A \in \mathfrak{A})$ or experiment $\mathscr{P} (\mathscr{P} \in \mathbf{P})$.

Additional problems arise when we want to study functions of information and entropy measures (that is, functions defined on the set of information and entropy values) by finding the proper qualitative relation for them.

Stanford University

BIBLIOGRAPHY

Aczél, J.: 1966, *Lectures on Functional Equations*, Academic Press, New York.

Aczél, J., Pickert, G., and Rado, F.: 1960, 'Nomogramme, Gewebe, und Quasigruppen', *Mathematica* (Cluj) **2**, 5–24.

Adler, R. L., Konheim, A. G., and McAndrew, M. H.: 1965, 'Topological Entropy', *Trans. Amer. Math. Soc.* **114**, 309–319.

Bar-Hillel, Y. and Carnap, R.: 1953, 'Semantic Information', *British Journal for the Philosophy of Science* **4**, 147–157

Belis, M. and Guiasu, S.: 1967, 'A Quantitative-Qualitative Measure of Information in Cybernetic Systems', *IEEE Transactions on Information Theory* **IT-14** (1967) 593–594.

De Fériet, J. K. and Forte, B.: 1967, 'Information et Probabilité', *C.R. Acad. Sci. Paris* **265A**, 110–114.

De Finetti, B.: 1937, 'La Prévision: Ses Lois Logiques, Ses Sources Subjectives', *Ann. Inst. Poincaré* **7**, 1–68. English translation in *Studies in Subjective Probability* (ed. by H. E. Kyburg, Jr. and H. E. Smokler), Wiley, New York, 1964, pp. 93–158.

Domotor, Z.: 1969, 'Probabilistic Relational Structures and Their Applications', Technical Report No. 144, May 14, Institute for Mathematical Studies in the Social Sciences, Stanford University, Stanford, Calif.

Erdös, P.: 1946, 'On the Distribution Function of Additive Functions', *Ann. Math.* **47**, 1–20.

Fadeev, D. K.: 1956, 'On the Concept of Entropy of a Finite Probability Scheme', *Uspechy Mat. Nauk* **11**, 227–231. (In Russian.)

Forte, B. and Pintacuda, N.: 1968a, 'Sull' Informazione Associata Alle Esperienze Incomplete', *Annali di Mathematica Pura ed Applicata* **80**, 215–234.

Forte, B. and Pintacuda, N.: 1968b, 'Information Fourn e par Une Expérience', *C.R. Acad. Sci. Paris* **266A**, 242–245.

Hintikka, K. J.: 1968, 'The Varieties of Information and Scientific Explanation', in *Logic, Methodology and Philosophy of Science, III* (ed. by B. van Rootselaar and J. F. Staal), North-Holland Publishing Company, Amsterdam, pp. 151–171.

Hintikka, K. J. and Pietarinen, J.: 1966, 'Semantic Information and Inductive Logic', in *Aspects of Inductive Logic* (ed. by K. J. Hintikka and P. Suppes), North-Holland Publishing Company, Amsterdam, pp. 81–97.

Ingarden, R. S.: 1963, 'A Simplified Axiomatic Definition of Information', *Bull. Acad. Sci.* **11**, 209–212.

Ingarden, R. S.: 1965, 'Simplified Axioms for Information Without Probability', *Roczniki Polskiego Tow. Matematycznego* **9**, 273–282.

Ingarden R. S. and Urbanik, K.: 1962, 'Information Without Probability', *Colloquium Mathematicum* **9**, 131–150.

Kendall, D. G.: 1964, 'Functional Equations in Information Theory', *Zeitschr. für Wahrsch. und Verw. Geb.* **2**, 225–229.

Khinchin, A. I.: 1957, *Der Begriffe der Entropien in der Wahrscheinlichkeitsrechnung.* (Arbeiten zur Informationstheorie, I: Mathematische Forschungsberichte), VEB, Berlin.

Kolmogorov, A. N.: 1965, 'Three Approaches to the Definition of the Concept "Quantity of Information"', *Problemy Peredaci Informacii* **1**, 3–11.

Kolmogorov, A. N.: 1967, 'Logical Basis for Information Theory and Probability Theory', *IEEE Transactions on Information Theory* **IT-14**, 662–664.

Kraft, C. H., Pratt, J. W., and Seidenberg, A.: 1959, 'Intuitive Probability on Finite Sets', *Ann. Math. Stat.* **30**, 408–419.

Lee, P. M.: 1964, 'On the Axioms of Information Theory', *Ann. Math. Stat.* **35**, 415–418.

Luce, R. D.: 1967, 'Sufficient Conditions for the Existence of a Finitely Additive Probability Measure', *Ann. Math. Stat.* **38**, 780–786.

Luce, R. D. and Tukey, J. W.: 1964, 'Simultaneous Conjoint Measurement: A New Type of Fundamental Measurement', *Journal of Mathematical Psychology* **1**, 1–27.

Maeda, S.: 1960, 'On Relatively Semi-orthocomplemented Lattices', *J. Sci. Hiroshima Univ.* **24**, 155–161.

Maeda, S.: 1963, 'A Lattice Theoretic Treatment of Stochastic Independence', *J. Sci. Hiroshima Univ.* **27**, 1–5.

Marczewski, E.: 1958, 'A General Scheme of the Notions of Independence in Mathematics', *Bull. de l'Acad. Polonaise des Sci.* **6**, 731–736.

Onicescu, O.: 1966, 'Energie Informationelle', *C.R. Acad. Sci. Paris* **263A**, (1966) 841–863.

Raiffa, H., Pratt, J. W., and Schlaifer, R.: 1964, 'The Foundations of Decision Under Uncertainty: An Elementary Exposition', *J. Amer. Stat. Assoc.* **59**, 353-375.

Rényi, A.: 1961, 'On Measures of Entropy and Information', in *Proc. of the Fourth Symp. on Math. Stat. and Probability* (ed. by J. Neyman), vol. I, University of California Press, Berkeley, Calif, pp. 457–561.

Savage, L. J.: 1954, *The Foundations of Statistics*, Wiley, New York.

Scott, D.: 1964a, 'Linear Inequalities and Measures on Boolean Algebras', unpublished paper.

Scott, D.: 1964b, 'Measurement Structures and Linear Inequalities', *Journal of Mathematical Psychology* **1**, 233–247.

Skornyakov, L. A.: 1964, *Complemented Modular Lattices and Regular Rings*, Oliver and Boyd, London.

Suppes, P.: 1961, 'Behavioristic Foundations of Utility', *Econometrica* **29**, 186–202.

Suppes, P. and Zinnes, J.: 1963, 'Basic Measurement Theory', in *Handbook of Mathematical Psychology* (ed. by R. D. Luce, R. R. Bush, and E. Galanter), vol. I, Wiley, New York, pp. 1–76.

Tveberg H.: 1958, 'A New Derivation of Information Functions', *Math. Scand.* **6**, 297–298.

Varma, S. and Nath, P.: 1967, 'Information Theory – A Survey', *J. Math. Sci.* **2**, 75–109.

Von Neumann, J.: 1960, *Continuous Geometry*, University Press, Princeton, N.J.

Weiss, P.: 1968, 'Subjektive Unsicherheit und Subjektive Information', *Kybernetik* **5**, (1968) 77–82.

REFERENCES

[1] The symbol $\circ$ denotes functional composition; that is, $I \circ P(A) = I(P(A))$, if $A \in \mathfrak{A}$. In general, we shall use the standard notation $M \xrightarrow{f} N$ or $f: M \to N$ for a function f which maps the set M into the set N. Complicated situations will be represented by diagrams in a well-known way.

[2] For simplicity we shall keep the same notation, even though we are working now with the algebra $\mathfrak{A}$ and not **P**.

[3] For typographical simplicity we use the same symbol that was used in Section 1.2 for different ordering.

[4] *Re* denotes the set of real numbers.

[5] *FQCP*-structure = finite qualitative conditional probability structure (see Domotor, 1969).

[6] $\mathfrak{A}[\mathscr{P}]$ denotes the smallest Boolean algebra containing $\mathscr{P}$.

PART III

INFORMATION AND LEARNING

DEAN JAMISON, DEBORAH LHAMON, AND PATRICK SUPPES

LEARNING AND THE STRUCTURE OF INFORMATION*

I. INTRODUCTION

The concept of information has received a great deal of attention in a number of disciplines ranging from mathematical statistics to experimental psychology over the past two decades, and the literature is now very extensive. In reviewing that literature, however, we were surprised to find how few studies are concerned with the theoretical relations between the concept of reinforcement in learning theory and the structure of information built into the structure of reinforcing events. For example, little attention is given to the systematic structure of reinforcement from the standpoint of information content in the traditional learning theories of Hull and Skinner. This can be regarded as a criticism of these theories, but the detailed analyses given in this paper show clearly enough that relating detailed learning-theoretic considerations to detailed information structures is not a simple matter. Any of the standard mathematical models of learning soon become difficult and awkward to handle once any but the simplest information structures on reinforcing events are used, though a number of analytic results can be obtained.

From a broad conceptual standpoint, this paper may be regarded as having two general aims. One is to bring together in a detailed way the ideas of contemporary mathematical learning theory and the concept of information structure. The other and subsidiary aim is to show by example how difficult the analysis of these concepts becomes if our objective is to carry the analysis through to the point of yielding explicit analytical results that can be tested against experimental data. Although this paper is long and many examples are considered, it clearly is only a kind of prolegomenon to a more systematic and more general theory of these matters. In originally planning the paper, we had hoped to include a final section on concept learning, but for a variety of reasons, including the evident length of the paper in its present form, we restricted ourselvez to a few remarks here and there on this subject. The study of conceps

J. Hintikka and P. Suppes (eds.), Information and Inference, 197–259.

learning differs from the topics we considered since the information structure in concept learning is placed on the stimulus set rather than the reinforcement set.

II. PAIRED-ASSOCIATE LEARNING

1. *Paired-Associate Learning with Complete Information*

In the experimental paradigm for the theories discussed in this section, the experimenter presents the subject with stimuli in random order. Each stimulus is paired to exactly one of N response alternatives. After seeing a stimulus, the subject chooses the response he believes is correct. After the subject has made a choice, the experimenter tells him what the correct response was. The subject then proceeds to the next stimulus. This *correction* procedure is distinguished from *noncorrection* procedures in which the subject is told only whether he was correct or incorrect. Noncorrection procedures are discussed briefly in Part II, Section 2 with other theories of incomplete information. Certain of our proposed models for the correction procedure bear mild resemblance to models for the noncorrection procedure presented by Millward (1964) and Nahinsky (1967).

The objective of a theory of PAL (paired-associate learning) is to predict the detailed statistical structure of subjects' response data in the type of experimental paradigm just described. Theories of PAL have the following general structure. For each of the (homogeneous) stimulus items there exists a set $\mathfrak{T}$ of states that the subject may be in on any trial and a set $\mathfrak{R}$ of response alternatives that he may choose from. There further exists a set $\mathfrak{E}$ of reinforcing events. Finally, there exist two functions: a function f that maps $\mathfrak{T} \times \mathfrak{R}$ into $[0, 1]$ and a function g that maps $\mathfrak{T} \times \mathfrak{E} \times \mathfrak{T}$ into $[0, 1]$. (Here $\mathfrak{T} \times \mathfrak{R}$ denotes the cross product of the sets $\mathfrak{T}$ and $\mathfrak{R}$.) The function f gives the probabilities of the various responses for each state; the function g gives the probabilities of state transitions for various reinforcements. In this way, a model of PAL may be considered an ordered quintuple, $\langle \mathfrak{E}, \mathfrak{R}, \mathfrak{T}, f, g \rangle$. A particular theory specifies precisely the members of the three sets and the form of the two functions.

The remainder of this section is divided into two parts. In the first part we give a brief review of eight existing theories of PAL. In the second part we present several new theories. For each new theory we present informally its assumptions, its basic mathematical structure, a few derivations, and its relations to other theories.

A. *Existing Theories of Paired-Associate Learning*

The linear model. Let $P(e_n)$ denote the probability of an error occurring on trial n. The basic assumption of the linear model is that $P(e_{n+1})$ is a fixed fraction of $P(e_n)$, specifically:

$$P(e_{n+1}) = \theta P(e_n) \tag{1}$$

If we make the natural assumption that $P(e_1)$ is equal to $(N-1)/N$, then

$$P(e_n) = \frac{N-1}{N}\,\theta^{n-1}. \tag{2}$$

Bush and Mosteller (1955) described the linear model in some detail.

The one-element model. The principal assumption of the one-element model is that for each stimulus element the subject is in one of two states – conditioned to the correct response or not conditioned to it. If he is not conditioned, then with probability c on any trial he becomes conditioned; once he becomes conditioned, he remains so. If the subject is conditioned, he responds correctly; if he is not, he guesses, responding correctly with probability $1/N$. The following transition matrix and error response probability vector summarize the one-element model, where C and $\bar{C}$ represent the conditioned and unconditioned states:

$$\begin{array}{c} \\ C \\ \bar{C} \end{array}\begin{array}{c} \begin{array}{cc} C & \bar{C} \end{array} \\ \begin{bmatrix} 1 & 0 \\ c & 1-c \end{bmatrix} \end{array} \begin{bmatrix} 0 \\ \dfrac{N-1}{N} \end{bmatrix}. \tag{3}$$

This matrix gives the probabilities of transition from one state to the next on each trial; the vector gives the probability of making an *incorrect* response in each state. The probability of error on trial n is easily shown to be given by:

$$P(e_n) = \frac{N-1}{N}(1-c)^{n-1}. \tag{4}$$

Bower (1961) compared the linear and one-element models on a wide variety of statistics for experiments with $N=2$. The one-element model fits much better than the linear model. But when $N>2$, the one-element model performs less well, although still better than the linear model.

The two-phase model. Norman (1964b) proposed a two-phase model for which he assumes that no learning occurs up to some trial k; after trial k, learning proceeds linearly with parameter θ. The trial of first learning, k, is geometrically distributed with parameter c. The probability of error on trial n is given by:

$$(5) \qquad P(e_n) = \begin{cases} (N-1)/N & \text{for} \quad n \leqslant k \\ \dfrac{N-1}{N}(1-\theta)^{n-k} & \text{for} \quad n > k \, . \end{cases}$$

When $\theta=1$, the equation reduces to the one-element model; when $c=1$, the equation reduces to the linear model.

The random-trial incremental model. In Norman (1964), the RTI (random-trial incremental) model postulated that on each trial learning occurs with probability c; if it does occur, it does so linearly with learning parameter θ. The following equation summarizes the model:

$$(6) \qquad P(e_{n+1}) = \begin{cases} (1-\theta)\, P(e_n) \text{ with probability } c \\ P(e_n) \text{ with probability } 1-c \, . \end{cases}$$

As with the two-phase model, if $\theta=1$, the RTI model reduces to the one-element model and if $c=1$, it reduces to the linear model.

The two-element model. Both the two-phase and the RTI models primarily represent extensions of the linear model; Suppes and Ginsberg (1963) suggested an extension of the one-element model to a two-element model. The subject is in any one of three states – C_0, C_1, and C_2; the subscript refers to the number of stimulus elements conditioned to the correct response. Those not conditioned to the correct response are unconditioned. The transition matrix and error probability vector given below summarize the model:

$$(7) \qquad \begin{array}{c} \\ C_2 \\ C_1 \\ C_0 \end{array} \begin{array}{c} \begin{array}{ccc} C_2 & C_1 & C_0 \end{array} \\ \begin{bmatrix} 1 & 0 & 0 \\ b & 1-b & 0 \\ 0 & a & 1-a \end{bmatrix} \end{array} \begin{bmatrix} 0 \\ 1-g \\ \dfrac{N-1}{N} \end{bmatrix}.$$

The model has three parameters: the conditioning probabilities a and b and the guessing probability g for when the subject is in state C_1. Predicting a stationary probability of success prior to last error is one of the major shortcomings of the one-element model; the two-element model avoids this shortcoming.

The long-short model. In their comprehensive overview of paired-associate learning models, Atkinson and Crothers (1964) proposed a model based on the distinction between long- and short-term stores. In state L the subject has the S-R association in long-term store and remembers it. In state S the subject always responds correctly, but may forget the association and drop back to a guessing state F. State F is initially reached by 'coding' the stimulus element from an uncoded state U; this coding occurs with probability c. The other parameters of the model are the probability a that when reinforcement occurs the subject goes into state L, and the probability f that an item in state S will move back to F. The transition matrix and error probability vector of the model are given below:

$$(8)\qquad \begin{array}{c} \\ L \\ S \\ F \\ U \end{array} \begin{array}{c} \begin{array}{cccc} L & S & F & U \end{array} \\ \begin{bmatrix} 1 & 0 & 0 & 0 \\ a & (1-a)(1-f) & (1-a)f & 0 \\ a & (1-a)(1-f) & (1-a)f & 0 \\ ca & c(1-a)(1-f) & c(1-a)f & 1-c \end{bmatrix} \end{array} \begin{bmatrix} 0 \\ 0 \\ \dfrac{N-1}{N} \\ \dfrac{N-1}{N} \end{bmatrix}.$$

The three-parameter version of this model is referred to as LS-3; a two-parameter version, LS-2, is obtained by setting $c=1$. Atkinson and Crothers point out that this model was constructed with an emphasis on reproducing specific psychological processes, though the reason the transition from S to S should have the same probability as the one from F to S remains unclear. Both the LS-3 and LS-2 models fit the data very well. Extensions of the LS-3 model and a trial-dependent forgetting (TDF) model to account for variations in list length are presented in the Atkinson and Crothers paper and extended by Calfee and Atkinson (1965). Rumelhart (1967) presented an illuminating overview and extensions of these models. However, we will discuss these variations no further.

A forgetting model. Bernbach (1965) proposed a three-parameter forgetting model for paired-associate learning. In state C the subject is always correct, and in state G he is correct with probability $1/N$. Immediately after reinforcement the subject is in state C; presumably if he were immediately tested he would always be correct, but before the next presentation of the stimulus there is a probability δ that he will forget. If the subject is in state C with probability θ, he permanently acquires the S-R association and moves to state C'. Finally, there is a probability β that if the subject guesses incorrectly, he learns the incorrect response he guessed. If so, he goes to state E in which his probability of success is zero. The forgetting model is represented by the following transition matrix and error probability vector:

(9)

$$\begin{array}{c} \\ C' \\ C \\ G \\ E \end{array}\begin{array}{c} \begin{array}{cccc} C' & C & G & E \end{array} \\ \begin{bmatrix} 1 & 0 & 0 & 0 \\ \theta & (1-\theta)(1-\delta) & (1-\theta)\,\delta & 0 \\ 0 & \left[1-\beta\left(\frac{N-1}{N}\right)\right](1-\delta) & \delta & \beta\left(\frac{N-1}{N}\right)(1-\delta) \\ 0 & (1-\beta)(1-\delta) & \delta & \beta(1-\delta) \end{bmatrix} \end{array} \begin{bmatrix} 0 \\ 0 \\ 1-1/N \\ 1 \end{bmatrix}.$$

Bernbach performed some experiments in which the forgetting model does rather better than the one-element model.

This completes our discussion of a number of existing models for paired-associate learning. We now turn to some new models.

B. *New Theories of Paired-Associate Learning*

The Dirichlet model. The name 'Dirichlet' is applied to this model since the generalization developed in Part II, Section 2 uses the general Dirichlet density. The model we shall now consider uses the one-dimensional version of the Dirichlet family known as the beta density. The intuitive idea of the model is that the subject can be in any state indexed by numbers on the interval $[0, 1]$. If the subject is in state $r (0 \leqslant r \leqslant 1)$ on trial n, he responds correctly with probability r, and his state on trial $n+1$ is drawn from a beta density on the interval $[r, 1]$. Figure 1 illustrates this.

Let us state the assumption more explicitly:

1. The state the subject is in on trial n is indexed by a real number r_n such that $0 \leqslant r_n \leqslant 1$.

2. If the subject is in state r_n, he responds correctly with probability r_n.

3. Let $f(r_{n+1}|r_n)$ be the density for r_{n+1} given r_n. Then

$$(10) \qquad f(r_{n+1} \mid r_n) = \begin{cases} 0 \text{ if } r_{n+1} < r_n \text{ or if } r_{n+1} > 1, \text{ and} \\ \dfrac{1}{B(\alpha, \beta)(1 - r_n)} \cdot \left(\dfrac{r_{n+1} - r_n}{1 - r_n}\right)^{\alpha - 1} \\ \qquad \cdot \left(1 - \dfrac{r_{n+1} - r_n}{1 - r_n}\right)^{\beta - 1} \\ \text{if } r_n \leqslant r_{n+1} \leqslant 1, \alpha > 0, \text{ and } \beta > 0. \end{cases}$$

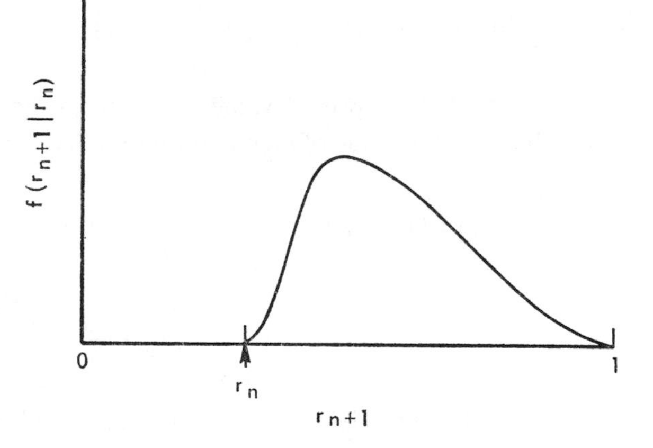

Fig. 1. The density for r_{n+1}.

The function $B(\alpha, \beta)$ is the beta function of α and β and is defined to equal $\int_0^1 x^{\alpha-1}(1-x)^{\beta-1}dx$.

4. On the first trial $r_1 = 1/N$, where N is the number of response alternatives.

THEOREM 1. *The learning curve for the Dirichlet model is given by*:

$$P(e_n) = \frac{N-1}{N}\left(\frac{\alpha}{\alpha + \beta}\right)^{n-1}.$$

PROOF. Denote the expected value of r_{n+1} given r_n by $E(r_{n+1}|r_n)$. It is an elementary property of beta densities that the density

$$\frac{1}{B(\alpha, \beta)} x^{\alpha-1}(1-x)^{\beta-1}$$

for $0 \leqslant x \leqslant 1$ has expectation $\alpha/\alpha+\beta$. Hence,

$$E(r_{n+1} \mid r_n) = r_n + \frac{\alpha}{\alpha+\beta}(1-r_n). \tag{11}$$

Now r_n is itself a random variable. The expected value of r_{n+1} given r_n is a linear function of r_n. But the expected value of a linear function of a random variable is simply equal to that same linear function of the expected value of the random variable, i.e.,

$$E[r_{n+1} \mid E(r_n)] = E(r_n) + \frac{\alpha}{\alpha+\beta}[1-E(r_n)]. \tag{12}$$

Thus from r_1 we can find the expected value of r_2; from the expected value of r_2 we find the expecte d value of r_3, etc. It follows that

$$\begin{aligned} E[(1-r_{n+1}) \mid E(1-r_n)] &= \left(1-\frac{\alpha}{\alpha+\beta}\right)E(1-r_n) \\ &= \frac{\beta}{\alpha+\beta}E(1-r_n). \end{aligned} \tag{13}$$

Since $1-r_1=(N-1)/N$, by recursion on (13) it follows that $E(1-r_n)=(N-1)/N[\beta/(\alpha+\beta)]^{n-1}$. But $P(e_n)$ is simply equal to $E(1-r_n)$; hence,

$$P(e_n) = \frac{N-1}{N}\left(\frac{\beta}{\alpha+\beta}\right)^{n-1}. \tag{14}$$

Q.E.D.

The quantity $\alpha/(\alpha+\beta)=1-\beta/(\alpha+\beta)$ represents the learning rate in this model; the learning curve generated is the same as for the linear and one-element models. *In fact, both the linear and one-element models are special cases of the Dirichlet.* The linear model results from setting $\theta=\beta/(\alpha+\beta)$ and allowing α and β to approach infinity.

The one-element model results from setting $c=\alpha/(\alpha+\beta)$ and letting α and β approach zero. The behavior of $f(r_{n+1})$ for various values of α is shown in Figure 2, where $\alpha/(\alpha+\beta)=.25$ and $r_n=.2$.

We assume that the subject fails to learn on each trial with some fixed probability, $1-\gamma$, but when he does learn, r_{n+1} is given by (10), which results in a three-parameter generalization of the Dirichlet model. Letting $\gamma=1$ gives the two-parameter Dirichlet. If $\gamma=c\neq 1$ and $\beta/(\alpha+\beta)=\theta$, letting α approach infinity gives Norman's RTI model as a special case

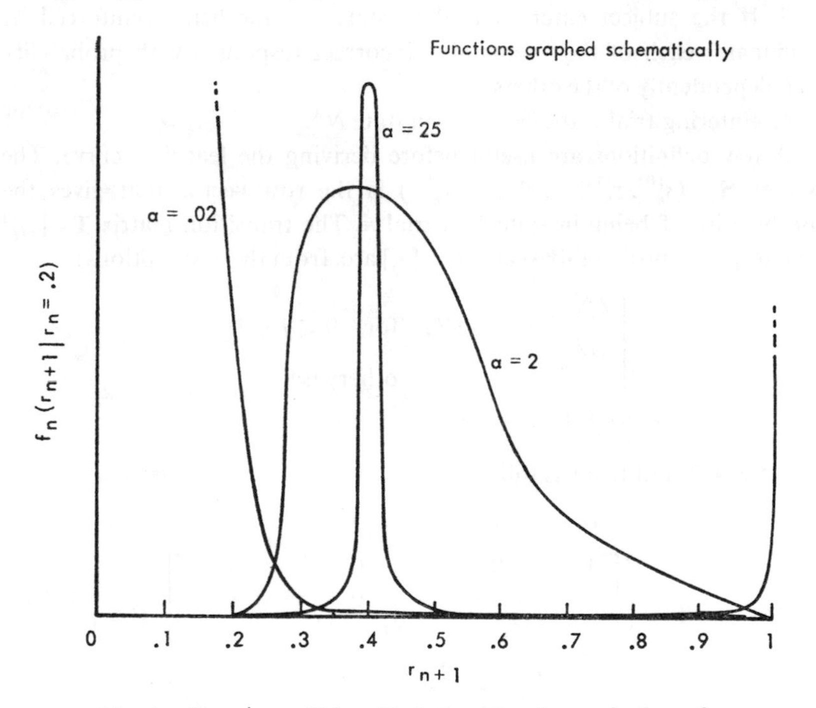

Fig. 2. $f(r_{n+1}|r_n=.2)$ for $\alpha/(\alpha+\beta)=.25$ and several values of α.

of the three-parameter Dirichlet model. The three-parameter Dirichlet model is an example of what Howard[1] calls a 'Markovian dynamic inference' model, with a continuous-state Markov chain.

The elimination model. The basic assumption of this model is that the subject learns by eliminating responses known to be incorrect. He eliminates each response possible on a given trial with a fixed probability, ε, independently of whether he eliminates other incorrect responses. More explicitly, the assumptions of the model are:

1. If there are N response alternatives, the subject can be in any of N

states labeled from 0 to N^*, where N^* is the number of wrong responses ($N^* = N - 1$). If the subject is in state i ($0 \leqslant i \leqslant N^*$), he has i possible wrong responses left to eliminate.

2. If the subject is in state i, the probability that he will make a correct response is $1/i+1$.

3. If the subject enters a trial in state i, after being reinforced he eliminates each of the i remaining incorrect responses with probability ε, independently of the others.

4. Entering trial 1, the subject is in state N^*.

A few definitions are useful before deriving the learning curve. The vector $\mathbf{S}_n = (s_n^{(0)}, s_n^{(1)}, \ldots, s_n^{(i)}, \ldots, s_n^{(N^*)})$ is the row vector that gives the probability of being in state i on trial n. The transition matrix $\mathbf{T} = [t_{ij}]$ and response probability vector $\mathbf{E} = [e_j]$ are, from the assumptions:

$$(15) \qquad t_{ij} = \begin{cases} \binom{i}{j} \varepsilon^{i-j} (1-\varepsilon)^j, & \text{for } 0 \leqslant j \leqslant i \\ 0 & , \text{otherwise} \end{cases};$$

$$e_j = j/j + 1 .$$

For $N=4$, $\mathbf{T}$ and $\mathbf{E}$ are as follows:

$$(16) \qquad \mathbf{T} = \begin{array}{c} \\ 0 \\ 1 \\ 2 \\ 3 \end{array} \begin{array}{c} \begin{array}{cccc} 0 & 1 & 2 & 3 \end{array} \\ \begin{bmatrix} 1 & 0 & 0 & 0 \\ \varepsilon & 1-\varepsilon & 0 & 0 \\ \varepsilon^2 & 2\varepsilon(1-\varepsilon) & (1-\varepsilon)^2 & 0 \\ \varepsilon^3 & 3\varepsilon^2(1-\varepsilon) & 3\varepsilon(1-\varepsilon)^2 & (1-\varepsilon)^3 \end{bmatrix} \end{array}, \quad \mathbf{E} = \begin{bmatrix} 0 \\ 1/2 \\ 2/3 \\ 3/4 \end{bmatrix}.$$

THEOREM 2. *The learning curve for the elimination model is given by*

$$P(e_n) = 1 - \frac{1 - [1 - (1-\varepsilon)^{n-1}]^N}{N(1-\varepsilon)^{n-1}} .$$

PROOF. First, it is evident that $P(e_n) = \mathbf{S}_n \mathbf{E}$ where $\mathbf{S}_n$ is the state probability vector on trial n, given by $\mathbf{S}_n = \mathbf{S}_1 \mathbf{T}^{n-1}$. $\mathbf{S}_1$ is simply equal to (0, 0,..., 0, 1). To proceed we must prove that $\mathbf{T}^n = [t_{ij}^{(n)}]$ is given by $t_{ij}^{(n)} = \binom{i}{j} [1 - (1-\varepsilon)^n]^{i-j} (1-\varepsilon)^{nj}$ if $j \leqslant i$, and $t_{ij}^{(n)} = 0$ otherwise. The proof is inductive. Clearly the assertion is true for $n=1$, where n is the power of the matrix. Let us assume that it is true for $k = n-1$; that is, assume $t_{i,j}^{(n-1)} =$

$\binom{i}{j}\gamma^j(1-\gamma)^{i-j}$ for $i \leqslant j$ with $\gamma=(1-\varepsilon)^{n-1}$. Henceforth it is understood that for $j>i$, t_{ij} equals zero. Then, multiplying $\mathbf{T}^{n-1}$ by $\mathbf{T}$ we obtain the general expression for $\mathbf{T}^n$: $\mathbf{T}^{n-1}\,\mathbf{T}=\mathbf{T}^n=[t_{ij}^{(n)}]$ where:

$$(17) \qquad t_{ij}^{(n)} = \sum_{1=0}^{N^*} \binom{i}{k} (1-\gamma)^{i-k} \gamma^i \binom{k}{j} \varepsilon^{k-j} (1-\varepsilon)^j .$$

Since $\binom{i}{k}\binom{k}{j} = \binom{i}{j}\binom{i-j}{k-j}$, and since the limits of the sum may be changed to j and i because the matrix is triangular,

$$(18) \qquad t_{ij}^{(n)} = \sum_{k=j}^{i} \binom{i}{j} \binom{i-j}{k-j} [\gamma(1-\varepsilon)]^j (1-\gamma)^{i-k} (\gamma\varepsilon)^{k-j} .$$

We now change the index of summation to $a=k-j$ and let d represent $i-j$. Therefore,

$$\begin{aligned}
(19) \qquad t_{ij}^{(n)} &= \binom{i}{j} [\gamma(1-\varepsilon)]^j \sum_{a=0}^{d} \binom{d}{a} (1-\gamma)^{d-a} (\gamma\varepsilon)^a \\
&= \binom{i}{j} [\gamma(1-\varepsilon)]^j [(1-\gamma)+\gamma\varepsilon]^d \\
&= \binom{i}{j} [\gamma(1-\varepsilon)]^j [1-\gamma(1-\varepsilon)]^{i-j} \\
&= \binom{i}{j} (1-\varepsilon)^{nj} [1-(1-\varepsilon)^n]^{i-j} .
\end{aligned}$$

This completes the subsidiary proof that $t_{ij}^{(n)}$ is given by

$$(20) \qquad t_{ij}^{(n)} = \begin{cases} \binom{i}{j} [1-(1-\varepsilon)^n]^{i-j} (1-\varepsilon)^{nj} & \text{if} \quad 0 \leqslant j \leqslant i \\ 0, \quad \text{otherwise}. \end{cases}$$

Multiplying $\mathbf{S}_1$ by $\mathbf{T}^{n-1}$ gives $\mathbf{S}_n=\mathbf{S}_1\mathbf{T}^{n-1}$, where

$$(21) \qquad \mathbf{S}_n^{(j)} = \binom{N^*}{j} [1-(1-\varepsilon)^{n-1}]^{N^*-j} (1-\varepsilon)^{(n-1)j} .$$

Multiplying this row vector by the column v ector **E**, we obtain:

$$(22) \qquad P(e_n) = \mathbf{S}_n\mathbf{E} = \sum_{j=0}^{N^*} j/j+1 \binom{N^*}{j} [1-(1-\varepsilon)^{n-1}]^{N^*-j} \times (1-\varepsilon)^{(n-1)j},$$

which can be transformed to:

$$(23) \qquad P(e_n) = \sum_{j=0}^{N^*} \binom{N^*}{j} [1-(1-\varepsilon)^{n-1}]^{N^*-j} (1-\varepsilon)^{(n-1)j} - \sum_{k=1}^{N^*+1} \frac{1}{(1-\varepsilon)^{n-1}(N^*+1)} \binom{N^*+1}{k} \times [1-(1-\varepsilon)^{(n-1)}]^{N^*+1-k} (1-\varepsilon)^{(n-1)k} = 1 - \frac{[1-(1-\varepsilon)^{n-1}]^{N^*+1}}{(1-\varepsilon)^{n-1}(N^*+1)}. \qquad \text{Q.E.D.}$$

The learning curve is the only statistic we shall derive for the elimination model. Before going on to extensions of this model, we should point out the following: First, when $N=2$, the elimination model is formally

TABLE I

Minimum χ^2 values for four one-parameter models[c]

Experiment	One-element[d]	Linear[d]	Elimination	Conditioning strength
Ia[a]	30.30	50.92	15.03	8.11
Ib[a]	39.31	95.86	17.63	14.41
III[a]	62.13	251.30	32.71	31.80
III[b]	150.66	296.30	101.11	95.26
IV[b]	44.48	146.95	31.76	39.37
Va[b]	102.02	201.98	56.52	53.74
Vb[b]	246.96	236.15	97.50	85.69
Vc[b]	161.03	262.56	117.76	90.26
Total	836.89	1542.02	470.02	418.64

[a]Three-response alternatives.
[b]Four-response alternatives.
[c]Total χ^2 for other models: 2-parameter: RTI, 284.39; 2-phase, 493.59; LS-2, 147.16; 3-parameter: LS-3, 137.26; 2-element, 259.56.
[d]Data from Atkinson and Crothers (1964).

identical to the one-element model, and, second, when $N>2$, the model predicts increasing probability of success prior to the trial of last error. This model is compared against data presented by Atkinson and Crothers in Table I.

The acquisition/elimination models. These models are two- and three-parameter generalizations of the elimination model. The basic notion behind the two-parameter acquisition/elimination model (AE-2) is that there is some probability c that the subject learns the correct response on any particular trial. If he fails to do so, he eliminates incorrect responses with probability ε as in the elimination model. More explicitly, AE-2 makes the same assumptions as the elimination model except that Assumption 3 is changed to:

3'. If the subject is in any state i, then after he is reinforced he acquires the correct response with probability c. If he fails to acquire the correct response then, with probability ε, he eliminates each of the i remaining incorrect responses, independently of others.

The following transition matrix, $\mathbf{T}'=[t'_{ij}]$ characterizes AE-2:

$$t'_{ij} = \begin{cases} c+(1-c)(\varepsilon)^i, & \text{for } j=0 \\ (1-c)\binom{i}{j}\varepsilon^{i-j}(1-\varepsilon)^j, & \text{for } 1\leqslant j\leqslant i\leqslant N^* \\ 0, & \text{otherwise}. \end{cases} \tag{24}$$

For $N=4$, the matrix is:

$$\mathbf{T}' = \begin{matrix} & 0 & 1 & 2 & 3 \\ 0 & 1 & 0 & 0 & 0 \\ 1 & c+(1-c)\varepsilon & (1-c)(1-\varepsilon) & 0 & 0 \\ 2 & c+(1-c)\varepsilon^2 & 2(1-c)(\varepsilon)(1-\varepsilon) & (1-c)(1-\varepsilon)^2 & 0 \\ 3 & c+(1-c)\varepsilon^3 & 3(1-c)\varepsilon^2(1-\varepsilon) & 3(1-c)\varepsilon(1-\varepsilon)^2 & (1-c)(1-\varepsilon)^3 \end{matrix}. \tag{25}$$

Model AE-2 reduces to the elimination model if $c=0$; it reduces to the one-element model if $\varepsilon=0$ or $N=2$. It can be extended to three parameters (AE-3) by assuming that when the subject learns an association (with probability c) he may pick up several more than just the correct one. The number he acquires is binomially distributed with parameters α and i, i being his state index. For example, if the subject is in state i and it is given that he learns on a particular trial, then with probability α^i he acquires just the correct response. Intuitively, α should be close to one. The assumptions of AE-3 are the same as those of the elimination model and AE-2 except that we substitute 3″ for 3′:

3″. If the subject is in state i at the beginning of a trial then, when reinforced, with probability c he acquires the correct response and up to i incorrect responses. He selects the number acquired with a binomial distribution with parameters α and i. With probability $1-c$ the subject acquires nothing, but he eliminates incorrect responses independently, each with probability ε.

The transition matrix for AE-3, $\mathbf{T}''=[t''_{ij}]$, is given in component form by

$$t''_{ij} = \begin{cases} c\binom{i}{j}\alpha^{i-j}(1-\alpha)^j + (1-c)\binom{i}{j}(\varepsilon)^{i-j}(1-\varepsilon)^j & \text{for } 0 \leqslant j \leqslant i \leqslant N^* \\ 0, & \text{otherwise}. \end{cases} \tag{26}$$

If $\alpha=1$, AE-3 reduces to AE-2; if $c=0$ or $c=1$, AE-3 reduces to the simple elimination model. The chief motivation for the AE-3 model is that it can give a bimodal transition distribution, which the binomial distribution in AE-2 cannot do.

An elimination model with forgetting. In the incorrect-response elimination models discussed so far, there has been no provision for regressing to a state in which the subject responds from *more* wrong responses, that is, for forgetting. It is plausible to assume that during the intertrial interval, after the subject has eliminated perhaps several incorrect responses, he might forget which ones he had eliminated, thus introducing some more wrong responses. The basic assumption of this forgetting model is that the responses learned previously to be incorrect are reintroduced, independently of one another, with some probability δ. More explicitly, the assumptions are:

1. If there are N response alternatives, the subject can be in any of N states labeled from 0 to N^*, where $N^*=N-1$. If the subject is in state $i(0 \leqslant i \leqslant N^*)$, he has i possible wrong responses left to eliminate.

2. If the subject is in state i, the probability that he will make a correct response is $1/i+1$.

3. If the subject enters a trial in state i, after being reinforced he eliminates each of the i remaining incorrect responses with probability ε, independently of the others.

4. Unless the subject is in state 0, between trials he forgets each response previously learned to be incorrect with probability δ, independently of the others. If the subject is in state 0, he stays there.

5. When the subject enters trial 1, he is in state N^*.

The subject enters trial 1 with state probability vector $\mathbf{S}_1=(0, 0,\ldots, 0,\ldots, 1)$ by Assumption 5. Shortly after reinforcement, the subject has state probability vector $\mathbf{S}_1'$ given by:

(27) $$\mathbf{S}_1' = \mathbf{S}_1\mathbf{T},$$

where $\mathbf{T}$ is the transition matrix given by (15). During the intertrial interval the subject may forget; his forgetting or reintroduction is represented by a matrix $\mathbf{F}$ that operates on $\mathbf{S}_1'$. $\mathbf{F}=[f_{ij}]$ is given by:

(28) $$\begin{cases} f_{00} = 1 \\ f_{0j} = 0 \quad \text{for} \quad 1 \leqslant j \leqslant N^* \\ f_{ij} = 0 \quad \text{for} \quad j < i \\ f_{ij} = \binom{N^* - 1}{j - i} \sigma^{j-i}(1-\delta)^{N^*-j} \quad \text{for} \quad 0 < i \leqslant j \leqslant N^*. \end{cases}$$

For $N=4$, $\mathbf{F}$ is:

(29) $$\mathbf{F} = \begin{array}{c} \\ 0 \\ 1 \\ 2 \\ 3 \end{array} \begin{array}{c} \begin{array}{cccc} 0 & 1 & 2 & 3 \end{array} \\ \begin{bmatrix} 1 & 0 & 0 & 0 \\ 0 & (1-\delta)^2 & 2\delta(1-\delta) & \delta^2 \\ 0 & 0 & (1-\delta) & \delta \\ 0 & 0 & 0 & 1 \end{bmatrix} \end{array}.$$

Thus $\mathbf{S}_2=\mathbf{S}_1'\mathbf{F}=\mathbf{S}_1\mathbf{TF}$. Or, more generally,

(30) $$\mathbf{S}_n = \mathbf{S}_1(\mathbf{TF})^{n-1},$$

and

(31) $$P(e_n) = \mathbf{S}_1(\mathbf{TF})^{n-1}\mathbf{E}.$$

Clearly this forgetting model could be generalized by replacing $\mathbf{T}$ with $\mathbf{T}'$ (24) or by $\mathbf{T}''$ (26).

A conditioning strength model. Atkinson (1961) suggested a generalization of stimulus-sampling theory that embodies the notion of 'conditioning strength'. Each response alternative has associated with it a conditioning strength; the total available amount of conditioning strength remains constant over trials. The probability that any given response will be made is its conditioning strength divided by the total available. Our model specializes Atkinson's work to paired-associate learning and generalizes it to include richer ways of redistributing conditioning strength after reinforcement. The assumptions of our model are:

1. If there are N response alternatives, the subject can be in any of N states. If the subject is in state i, $(0 \leqslant i \leqslant N-1)$, the conditioning strength of the correct response is $N-i$. The total available conditioning strength is N.

2. The probability of a correct response is equal to the response strength of the correct response divided by total response strength. That is to say, if the subject is in state i, his probability of being correct is $(N-i)/N$, and the probability of being incorrect is i/N.

3. If the subject is in state i on trial n, on trial $n+1$ he can be in any state between i and 0; which state he enters is given by a binomial distribution with parameters i and α.

4. On trial 1, $i=N-1$.

The transition matrix of this model is identical to that of the elimination model; all that differs is the response probability vector. The matrix and response probability vector are shown below.

(32)

$$\mathbf{T}^* = \begin{array}{c} \\ 0 \\ 1 \\ 2 \\ 3 \\ \vdots \\ N-1 \end{array} \begin{array}{c} \begin{array}{ccccc} 0 & 1 & 2 & 3 & \ldots N-1 \end{array} \\ \begin{bmatrix} 1 & 0 & 0 & 0 & \ldots 0 \\ \alpha & (1-\alpha) & 0 & 0 & \ldots 0 \\ (\alpha)^2 & 2\alpha(1-\alpha) & (1-\alpha)^2 & 0 & \ldots 0 \\ (\alpha)^3 & 3\alpha^2(1-\alpha) & 3\alpha(1-\alpha)^2 & (1-\alpha)^3 & \ldots 0 \\ & & & & \vdots \\ (\alpha)^{N-1} & \ldots & & & \ldots (1-\alpha)^{N-1} \end{bmatrix} \end{array} \quad \mathbf{E}^* = \begin{bmatrix} 0/N \\ 1/N \\ 2/N \\ \vdots \\ \frac{N-1}{N} \end{bmatrix}.$$

The learning curve is given by:

$$(33) \qquad P(e_n) = \mathbf{S}_1 \mathbf{T}^{*(n-1)} \mathbf{E}^* .$$

$\mathbf{T}^{*(n-1)}$ is given by (20) and $\mathbf{S}_1\mathbf{T}^{*(n-1)}$ by (21) where N^* must be replaced by $N-1$. Multiplying $\mathbf{S}_1\mathbf{T}^{*(n-1)}$ by $\mathbf{E}^*$, we obtain

$$(34) \qquad P(e_n) = \sum_{j=0}^{N-1} j/N \binom{N-1}{j} \alpha^{j(n-1)}(1-\alpha)^{(n-1)(N-1-j)} .$$

Ignoring N in the denominator, what remains is the expression for the expectation of a binomial density with parameters $N-1$ and $(1-\alpha)^{n-1}$. As this expectation is $(N-1)(1-\alpha)^{n-1}$,

$$(35)\qquad P(e_n) = \frac{N-1}{N}(1-\alpha)^{n-1},$$

which is the same learning curve as that for the linear and one-element models.

Clearly two- and three-parameter generalizations of the conditioning strength model are obtained by using the matrices given in (24) and (26) instead of **T***.

C. *Comparison of the One-Parameter Elimination and Conditioning Strength Models*

Atkinson and Crothers (1964) presented results from eight PAL experiments, in which three have three response alternatives and five have four response alternatives. Parameters are estimated by a minimum χ^2 technique from the 16 possible sequences in the data of correct and incorrect responses on trials 2 to 5. Atkinson and Crothers give results for many models; their results for the linear and one-element models are shown in Table I (see p. 209). Also shown in Table I are the results we

TABLE II

Parameter estimates for four one-parameter models

Experiment	One-element[a]	Linear[a]	Elimination	Conditioning strength
	c	θ	ε	α
Ia	.383	.414	.50	.55
Ib	.328	.328	.56	.60
II	.273	.289	.59	.69
III	.203	.258	.61	.70
IV	.281	.297	.52	.66
Va	.125	.164	.74	.84
Vb	.172	.250	.62	.70
Vc	.289	.336	.52	.66

[a]Data from Atkinson and Crothers (1964).

obtained for the one-parameter elimination and conditioning strength models. Table II shows the parameter estimates. Our theoretical predictions were obtained by computer simulation.

2. *Paired-Associate Learning with Incomplete Information: Noncontingent Case*

The general structure considered in this subsection is paired-associate learning with multiresponse reinforcement. We deal here with noncontingent reinforcement, and then, in the next subsection, we deal very briefly with reinforcement contingent on the subject's response. On each trial the subject responds with one of N alternatives. He is then reinforced with a subset of these N alternatives consisting of the one correct response and D distractors, of cardinality A in all (where $A = D + 1$). If A is one, then the paradigm is exactly that of determinate reinforcement just considered. If A is greater than one, then on any single trial the subject cannot rationally determine the correct response. On each trial the correct response is reinforced. The D distractors are selected randomly on each trial from the N^* possible wrong responses. Thus over trials the correct response will be the one response which is always reinforced. The subject's task is to make as many correct responses as he can and to learn the correct response as quickly as he can.

Normative model. Given the above paradigm, for some of the extensions, it is necessary to make predictions about the optimal behavior of a subject with perfect memory. Perfect memory of the entire reinforcement history is not required for normative behavior. If on each trial the reinforcement sets are intersected, then only the resulting intersection needs to be remembered. Thus if on trial 1 the subject is told that the correct response is among *a, b, c*, and *d*, where A is 4, and on trial 2 that the correct response is among *a, b, c*, and *e*, he need only remember *a, b*, and *c*, the members of the intersection as he begins trial 3. He then intersects this set with the next reinforcement set. Successive reinforcements and intersections eventually will lead to the correct response. The task now is to describe the 'eventually'.

Let N states $0, 1, \ldots, N^*$ be defined as for the elimination models in Part II, Section 1. Thus state i is the state of having i wrong responses, plus the correct one, that remain in the intersection on a given trial immediately before making a response. The subject responds from this set of $i+1$ responses, and then is shown A reinforcers. Since the subject is assumed to be acting normatively he intersects the new reinforcement set with the old intersection and remembers the resulting intersection until the next trial. The number of wrong responses now in memory is the cardinality of the intersection minus 1, and this is the number of the state in which the subject enters the next trial. Obviously j, the index of this new state, cannot be greater than i, which after the first reinforcement cannot be greater than D.

Letting $\mathbf{NN}=[nn_{ij}]$ be the transition matrix for the normative model, the general expression follows immediately by considering the transition from state i to j as the event of exactly j out of the D reinforced distractors being among the i distractors in the previous intersection.

Thus we obtain

$$(36) \qquad nn_{ij} = \begin{cases} \binom{D}{j} \dfrac{(i)_j (N^* - i)_{D-j}}{(N^*)_D} & \text{for } 0 \leqslant j \leqslant i \leqslant N^* \text{ and } j \leqslant D \\ 0, & \text{otherwise} \end{cases}$$

where

$$(a)_b = \binom{a}{b} \cdot b! = a(a-1)\ldots(a-b+1).$$

The normative transition matrix and error vector for $A=2$ are given as an example.

$$(37) \qquad \mathbf{NN}_{A=2} = \begin{array}{c} \\ \end{array} \begin{matrix} & 0 & 1 & 2\ldots N^* \end{matrix}$$

$$\mathbf{NN}_{A=2} = \begin{bmatrix} 0 & 1 & 0 & 0\ldots0 \\ 1 & \dfrac{N^*-1}{N^*} & \dfrac{1}{N^*} & 0\ldots0 \\ \vdots & \vdots & \vdots & \vdots \quad \vdots \\ i & \dfrac{N^*-i}{N^*} & \dfrac{i}{N^*} & 0\ldots0 \\ \vdots & \vdots & \vdots & \vdots \quad \vdots \\ N^* & 0 & 1 & 0 \quad 0 \end{bmatrix}, \quad \mathbf{E} = \begin{bmatrix} 0 \\ \frac{1}{2} \\ \vdots \\ \dfrac{i}{i+1} \\ \vdots \\ \dfrac{N^*}{N} \end{bmatrix}.$$

If S_n is the state probability vector as before, and the subject again enters trial 1 in state N^*, by virtue of intersecting the reinforcement subset with the entire set, the subject must enter trial 2 in state D. Thus

$$s^{(j)} = \begin{cases} 1, & \text{if } j = D \\ 0, & \text{otherwise.} \end{cases} \tag{38}$$

This equation also can be obtained directly from the transition matrix in (37).

Although states A through N^* are irrelevant except for entering state N^* on the first trial, they will be needed later, and thus for convenience are introduced here.

The equation for the state vector is given below:

$$\mathbf{S}_n = \mathbf{S}_1 \mathbf{NN}^{n-1}. \tag{39}$$

Letting $\mathbf{S}_n = [\mathbf{S}'_n, \mathbf{S}''_n]$ with the partition after column D, and letting $\mathbf{NN} = \begin{bmatrix} \mathbf{NN}' & \mathbf{0} \\ \mathbf{NN}'' & \mathbf{0} \end{bmatrix}$, with the partition after column and row D, we obtain

$$\mathbf{S}'_n = \mathbf{S}'_2 (\mathbf{NN}')^{(n-2)}. \tag{40}$$

We now derive the normative learning curve. As before, the probability of an error on trial n is found by multiplying $\mathbf{S}_n$ and $\mathbf{E}$; thus

$$P^*(e_n) = \mathbf{S}_n \mathbf{E} = \sum_{j=0}^{D} nn^{n-2} D_j \cdot \frac{j}{j+1}. \tag{41}$$

The powers of the $\mathbf{NN}$ matrix for $A=2$ given in (37) are readily found, and an explicit solution to the learning curve is possible. The power of the matrix with the extra states eliminated is given below.

$$(\mathbf{NN}')^n = \begin{array}{c} \\ 0 \\ 1 \end{array}\!\!\begin{array}{c} \begin{array}{cc} \quad 0 & \qquad 1 \end{array} \\ \begin{bmatrix} 1 & 0 \\ 1 - \left(\dfrac{1}{N^*}\right)^n & \left(\dfrac{1}{N^*}\right)^n \end{bmatrix} \end{array}. \tag{42}$$

Thus the learning curve and total errors are obtained:

$$P(e_n) = \begin{cases} \dfrac{N^*}{N}, & n = 1 \\ \dfrac{1}{2}\left(\dfrac{1}{N^*}\right)^{n-2}, & n = 2, 3, \ldots \end{cases} \tag{43}$$

$$E(\text{total errors}) = \sum_{n=1}^{\infty} P(e_n) = \frac{N^*(3N^* - 1)}{2N(N^* - 1)}. \tag{44}$$

This analytic solution for $A=2$ is given only as an example; numerical solutions for several specific N, A pairings are included in the Appendix. They are used there to compare real subject performance with the normative model.

At this point extensions of some models which do not reduce to the normative one are discussed. The normative model will be used later in extensions of other models.

One-element model. Several extensions of the one-element model outlined in Part II, Section 1 are possible and are considered here. An alternative generalization is discussed later as a special case of another model. The assumptions of this version of the one-element model are:

1. On each trial, the subject is either unconditioned or conditioned to exactly one of the N response alternatives. The unconditioned state will be denoted $\bar{C}$; the state of being conditioned to the correct response will be denoted C; and the state of being conditioned to any of the N^* incorrect responses will be denoted W.

2. If the subject is in state $\bar{C}$, he makes each response with a guessing probability, $1/N$. Otherwise he makes the response to which he is conditioned.

3. On any given trial, with probability $1-c$, the reinforcement is ineffective and the state of conditioning is unchanged. With probability c the reinforcement is effective. With effective reinforcement, if the subject is in state $\bar{C}$, he conditions with equal likelihood to any one, but exactly one, of the A reinforcers. If he is in a conditioned state and the response to which he is conditioned appears in the reinforcement set, he remains conditioned to that response. If the response does not appear, and if the reinforcement is effective, he rejects the response to which he was conditioned and becomes conditioned to exactly one of the responses reinforced on that trial.

4. Entering trial one, the subject is in state $\bar{C}$. Thus for the one-element model the transition matrix and error vector are:

$$(45)\qquad \begin{array}{c} \\ C \\ W \\ \bar{C} \end{array}\!\!\begin{array}{c} \begin{array}{ccc} C & W & \bar{C} \end{array} \\ \left[\begin{array}{ccc} 1 & 0 & 0 \\ c\cdot\dfrac{N^*-D}{N^*A} & 1-\dfrac{c(N^*-D)}{N^*A} & 0 \\ c\cdot\dfrac{1}{A} & c\cdot\dfrac{D}{A} & 1-c \end{array}\right] \end{array}, \quad \mathbf{E}=\left[\begin{array}{c} 0 \\ 1 \\ \dfrac{N^*}{N} \end{array}\right].$$

By raising the transition matrix to the $(n-1)$st power, the learning curve and expectation for total errors are found to be as follows:

$$(46) \qquad P(e_n) = \frac{N^*}{N} 1 - \frac{c}{A} \frac{(N^* - D)^{n-1}}{N^*},$$

and

$$(47) \qquad E(\text{total errors}) = \frac{N^*}{N} \frac{AN^*}{c(N^* - D)}.$$

No other statistics will be derived. The most obvious test, however, is not the learning curve itself, but the prediction of the run of errors while the subject is in state W. Once the subject moves out of state $\bar{C}$, no successes are predicted until he learns.

It should be noted that the one-element model does not reduce to the normative model for any value of c. As c increases, the probability of conditioning wrongly increases at the same rate as the probability of conditioning correctly.

An interesting extension of the one-element model has been worked out for A that varies in size on each trial from 1 to N, with probability π_a that $A=a$. The basic assumptions of the model are the same, but the state transition probabilities are altered by the experimental change. Let

$$(48) \qquad B = P(W_{n+1} \mid W_n).$$

Then, if the learning curve is analogous to that with constant A, we should expect

$$(49) \qquad q_n = \frac{N^*}{N} B^{n-1}.$$

We now prove this. Let $M = P(W_{n+1} | \bar{C}_n)$. So $P(\bar{C}_{n+1} | \bar{C}_n)$ remains $1-c$. Then by raising the transition matrix to the $(n-1)$st power, and assuming the subject starts in state $\bar{C}$,

$$(50) \qquad P(W_n) = \frac{M\left[(1-c)^{n-1} - B^{n-1}\right]}{1 - c - B}.$$

But, since

$$(51) \qquad B = \sum_{a=1}^{N} \pi_a \left(1 - \frac{c(N-a)}{(N-1)a}\right) = 1 - \frac{c}{N-1}\left(N \sum_{a=1}^{N} \frac{\pi_a}{a} - 1\right),$$

and

$$(52)\qquad M = \sum_{a=1}^{N} \pi_a c\left(1 - \frac{1}{a}\right) = c\left[1 - \sum_{a=1}^{N} \frac{\pi_a}{a}\right],$$

$$(53)\qquad P(W_n) = \frac{1-N}{N}\left[(1-c)^{n-1} - B^{n-1}\right].$$

Thus,

$$(54)\qquad q_n = P(W_n) + \frac{N-1}{N} P(\bar{C}_n) = P(W_n) + \frac{N-1}{N}(1-c)^{n-1} = \frac{N-1}{N} B^{n-1}.$$

Linear models. Let $\mathbf{p}_n = (p_{1,n}, p_{2,n}, \ldots, p_{N,n})$ represent the response probability vector on trial n. That is, $p_{i,n}$ is the probability of making the ith response on trial n. N is the number of response alternatives. A *linear* model for learning asserts that $\mathbf{p}_{n+1}$ is a linear function of $\mathbf{p}_n$; the exact nature of that linear function depends on the reinforcement. Consider as an example a situation with the two response alternatives, a_1 and a_2, where a_1 is always correct. The linear model for this situation is represented by a transformation matrix, $\mathbf{L} = [l_{ij}] = \begin{bmatrix} 1-\alpha & \alpha \\ 1-\theta & \theta \end{bmatrix}$. The vector $\mathbf{p}_{n+1}$ is given by the following expression:

$$(55)\qquad (p_{1,n+1}, p_{2,n+1}) = (p_{1,n}, p_{2,n}) \begin{bmatrix} 1-\alpha & \alpha \\ 1-\theta & \theta \end{bmatrix}.$$

The elements of the matrix $\mathbf{L}$ clearly must be independent of $\mathbf{p}_n$ or the model would be nonlinear. For learning to occur, θ must be greater than α.

In the example above only one reinforcement is given (i.e., this is the situation considered in Part II, Section 1), hence, only one transition matrix. In general the transition matrix must be indexed by the reinforcement E. The class of all linear models corresponds to the class of all transition matrices $\mathbf{L}(E) = [l_{ij}(E)]$ such that:

$$(56)\qquad l_{ij}(E) \geqslant 0 \quad \text{for} \quad 1 \leqslant i, j \leqslant N$$

and

$$(57)\qquad \sum_{j=i}^{N} l_{ij}(E) = 1 \quad \text{for} \quad 1 \leqslant i \leqslant N,$$

where E is a particular reinforcement. A linear model specifies for each reinforcement E a matrix $\mathbf{L}(E)$ such that

$$(58) \qquad (\mathbf{p}_{n+1} \mid \mathbf{p}_n, E) = \mathbf{p}_n \mathbf{L}(E),$$

as well as a starting vector, $\mathbf{p}_1$. Without placing further constraints on $\mathbf{L}$, we have an $N \times (N-1)$ parameter model. To pare these down to a single parameter, we make four further assumptions. The first, third, and fourth assumptions seem indispensable; relaxing the second would give a somewhat more general model. The assumptions are these.

1. Relabeling the response alternatives in no way affects the predictions of response probabilities.
2. If $a_i \in E_n$, where E_n is the subset of the response alternatives in the reinforcement set on trial n, then $l_{ii}(E_n) = 1$. In the example of (55), this corresponds to assuming $\alpha = 0$ instead of simply assuming $\alpha < \theta$.
3. If $r_i \in E_n$ then $p_{i,\,n+1} > p_{i,\,n}$.
4. $\mathbf{p}_1 = (1/N, 1/N, \ldots, 1/N)$.

The preceding assumptions limit us to two distinct one-parameter models. To see this, consider the $N=4$ with $A=2$ case. For convenience we consider that the first response is correct, i.e., it is always in the reinforcement set. Each of the remaining three responses appears in the reinforcement set with probability $\frac{1}{3}$. The two possible reinforcement matrices for when the first (correct) response and the second response are reinforced are given by:

$$(59) \qquad \mathbf{L}^{(1)} = \begin{bmatrix} 1 & 0 & 0 & 0 \\ 0 & 1 & 0 & 0 \\ \alpha/2 & \alpha/2 & 1-\alpha & 0 \\ \alpha/2 & \alpha/2 & 0 & 1-\alpha \end{bmatrix} \quad \text{and}$$

$$\mathbf{L}^{(2)} = \begin{bmatrix} 1 & 0 & 0 & 0 \\ 0 & 1 & 0 & 0 \\ \alpha/3 & \alpha/3 & 1-\alpha & \alpha/3 \\ \alpha/3 & \alpha/3 & \alpha/3 & 1-\alpha \end{bmatrix}.$$

The values of the first two rows follow from Assumption 2. Since the models are linear, none of the $p_{n,\,i}$ can appear in the matrices. From Assumption 3, $0 < \alpha \leqslant 1$ and, from Assumption 1, the constant α appearing in row 3 must have the same value as the α in row 4. The transition matrix

$\mathbf{L}^{(1)}$ follows if we assume that the decrement in any response alternative not reinforced is spread evenly among those that are reinforced; $\mathbf{L}^{(2)}$ follows if we assume that the decrement is spread among all the rest.

Let us now derive the learning curve for the $\mathbf{L}^{(1)}$ transition matrix. As there are three equiprobable reinforcements, and again assuming that response 1 is correct,

$$(60)\qquad p_{1,n+1}^{(1)} = 1/3\left(1\cdot p_{1,n} + 0\cdot p_{2,n} + \frac{\alpha}{2}\cdot p_{3,n} + \frac{\alpha}{2}\cdot p_{4,n}\right) + 1/3\left(1\cdot p_{1,n} + \frac{\alpha}{2}\cdot p_{2,n} + 0\cdot p_{3,n} + \frac{\alpha}{2}\cdot p_{4,n}\right) + 1/3\left(1 + p_{1,n} + \frac{\alpha}{2}\cdot p_{2,n} + \frac{\alpha}{2}\cdot p_{3,n} + 0\cdot p_{4,n}\right)$$

or

$$(61)\qquad p_{1,n+1}^{(1)} = p_{1,n} + \frac{\alpha}{3}(1 - p_{1,n}).$$

From this recursion and Assumption 4, it follows that

$$(62)\qquad p_{1,n}^{(1)} = 1 - \left[\frac{3}{4}\left(1 - \frac{\alpha}{3}\right)^{n-1}\right].$$

Using similar arguments with the $L^{(2)}$ transition matrix, we find that

$$(63)\qquad p_{1,n}^{(2)} = 1 - \left[\frac{3}{4}\left(1 - \frac{2\alpha}{9}\right)^{n-1}\right].$$

These results generalize to arbitrary N and A. We continue to assume that response 1 is correct. The following recursion gives $p_{1,\,n+1}$

$$(64)\qquad p_{1,n+1} = p_{1,n} + \sum_{i=2}^{N} p_{i,n}\cdot\frac{\alpha}{J}\cdot K\cdot L,$$

where L is the probability of each reinforcement set, K is the number of times each $p_{i,\,n}$ appears in the generalization of the sum given in (60) and J is the number by which α must be divided. This number depends on whether the decrement is spread among all response alternatives or only among these reinforced. L is equal to $\binom{N-1}{A-1}^{-1}$; K equals $\binom{N-2}{A-1}$; and J equals A or $N-1$.

Under the assumption that the decrement is spread only among those reinforced, i.e., $J=A$, the learning curve is:

$$(65a)\qquad p_{i,n+1}^{(1)} = 1 - \left[\frac{N-1}{N}\left(1 - \frac{\alpha N - A}{AN - 1}\right)^{n-1}\right].$$

Under the alternative assumption, $J=N-1$, the learning curve is:

$$(65b)\qquad p_{1,n}^{(2)} = 1 - \left[\frac{N-1}{N}\left(1 - \frac{\alpha(N-A)}{(N-1)^2}\right)^{n-1}\right].$$

Before we leave the linear models, consider a geometric interpretation for the $N=3$, $A=2$ case (in which it makes no difference whether the decrement is spread to only those reinforced or to all). The triangle ABC in Figure 3 represents all possible values of p_n; one particular value is shown. Assume that responses 1 and 2 are reinforced. Let S be the point

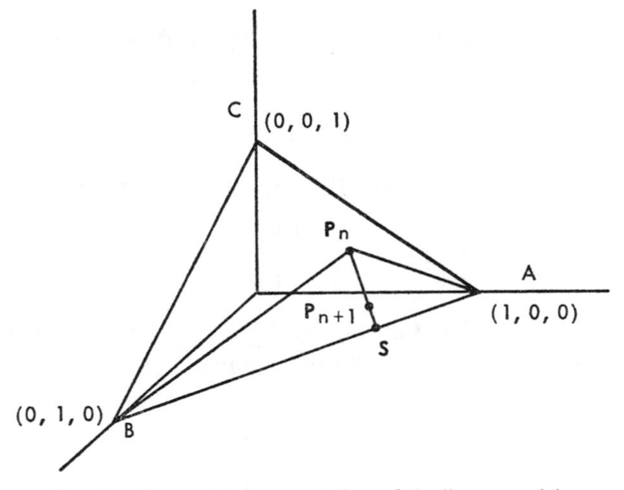

Fig. 3. Geometric interpretation of the linear models.

on the line AB such that the vectors $\mathbf{S}-\mathbf{p}_n$ are perpendicular to AB. Then the linear matrix models previously developed are equivalent to the geometric assertion that $\mathbf{p}_{n+1}=\mathbf{p}_n+\alpha(\mathbf{S}-\mathbf{p})$. Thus the *area* of triangle Ap_nB is decreased by a fixed fraction, whereas, in the determinate case, a *length* was decreased by a fraction α.

General Dirichlet model. As before, there are N response alternatives, A of which are reinforced on every trial. One of the A is correct; the remaining $A-1$ are chosen randomly from the $N-1$ incorrect responses. Let the vector $\mathbf{r}^{(n)}=(r_1^{(n)}, r_2^{(n)},..., r_N^{(n)})$ give the probabilities of making various responses on trial n. Clearly,

$$\sum_{j=1}^{N} r_j^{(n)} = 1 \quad \text{and} \quad r_j \geqslant 0 \quad \text{for} \quad 0 < j \leqslant N. \tag{66}$$

Let R be the set of all possible vectors $\mathbf{r}^{(n)}$; R is, then, a simplex in N-space. Our purpose first is to describe qualitatively the effect of reinforcement on $\mathbf{r}^{(n)}$. The vector $\mathbf{r}^{(n+1)}$ will be some point in the A-dimensional simplex in R whose points are linear combinations of $\mathbf{r}^{(n)}$ and the unit vectors corresponding to the responses reinforced. The simplex generated by $\mathbf{r}^{(n+1)}$ is denoted A^*. Figure 4 shows the case $N=3$, $A=2$ when responses 1 and 3 are reinforced.

The basic assumption of the general Dirichlet model is that the value of $\mathbf{r}^{(n+1)}$ given $\mathbf{r}^{(n)}$ is a random variable distributed according to an

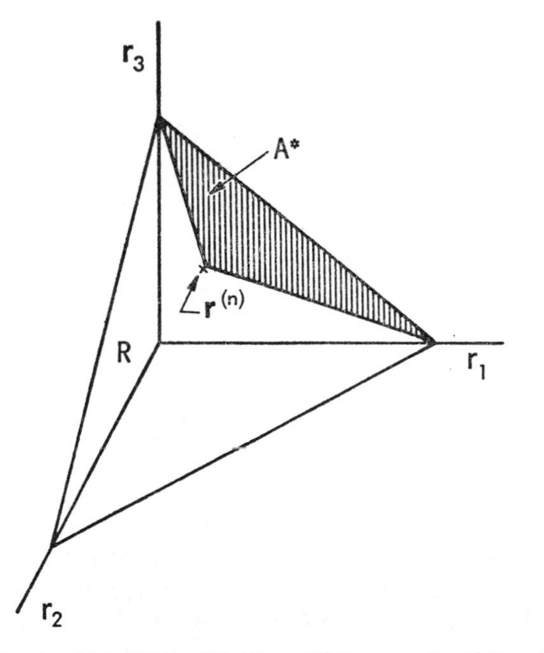

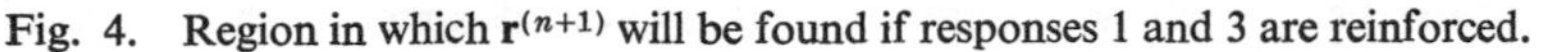
Fig. 4. Region in which $\mathbf{r}^{(n+1)}$ will be found if responses 1 and 3 are reinforced.

A-variate Dirichlet density over the region A^*. A further assumption is that this density is symmetric with respect to the responses reinforced. More explicitly, the assumptions of the theory are:

1. The state the subject is in on trial n is indexed by a vector $\mathbf{r}^{(n)}=(r_1^{(n)}, r_2^{(n)},..., r_N^{(n)})$ whose components are such that Equation (66) is satisfied.

2. If the subject is in state $\mathbf{r}^{(n)}$, he makes response i with probability $r_i^{(n)}$.

3. The density for $\mathbf{r}^{(n+1)}$ given $\mathbf{r}^{(n)}$ is an A-variate Dirichlet density over the previously defined region A^* with parameters $\alpha_1, \alpha_2,..., \alpha_A$, and β. (See Wilks, 1962 for a general discussion of the Dirichlet density.) Further, $\alpha_1=\alpha_j=\alpha$ for $1 \leqslant i, j \leqslant A$.

4. $\mathbf{r}^{(1)}=(1/N, 1/N,..., 1/N)$.

The A-variate Dirichlet density is defined on the standard region X such that $x_i \geqslant 0$ for $1 \leqslant i \leqslant A$ and $\sum_{i=1}^{A} x_i \leqslant 1$. The algebraic tangle involved in translating the region X into the region A^* may be avoided by considering only the marginal density for the probability of the correct response (which probability will be denoted $r_c^{(n)}$). Consider Figure 5. The region

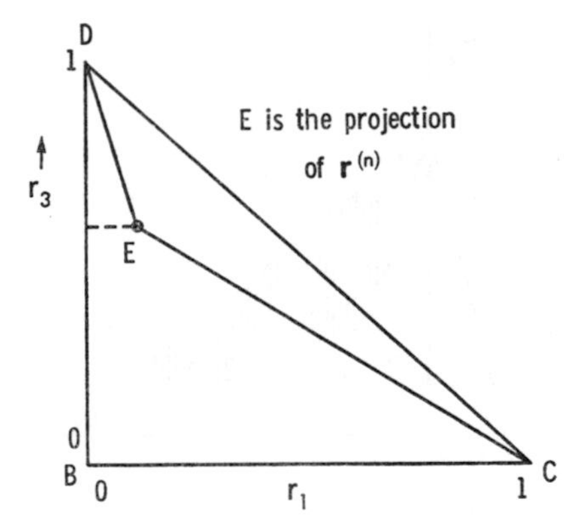

Fig. 5. The projection of A^* onto the $r_1 - r_3$ plane.

DEC is the straight on projection of A^* (from Figure 4) onto the $r_1 - r_3$ plane. The region X is the region BDC. Let the correct response be 3. All we need know is the marginal density along the line DE. From Wilks (1962, Th^{m} 7.7.2), we find that in this case, with $A=2$, the marginal is a beta density with parameters α and $\alpha+\beta$ and hence with expectation

$\alpha/(2\alpha+\beta)$. In general, the marginal distribution is a beta distribution with parameters α and $(A-1)\,\alpha+\beta$ and hence with expectation $\alpha/(A\alpha+\beta)$.

From here the derivation of the learning curve strictly parallels the development in Section 2.

$$(67) \qquad E(r_c^{(n+1)}) = E(r_c^{(n)}) + \frac{\alpha}{A\alpha+\beta}\,[1 - E(r_c^{(n)})].$$

Repeating the arguments of Part II, Section 1 we find, for $A<N$:

$$(68) \qquad P(e_{n+1}) = \left(\frac{(A-1)\,\alpha+\beta}{A\alpha+\beta}\right)^{n-1} N - 1/N.$$

Notice that for fixed α, β, and N, increasing A decreases the learning rate, as it should.

Elimination models. All elimination models generalize similarly from the models in Part II, Section 1. First the generalization of the one-parameter model is given, followed by the others in less detail.

The assumptions for the one-parameter elimination model are the same as those in Part II, Section 1 with the condition that the subject can eliminate only the responses on a trial that have been shown to be incorrect. With determinate reinforcement this condition could be introduced, but it would be inconsequential because on every trial it is possible to transit to state 0, that is, it is possible to eliminate all wrong responses. With multiresponse reinforcement the subject cannot always eliminate all wrong responses. If the subject narrows the correct response down to a, f, or g, and is shown a reinforcement set of a, b, e, and g, then the best he can do is to eliminate f. Thus the new transition probabilities are tied to the normative transition probabilities for the estimates of the best possible, or normative, move. More explicitly, letting $\mathbf{TT}=[tt_{ij}]$ be the multiple-response elimination transition matrix,

$$(69) \qquad \begin{aligned} tt_{ij} &= P(\text{state}_{n+1}=j \mid \text{state}_n = i) \\ &= \sum_{k=0}^{j} P(\text{state}_{n+1}=j \mid \text{state}_n = i,\ \text{normative} = k) \\ &\qquad\qquad \times P(\text{norm.} = k \mid \text{state}_n = i). \end{aligned}$$

The sum is only to j, as the subject can move no further than the normative move. The second term in the sum is obviously just the normative transition probability. The first term is the probability that if you start with i wrong responses and $i-k$ can be eliminated, j wrong responses remain. Thus, $j-k$ responses which might have been eliminated were not. This term is equivalent to the determinate reinforcement probability of transit from state $i-k$ to state $j-k$. Thus

$$(70) \qquad tt_{ij} = \sum_{k=0}^{j} t_{i-k,\, j-k} nn_{ik}\,.$$

Here the use of the 'extra' normative states is seen. If the subject by incompletely eliminating wrong responses is in a state between D and N^*, the normative probabilities for moves out of these states are needed. While $\mathbf{TT}(=[tt_{ij}])$ can be written in terms of N, A, and ε instead of as it was in (70), the terms do not reduce considerably, and we feel the above formulation is conceptually clearer.

The error probabilities, given the state, are the same as before, and thus the learning curve is directly analogous.

$$(71) \qquad P(e_n) = \mathbf{S}_1 \mathbf{TT}^{n-1}\mathbf{E} = \sum_{j=0}^{N^*} tt^{n-1}_{N^*,j} \cdot \frac{j}{j+1}\,.$$

The generalization of the AE-2 and AE-3 models is the same as that for the one-parameter elimination model. In (70), for tt substitute tt' or tt'', for t substitute t' or t'', and in (71) make the same substitutions. TT' and TT'' are then the new transition matrices for the multiresponse reinforcement version of the AE-2 and AE-3 models, respectively.

The AE-2 model with ε set equal to 0 (no elimination occurs) is an alternative extension of the one-element model. Here with probability c the subject acquires, or conditions to, the entire group of responses that were both in memory and in the current reinforcement set. This generalization reduces to the normative model if c is set equal to 1.

The generalization of the elimination-forgetting model is comparable to that for the other three elimination models. Since the incomplete information affects only the number of responses possible for elimination, not the forgetting given the state immediately following reinforcement, only the $\mathbf{T}$ matrix, i.e., the elimination matrix, is affected. The effect is precisely that of the elimination model. Thus if $\mathbf{TT}$ is defined as in (70) the formulation of the model is the same as for the determinate reinforcement case, substituting $\mathbf{TT}$ for $\mathbf{T}$.

Conditioning-strength model. The generalization of this model is precisely parallel to the generalization of the elimination model. It does not reduce to the normative model, because of the difference in response assumptions. Therefore as ε approaches 1, and the transition matrix approaches the normative matrix, the conditioning strength model predicts learning faster than that predicted by the normative theory. Needless to say, this prediction could not hold in practice, and the model needs investigation for more intermediate values of ε.

Degraded normative model. The elimination models assume that the subject never eliminates more responses than he logically can, although he may eliminate fewer. The one-element model requires that the reinforcement have no effect or that the subject select only one of the reinforced responses, thus he always eliminates too many.

The degraded normative model, (*DN*), derives from considerations of the normative model degraded in intuitive ways to human performance. Supposedly the subject keeps in memory a running set of responses which he thinks contains the correct response. He responds randomly from this set, or from the full set if his memory set is empty. When reinforced, he correctly intersects his memory set with the outcome set. If the intersection has only one member, he may settle on this as the correct response. The subject may eliminate some of the responses if his memory set remains in the former mode during the intertrial interval (ITI), and he may also introduce responses which were not there previously.

More explicitly and completely, the assumptions of this model are:

1. At the beginning of each trial the subject is in exactly one state. These states are C, $\bar{C}$, 0, 1, ..., N^*, 0^ω, 1^ω, ..., $N^{*\omega}$, where N^* is the number of wrong responses possible. If the subject is in state i, then he has i wrong responses and the correct response in the set of responses he is trying to remember. If he is in i^ω then he has only i wrong responses, without the correct response, in his memory set. If he is in state C or $\bar{C}$, then he has exactly one response in memory, the correct one if C, a wrong one if $\bar{C}$, and this memory singlet is unaffected by the ITI.

2. If the subject is in state i, he makes a correct response with probability $1/i+1$. If he is in state C, he makes a correct response with probability 1. If he is in any other state, he always makes an error.

3. If the subject enters a trial in an unconditioned state, he correctly

intersects his memory set with the outcome. If this intersection is of cardinality 1, then with probability c the subject conditions to that one response, entering C or $\bar{C}$ if the response is correct or incorrect, respectively. If the cardinality of the intersection is 0, the subject takes the entire outcome set as the intersection. If the cardinality of the intersection is not 1, or if the cardinality is 1 and the subject does not condition, then during the ITI the subject eliminates independently with probability ε' each response in the intersection and introduces independently with probability δ' each response not in the intersection. If the subject enters the trial in a conditioned state, and if the conditioned response is in the outcome, then he remains conditioned to that response, and this state is unaffected by the ITI. If the conditioned response is not in the outcome, then with probability c the subject exits from the conditioned state and returns to the elimination-introduction mode of behavior. With probability $1-c$ the subject remains conditioned.

4. When the subject enters trial 1, he is in state N^*.

The necessity for some sort of assumption such as the conditioning part of Assumption 3 is clear if the asymptotic probability of the correct response is 1. An alternative not discussed here is to let ε' and δ' be trial dependent and to decrease with time. It is more probable that the subject will select the correct response once he has narrowed down the possibilities to a single response. The assumption of perfect intersection, with degradation occurring only later, may well prove unrealistic with large A, but it permits the natural interpretation, also not discussed here, of having ε' and δ' dependent on the length or type of ITI. Thus the amount of memory degradation varies with the amount of ITI interference. He returns to the intersection mode from the conditioned mode. Since the subject's memory contains only the single wrong response, and since by assumption, this response was not in the outcome, the subject's immediate intersection is of cardinality 0, and he takes the full outcome of cardinality A for his intersection.

The state of the subject at the start of trial n is termed S_n as before. For convenience the intermediate states are also labeled. The state immediately following intersection is termed I_n. If I_n has cardinality 1 or 0, the state following transformation of this intersection is termed IT_n. The values of I_n and IT_n are specified as are those for S_n.

The transition matrix from S_n to S_{n+1} is more comprehensible and more readily written as the product of the intermediate transition matrices. The transition from S_n to I_n depends only on the value of S_n, the experimental variables, and the outcome. Thus, where applicable, the normative transition matrix provides the basis for the transition. As the correct response is always in the outcome, its presence or absence in the intersection is the same as in the memory before the intersection. The equations for the general term are given by:

$$(72)\quad (\text{trans } S_n \to I_n)_{xy} = \begin{cases} nn_{xy} & \text{for } x = 0, 1, \ldots, N^* \text{ and } y = 0, 1, \ldots, N^* \\ nn_{xy} & \text{for } x = 0^\omega, 1^\omega, \ldots, N^{*\omega} \text{ and } y = 0^\omega, 1^\omega, \ldots, N^{*\omega} \\ 1 & \text{for } x = y = C \\ \dfrac{D}{N^*} + (1-c)\left(1 - \dfrac{D}{N^*}\right) & \text{for } x = y = \bar{C} \\ c\left(1 - \dfrac{D}{N^*}\right) & \text{for } x = \bar{C} \text{ and } y = 0^\omega \\ 0, & \text{otherwise.} \end{cases}$$

The matrix for the transformation of I_n to IT_n follows immediately from the assumption of the model. The general term is given by

$$(73)\quad (\text{trans } I_n \to IT_n)_{xy} = \begin{cases} c & \text{for } x = 0 \text{ and } y = C \\ c & \text{for } x = 1 \text{ and } y = \bar{C} \\ 1-c & \text{for } x = y = 0, \text{ or } 1^\omega \\ 1 & \text{for } x = 0^\omega \text{ and } y = D \\ 1 & \text{for } x = y = C, \bar{C}, 1, 2, \ldots, N^*, 2^\omega, 3^\omega, \ldots, N^{*\omega} \\ 0, & \text{otherwise.} \end{cases}$$

The transition from IT_n to S_{n+1} is a function of the parameters ε' and δ'. The general term of this matrix is derived by considering the correct response separately from the wrong responses. In transiting from either i or i^ω to either j or j^ω, in each of the four cases, the subject starts with i wrong responses and ends with j wrong responses. The probability of this transition, disregarding momentarily the correct response, is given by $\mathbf{TTT} = [ttt_{ij}]$, where

(74) $ttt_{ij} = Pr$ (j wrong responses available on trial $n+1 \mid i$ wrong responses available at the end of trial n)

$$(74) \qquad = Pr\,(S_{n+1} = j \;\text{ or }\; j^{\omega} \mid IT_n = i \;\text{ or }\; i^{\omega})$$
$$= \sum_{k=0}^{N^*} \binom{i}{k} \varepsilon'^k (1-\varepsilon')^{i-k} \binom{N^* - i}{j+k-i} \delta'^{j+k-i} (1-\delta')^{N^*-j-k}.$$

The general matrix for the IT_n to S_{n+1} transition follows immediately when the correct response and the conditioned states are considered.

$$(75) \qquad (\text{trans}\,(IT_n \to S_{n+1}))_{xy} = \begin{cases} (1-\varepsilon')\,ttt_{xy}, & \text{if } x = 0, 1, \ldots, N^* \text{ and } y = 0, 1, \ldots, N^* \\ \varepsilon' ttt_{xy}, & \text{if } x = 0, 1, \ldots, N^* \text{ and } y = 0^{\omega}, 1^{\omega}, \ldots, N^{*\omega} \\ \delta' ttt_{xy}, & \text{if } x = 0^{\omega}, 1^{\omega}, \ldots, N^{*\omega} \text{ and } y = 0, 1, \ldots, N^* \\ (1-\delta')\,ttt_{xy}, & \text{if } x = 0^{\omega}, 1^{\omega}, \ldots, N^{*\omega} \text{ and } y = 0^{\omega}, 1^{\omega}, \ldots, N^{*\omega} \\ 1, & \text{if } x = y = C \text{ or } x = y = \bar{C} \\ 0, & \text{otherwise}. \end{cases}$$

Thus the components are computed, and the full matrix for the transition probabilities from S_n to S_{n+1} can be obtained by multiplying the three transition matrices. Using Assumption 4, the N^*th row of the transition matrix to the $(n-1)$st power is the state probability vector for the nth trial. The error probability vector, derived directly from the response assumption, is the same as that for the elimination models, with i in all additional states, except C, where the subject has probability 0 of an error.

The transition matrices have been worked out for the special case of $N=3$, $A=2$, and are given below.

$$(76) \qquad S_n \to I_n: \quad \begin{array}{c|cccccccc} & 0 & 1 & 2 & 0^{\omega} & 1^{\omega} & 2^{\omega} & C & \bar{C} \\ \hline 0 & 1 & 0 & 0 & 0 & 0 & 0 & 0 & 0 \\ 1 & \frac{1}{2} & \frac{1}{2} & 0 & 0 & 0 & 0 & 0 & 0 \\ 2 & 0 & 1 & 0 & 0 & 0 & 0 & 0 & 0 \\ 0^{\omega} & 0 & 0 & 0 & 1 & 0 & 0 & 0 & 0 \\ 1^{\omega} & 0 & 0 & 0 & \frac{1}{2} & \frac{1}{2} & 0 & 0 & 0 \\ 2^{\omega} & 0 & 0 & 0 & 0 & 1 & 0 & 0 & 0 \\ C & 0 & 0 & 0 & 0 & 0 & 0 & 1 & 0 \\ \bar{C} & 0 & 0 & 0 & \frac{c}{2} & 0 & 0 & 0 & 1 - c/2 \end{array}$$

$$(77) \qquad I_n \rightarrow IT_n: \quad \begin{array}{c} \\ 0 \\ 1 \\ 2 \\ 0^{\omega} \\ 1^{\omega} \\ 2^{\omega} \\ C \\ \bar{C} \end{array} \begin{array}{c} \begin{array}{cccccccc} 0 & 1 & 2 & 0^{\omega} & 1^{\omega} & 2^{\omega} & C & \bar{C} \end{array} \\ \begin{bmatrix} 1-c & 0 & 0 & 0 & 0 & 0 & C & 0 \\ 0 & 1 & 0 & 0 & 0 & 0 & 0 & 0 \\ 0 & 0 & 1 & 0 & 0 & 0 & 0 & 0 \\ 0 & 1 & 0 & 0 & 0 & 0 & 0 & 0 \\ 0 & 0 & 0 & 0 & 1-c & 0 & 0 & C \\ 0 & 0 & 0 & 0 & 0 & 1 & 0 & 0 \\ 0 & 0 & 0 & 0 & 0 & 0 & 1 & 0 \\ 0 & 0 & 0 & 0 & 0 & 0 & 0 & 1 \end{bmatrix} \end{array}$$

$$(78) \qquad \mathbf{TTT}: \quad \begin{array}{c} \\ 0 \\ 1 \\ 2 \end{array} \begin{array}{c} \begin{array}{ccc} 0 & 1 & 2 \end{array} \\ \begin{bmatrix} (1-\delta')^2 & 2(1-\delta')\,\delta' & \delta'^2 \\ \varepsilon'(1-\delta') & (1-\varepsilon')(1-\delta') + \varepsilon'\delta' & (1-\varepsilon')\,\delta' \\ \varepsilon'^2 & 2\varepsilon'(1-\varepsilon') & (1-\varepsilon')^2 \end{bmatrix} \end{array}$$

$$(79) \qquad IT_n \rightarrow S_{n+1}: \quad \begin{array}{c} \\ 0 \\ 1 \\ 2 \\ 0^{\omega} \\ 1^{\omega} \\ 2^{\omega} \\ C \\ \bar{C} \end{array} \begin{array}{c} \begin{array}{cccc} 0\;1\;2 & 0^{\omega}\;1^{\omega}\;2^{\omega} & C & \bar{C} \end{array} \\ \left[\begin{array}{cc|cc} (1-\varepsilon')\,TTT & \varepsilon' TTT & \multicolumn{2}{c}{0} \\ \delta' TTT & (1-\delta')\,TTT & & \\ \hline \multicolumn{2}{c}{0} & 1 & 0 \\ & & 0 & 1 \end{array}\right] \end{array}$$

The number of states postulated makes obvious the complexity of working out the model for cases with $N > 3$ and $A > 2$.

The relation of the degraded normative to the normative is obvious. If the degrading parameters ε' and δ' are both 0, the intersection is retained perfectly throughout the ITI, and the two models are equivalent. In this case the parameter c is irrelevant to the subject's behavior. Because the degraded normative is equivalent to the normative model in this extreme instance, it is also equivalent to the elimination models at the points where they are equivalent to the normative.

While ε and ε' are conceptually similar, they operate over disjoint domains. The parameter ε only allows elimination down to the correct intersection. The parameter ε' begins elimination at the correct intersection. The parameters δ and δ' are more similar, with the exception that δ' always begins introduction at the intersection and δ begins at whatever intermediate state the elimination operation left the subject. As δ' ranges over all responses not in the correct intersection, and ε ranges only over those not in the intersection but in the previous memory, δ' and ε are not comparable.

Because of the lack of an equivalent for ε, the degraded normative does not reduce to the elimination models, nor vice versa. The elimination-forgetting model reduces to the *DN* model only if elimination is discarded. That is, if ε is 1 in the *E-F* model, and ε' is 0 in the *DN* model, then δ corresponds to δ'. Similarly if in the 3-parameter version of the *E-F* model, α is 1 and ε' is again 0, then ε is irrelevant and the models are equivalent. In both cases the parameter c of the *DN* model must be 1, when Assumption 4 of the *E-F* models is considered.

The *DN* model reduces to the one-element model only if A is 1. If ε' is 1 and δ' is 0, or if δ' is 1 and ε' is 0, then the subject responds from the entire response set prior to conditioning, and the parameters c are exactly equivalent. If A is greater than 1, there is no way in the *DN* model as it now stands to approximate the one-element requirement that the subject move only to another conditioned state if he moves out of a wrongly conditioned state.

3. *Paired-Associate Learning with Incomplete Information: Contingent Case*

In right-wrong reinforcement, a subject is told 'right' when his response is correct and 'wrong' when his response is incorrect. Thus the reinforcement size is effectively 1 to $N-1$, contingent on the response. A slightly more general version of this paradigm is to tell the subject 'right' when he is correct, but to give him a subset of size A, including the correct response, when he is wrong. Given the paradigm, the constraint should probably be made that A alternatives presented when the response was wrong not include that response. Bower (1962) performed the relevant experiments with N equals 3 or 8, and $A=2$. He also reports right-wrong results. The two paradigms are experimentally distinct. In the right-wrong set-up, the 'wrong' reinforcements are not used. Instead of A being $N-1$, it is effectively N; that is, the subject uses no information from these reinforcements. On the other hand, in the (3:2) condition with the seemingly minor change of reporting that one of the remaining two is correct, learning is faster. For any theory, we must consider the effect of reactions to the experimental design as well as to the information to be learned.

Leaving the right-wrong problem, the performance on the partial correction experiment is of interest. Following the first success, when the subject sees the correct response by itself, there is a qualitative change in behavior. Bower postulates this is a recognition state in which the subject can pick the correct response from two reinforced responses with probability 1. Using the models discussed in this paper, we have no way to obtain this differential knowledge of the members of the reinforcement set at the time of reinforcement, given that the knowledge was not useful prior to reinforcement. Several variations of the one-element model were tested against Bower's data and failed relative to Bower's model, apparently because they were not able to handle properly the transition point of the first success. Because of its two-process nature, the acquisition-elimination model discussed above could handle this contingent reinforcement situation by making c, the acquisition parameter, contingent on having the correct response reinforced singly. Such a model has not been devised or tested, although the direction to take is clear.

4. *Paired-Associate Learning with a Continuum of Response Alternatives*

In the experimental paradigms discussed so far, subjects select their response from one of a finite (and usually small) set of alternatives. Linear and stimulus-sampling models for situations involving a continuum of response alternatives have been proposed by Suppes (1959, 1960). A brief description of experiments run by Suppes and Frankmann (1961) and by Suppes, Rouanet, Levine, and Frankmann (1964) give a feel for the type of experimental setup we shall now consider.

In these experiments subjects sat facing a large circular disk. After the subject responded by setting a pointer to a position on the circumference of the disk, he was reinforced by a light that appeared at some point on the circumference. As the subject saw exactly where the light flashed, i.e., what his response 'should' have been, reinforcement was determined. In these studies reinforcement was also noncontingent. The reinforcement density in the 1961 study was triangular on $0-2\pi$; in the 1964 study it was bimodal, consisting of triangular sections on $0-\pi$ and $\pi-2\pi$. By *reinforcement density* we mean the probability density function from which reinforcement is drawn. For example, if $f(y)$ is the reinforcement density, the probability that the reinforcement will appear between a and b is $\int_a^b f(y)dy$, and this probability is contingent on neither trial number nor the subject's previous response.

The experimental paradigm just described corresponds more fully to probability learning than to PAL and will be considered again later. Variations of it, however, correspond to PAL.

Complete information. We consider a list of length L of distinct stimuli (trigrams, for example). Each stimulus corresponds to a single, fixed region on the circumference of the experimental disk. The subject is shown the stimulus, indicates his response with his pointer, is shown the region considered correct, and then is shown the next stimulus. His response is considered correct if it falls in the reinforced region; otherwise, it is incorrect. We wish now to derive a learning curve for the subject.

Denote the center of the correct region by e and let the correct region extend a distance α on either side of e. The subject's response is given by a density $r_n(x)$ for trial n. If the subject is known to be conditioned to some point z, then the density for his response is a smearing density $k(x \mid z)$. The parameter z itself is a random variable, and we shall denote its density on trial n by $g_n(z)$. The conditioning assumption we shall make is that with probability $1-\theta$ the parameter of the subject's smearing distribution makes no change after reinforcement, and with probability θ, z is distributed by an 'effective reinforcement density' $f(y)$. Subsequently, we shall consider two candidates for $f(y)$. First, observe what happens to the reinforcement density $g(z)$. (All these matters are discussed in detail in Suppes (1959) with a different interpretation of the effective reinforcement density.) The density g changes in the following way:

$$g_{n+1}(z) = (1-\theta)\, g_n(z) + \theta f(z). \tag{80}$$

If we assume that $g_1(z)$ is uniform $(=1/2\pi)$, we find from the above recursion that

$$g_n(z) = (1-\theta)^{n-1}/2\pi + [1-(1-\theta)^{n-1}]\, f(z). \tag{81}$$

The probability of being correct on trial n, $P(S_n)$, is given by

$$(82) \qquad P(S_n) = \int_0^{2\pi} \int_{e-\alpha}^{e+\alpha} k(x \mid z)\, g_n(z)\, dx\, dz\,.$$

Two plausible assumptions concerning $f(y)$ are:

$$(83) \qquad f_1(y) = \delta(y - e)\,,$$ [2]

or

$$(84) \qquad f_2(y) = \begin{cases} 1/2\alpha & e - \alpha \leqslant y \leqslant e + \alpha \\ 0\,, & \text{elsewhere}\,. \end{cases}$$

If conditioning occurs, $f_1(y)$ asserts that z becomes e; $f_2(y)$ asserts that z becomes uniformly distributed over the correct region. The learning curves for $f_1(y)$ and $f_2(y)$ follow: For f_1,

$$(85) \qquad P(S_n) = (1-\theta)^{n-1}\frac{\alpha}{\pi} + [1-(1-\theta)^{n-1}] \int_{e-\alpha}^{e+\alpha} k(x \mid e)\, dx\,,$$

and for f_2,

$$(86) \qquad P(S_n) = (1-\theta)^{n-1}\frac{\alpha}{\pi} + [1-(1-\theta)^{n-1}]\frac{\pi}{\alpha} \int_{e-\alpha}^{e+\alpha} \int_{e-\alpha}^{e+\alpha} k(x \mid z)\, dx\, dz\,.$$

For the present, we shall derive no further statistics for these models.

Incomplete information. The experiment is organized so that a total of A regions of fixed width 2α are presented to the subject each time he is reinforced. One of these regions is fixed with center at y; the others have their centers uniformly distributed on $0-2\pi$ each trial. (Hence, there can be overlap among the reinforcers.) A list of stimuli is assumed. The subject starts with z uniformly distributed on the region $0-2\pi$. The conditioning assumptions are: (i) if the subject responds in a reinforced region, conditioning remains unchanged; and (ii) if he does not, with probability $(1-\theta)$ his conditioning remains unchanged, and with probability θ, it is spread uniformly over the reinforced regions. Let us start with some definitions. The total area expected to be covered by reinforcers on any given trial, is γ, where

$$(87)\qquad 1-\gamma=\int_{0}^{2\pi}(2\pi-2\gamma)^{A}\,dt\,,$$

and hence,

$$(88)\qquad \gamma=2\pi\left[1-(2\pi-2\gamma)^{A}\right].$$

Let s_n denote the event of responding in a reinforced region on trial n, W the event of being wrongly conditioned on trial n; C_n the event of being correctly conditioned on trial n (i.e., z is in the one 'correct' region). Then,

$$(89)\qquad P(s_n \mid C_n)\approx 1/2\alpha\int_{y-\alpha}^{y+\alpha}\int_{y-\alpha}^{y+\alpha}k(x\mid z)\,dx\,dz=\beta,\text{ by definition, and}$$

$$(90)\qquad P(s_n \mid W_n)=1-(2\pi-2\alpha)^{A}=\gamma/2\pi\,.$$

Equation (89) is an approximation, because there is some (small) probability that the subject will guess outside the correct region *and* be reinforced by one of the distractors. Also, we can write the transition probabilities:

$$(91)\qquad \begin{aligned} P(C_{n+1}\mid C_n) &= P(C_{n+1}\mid C_nS_n)\,P(S_n)+P(C_{n+1}\mid C_n\bar{S}_n)\\ &=\beta+(1-\theta)(1-\beta)+(1-\beta)\,\theta\,\frac{2\alpha}{\gamma}=m,\text{ by definition}\end{aligned}$$

and

$$P(C_n\mid W_n)=\frac{\gamma}{2\pi}+\left(1-\frac{\gamma}{2\pi}\right)\theta\,\frac{2\alpha}{\gamma}=n,\text{ by definition}\,.$$

The transition matrix between states and error probability vector are, therefore,

$$(92)\qquad \mathbf{T}=\begin{bmatrix} m & 1-m\\ n & 1-n\end{bmatrix}\quad\begin{bmatrix}\beta\\ \gamma/2\pi\end{bmatrix}.$$

If $\mathbf{S}_n$ is the vector that represents the probabilities of being in the 2 states on trial n, then $\mathbf{S}_1=\left(\alpha\,/\,\pi,\dfrac{\pi\quad\alpha}{\pi}\right)$. The learning curve is:

$$(93)\qquad P(x_n E\,\text{correct region})=\mathbf{S}_1\mathbf{T}^{n-1}\begin{bmatrix}\beta\\ \gamma/2\pi\end{bmatrix}.$$

We shall complete this discussion by deriving the expression for the powers of **T**. The eigenvalues of **T** can be shown to be: $\lambda_1 = 1$ and $\lambda_2 = m - n$. Let **Q** be the matrix of the eigenvectors generated from λ_1 and λ_2. Then,

$$(94) \qquad \mathbf{Q} = \begin{bmatrix} 1 & \dfrac{m-1}{n} \\ 1 & 1 \end{bmatrix} \quad \text{and} \quad \mathbf{Q}^{-1} = \frac{n}{n-m+1} \begin{bmatrix} 1 & \dfrac{1-m}{n} \\ -1 & 1 \end{bmatrix}.$$

It is a theorem of matrix analysis that

$$(95) \qquad \mathbf{T}^n = \mathbf{Q}\Lambda^n\mathbf{Q}^{-1} \quad \text{where} \quad \Lambda = \begin{bmatrix} \lambda_1 & 0 \\ 0 & \lambda_2 \end{bmatrix}.$$

By multiplying and simplifiying as much as possible, we find

(96)

$$\mathbf{T}^n = \begin{bmatrix} n + (1-m)(m-n)^n & 1 - m + (1-m)(m-n)^n \\ n + (m-n)^n & 1 - m + (m-n)^n \end{bmatrix} \frac{1}{n-m+1}.$$

III. PROBABILITY LEARNING

If an experiment is constructed so that the only reward a subject receives is that of being correct, the reinforcements can be characterized by the amount of information he receives concerning the correct response. More specifically, if $\mathfrak{R}$ is the set of response alternatives and $\mathfrak{C}$ is the set of possible reinforcements, then $\mathfrak{C}$ is the set of all subsets (power set) of $\mathfrak{R}$. The notion here is that after responding on a given trial the subject is shown some $e_i \in \mathfrak{C}$ and told that the correct response for that trial is included in e_i. In the general noncontingent case (i.e., the reinforcement is not contingent on the subject's response), each e_i will be shown with a probability π_i independent of the subject's prior responses and the trial number.

We now consider the experimental paradigm in which the number of responses in the reinforcement set is a constant, $j (1 \leqslant j \leqslant N$, where N is the cardinality of $\mathfrak{R}$), but no one response is necessarily always present. Thus, the paradigm is that of probability learning.

Previous theories of probability learning have dealt primarily with the case $j = 1$. We shall present theories for arbitrary j. The first theory presented is attractive since it implies a natural generalization of the well-known probability matching theorem. Unfortunately, this theory is intuitively unacceptable for extreme values of the π's. The second theory gives the probability matching theorem for $j=1$, but unless $j=1$, or $N-1$, it is mathematically untractable. These two theories are essentially all or none; we shall also discuss a third, linear theory.

1. *Probability Learning Without Permanent Conditioning*

The assumptions of this theory are:

1. On every trial the stimulus element is conditioned to exactly one of the N responses, or it is unconditioned. At the outset it is unconditioned.
2. After reinforcement, the stimulus-element conditioning remains unaltered with probability $1-\theta$. The stimulus element becomes conditioned to any one of the A members of the reinforcement set with probability θ/A.
3. If unconditioned, the subject makes each response with a guessing probability of $1/N$; if the subject is conditioned, he makes the response he is conditioned to.

We shall designate the set of possible responses by $A=\{a_1, a_2, \ldots, a_N\}$. The probability of response a_i on trial is denoted by $p_{i,n}$. The asymptotic probability of a_i, i.e., $\lim_{n\to\infty} p_{i,n}$, is denoted p_i. By relabeling, any response can be denoted 'a_1'; hence, we shall derive only p_1. As each reinforcement set has A members, there are a total of $\binom{N}{A}=N!/A!$ $(N-A)!$ different reinforcement sets. Of these reinforcement sets a number $k=\binom{N-1}{A-1}$ will contain a_1. We shall denote by $e_1, e_2, \ldots, e_k$ those reinforcement sets that contain a_1; the probabilities that these reinforcement sets will occur are $\pi_1, \pi_2, \ldots, \pi_k$.

THEOREM 3 (*probability matching*). Assumptions 1 to 3 *imply that*

$$p_1 = \sum_{i=1}^{k} \pi_i/A. \tag{97}$$

PROOF. Let $C_{i,n}$ be the event of being conditioned to a_i on trial n, and let $p(C_{i,n})$ be the probability of this event. By the theorem of total probability and by assuming that n is sufficiently large, we can neglect the possibility of being unconditioned. Thus,

$$(98)\qquad P(C_{1,n+1}) = P(C_{1,n+1} \mid C_{1,n})\, P(C_{1,n}) + \sum_{j=2}^{N} P(C_{1,n+1} \mid C_{j,n})\, P(C_{j,n}).$$

The value of $p(C_{1,n+1}|C_{1,n})$ is obtained by noting that one can be in state C_1 on $n+1$ after being in state C_1 on n if either the subject's conditioning is unaltered (with probability $1-\theta$) or if a_1 is in the reinforcement set shown, and he becomes conditioned to it (with probability $\theta/A\sum_{i=1}^{k}\pi_i$). Thus,

$$(99)\qquad P(C_{1,n+1} \mid C_{1,n}) = (1-\theta) + \theta/A \sum_{i=1}^{k} \pi_i .$$

If $j \neq 1$, the subject can be in state C_1 on n only if A_1 is in the reinforcement set shown and he becomes conditioned to it. Thus,

$$(100)\qquad P(C_{1,n+1} \mid C_{j,n}) = \theta/A \sum_{i=1}^{k} \pi_i .$$

For large n, $P(C_{i,n+1}) = P(C_{i,n}) = p_i$; hence, (98) can be written in the following way:

$$(101)\qquad p_i = \left[(1-\theta) + \theta/A \sum_{i=1}^{k} \pi_i\right] p_i + \sum_{j=2}^{N} \left(\theta/A \sum_{i=1}^{k} p_j\right)$$

or,

$$(102)\qquad p_1 = p_1 - \theta p_1 + \left(\theta/A \sum_{i=1}^{k} \pi_i\right) \sum_{j=1}^{N} p_j .$$

Since $\sum_{j=1}^{N} p_j = 1$,

$$(103)\qquad \theta p_1 = \theta/A \sum_{i=1}^{k} \pi_i ,$$

which, by cancelling θ, gives the desired result:

$$(104)\qquad p_1 = \sum_{i=1}^{k} \pi_i/A . \qquad \text{Q.E.D.}$$

Some special cases of the above are: $N=2$ and $A=1$; here $p_1 = \pi_1$. For $N=3$ and $A=2, p_1=(\pi_1+\pi_2)/2$; for $N=6$ and $A=3, p_1=(\pi_1+\pi_2 + \cdots + \pi_{10})/3$.

Let us look at the case $N=3$ and $A=2$ in a little more detail: $e_1 = \{a_2, a_3\}$,

$e_2=\{a_1, a_3\}$, and $e_3=\{a_2, a_3\}$. Assume that $\pi_1=\pi_2=.5$ and $\pi_3=0$. Clearly, then, $p_1=.5$ and $p_2=p_3=.25$. Notice that since $\pi_3=0$, a_1 is *always* in the reinforcement set. Data from the experiment reported in the Appendix show that when one response is always reinforced (paired-associate learning), subjects learn to select it only. Hence the empirical value of p_1 is 1. It is obvious, then, that the theory just presented will break down if one or more of the π_is tends to zero; how well it will do for nonextreme values of the π_is remains to be seen.

2. *Probability Learning With Permanent Conditioning*

Assumptions 1 and 3 of this model are the same as for probability learning without permanent conditioning. Assumption 2 is changed to:

2′. (i) if the stimulus element is conditioned to one of the responses reinforced, it remains so conditioned; and

(ii) if the stimulus element is not conditioned to one of the responses reinforced, then with probability $1-\theta$ its conditioning remains unchanged, and with probability θ/A, it becomes conditioned to any one of the A members of the reinforcement set.

Unfortunately, this model is less mathematically tractable than the preceding one and asymptotic response probabilities were obtained only for the special cases $A=1$, $A=N-1$, and $N=4$ with $A=2$. As before, the subject's being in state i on trial n will be denoted by $C_{i,n}$. Let us first derive the asymptotic response probabilities for $A=1$.

The reinforcement sets are $e_1=\{a_1\}$, $e_2=\{a_2\}$, etc., and appear with probabilities $\pi_1, \pi_2, \ldots, \pi_N$. Thus,

$$(105) \qquad P(C_{i,n+1} \mid C_{i,n}) = (1-\theta) + \pi_i\theta\,,$$

since with probability $1-\theta$ the subject's conditioning undergoes no change and with probability $\pi_i\theta$ he is reinforced with a_i and conditions to it. If $j\neq 1$, $P(C_{i,n+1}|C_{j,n})=\pi_i\theta$. By the theorem on total probability,

$$(106) \qquad p_i = ((1-\theta)+\pi_i\theta)\, p_i + \left(\sum_{j=1}^{N} \pi_i\theta p_j - \pi_i\theta p_i\right).$$

But this is equivalent to:

$$(107) \qquad p_i = (1-\theta)\, p_i + \theta\pi_i \sum_{j=1}^{N} p_j\,,$$

so we obtain, for $A=1$, the probability matching result:

(108) $$p_i = \pi_i.$$

For $A=N-1$ let us denote the reinforcement sets in the following way: $e_i=\{a_j:j\neq i\}$. That is, e_i contains all the responses *except* a_i. Clearly there are a total of N reinforcement sets whose probabilities will be given by $\pi_1, \pi_2, \ldots, \pi_N$. At this point it may be helpful to look at the transition matrix from state C_i to the other states. The notation C_ie_j means that the subject was in state C_i and received reinforcing set e_j.

(109)

	C_1	C_2	$\ldots$	C_i	$\ldots$	C_N
C_ie_1	0	0	$\ldots$	1	$\ldots$	0
C_ie_2	0	0	$\ldots$	1	$\ldots$	0
$\vdots$						
C_ie_i	$\theta/(N-1)$	$\theta/(N-1)$	$\ldots$	$1-\theta$	$\ldots$	$\theta/(N-1)$
$\vdots$						
C_ie_N	0	0	$\ldots$	1	$\ldots$	0

Thus we see that $P(C_{i,\,n+1}|C_{i,\,n})$ is equal to $(1-\pi_i)+(1-\theta)\,\pi_i$. For $j\neq i$, $P(C_{i,\,n+1}|C_{j,\,n})=\pi_i\theta/(N-1)$. By the theorem on total probability, we see that:

(110) $$p_i = (1 - \pi_i + \pi_i - \theta\pi_i)\; p_i + \theta/(N-1)\left\{\left(\sum_{k=1}^{N} \pi_k p_k\right) - \pi_i p_i\right\},$$

or

(111) $$\pi_1 p_1 + \pi_2 p_2 + \cdots + \pi_i p_i + \cdots + \pi_N p_N = N\pi_i p_i.$$

As this is true for all i,

(112) $$p_i/p_j = \pi_j/\pi_i.$$

Since $p_1+p_2+\cdots+p_i+\cdots+p_N=1$,

(113) $$\frac{1}{p_i} = \frac{p_1}{p_i} + \frac{p_2}{p_i} + \cdots + \frac{p_i}{p_i} + \cdots + \frac{p_N}{p_i}.$$

Substituting (112) into (113) we obtain

(114) $$p_i = 1/\sum_{k=1}^{N} \frac{\pi_i}{\pi_k},$$

which is equivalent to

$$(115)\qquad p_i = \frac{\prod_{j=1}^{n} \pi_j \cdot \frac{1}{\pi_i}}{\sum_k \left(\prod_{j=1}^{n} \pi_j\right)/\pi_k}.$$

The derivation of asymptotic response probabilities for $N=4$, $A=2$ is both tedious and unilluminating; we shall state only the results. The six reinforcing events are labeled as follows: $e_i=\{a_1, a_2\}$, $e_2=\{a_1, a_3\}$, $e_3=\{a_1, a_4\}$, $e_4=\{a_2, a_3\}$, $e_5=\{a_2, a_4\}$, and $e_6=\{a_3, a_4\}$. The response probabilities are given by:

$$(116)\qquad \mathbf{p} = \mathbf{B}^{-1}\mathbf{r}$$

where

(117)

$$\mathbf{p} = \begin{bmatrix} p_1 \\ p_2 \\ p_3 \\ p_4 \end{bmatrix} \quad \mathbf{r} = \begin{bmatrix} 1 \\ 0 \\ 0 \\ 0 \end{bmatrix} \quad \text{and}$$

$$\mathbf{B} = \begin{bmatrix} 1 & 1 & 1 & 1 \\ -2(\pi_4+\pi_5+\pi_6) & \pi_4+\pi_5 & \pi_4+\pi_6 & \pi_5+\pi_6 \\ \pi_4+\pi_5 & -2(\pi_2+\pi_3+\pi_6) & \pi_1+\pi_5 & \pi_1+\pi_4 \\ \pi_1+\pi_3 & \pi_2+\pi_6 & -2(\pi_1+\pi_3+\pi_5) & \pi_2+\pi_4 \end{bmatrix}.$$

When $A=N=1$, if π_i is equal to zero, a_i will appear in every reinforcement set. As we have seen, the theory of probability learning without permanent conditioning fails to predict the empirical result that in this case p_i equals one. The model just described does predict that $p_i=1$ on the assumptions that $\pi_i=0$ and for $j\neq i\ \pi_j>0$. To see this, let us write out (115):

$$(118)\qquad p_i = \frac{\pi_1\pi_2\ldots\pi_{i-1}\pi_{i+1}\ldots\pi_N}{\pi_1\pi_2\ldots\pi_{N-1}+\cdots+\pi_1\pi_2\ldots\pi_{i-1}\pi_{i+1}+\cdots+\pi_2\pi_3\ldots\pi_N}.$$

Now all the terms in the denominator but one contain π_i; therefore, they vanish. The one that does not contain π_i must be unequal to zero since for $j\neq i$, $\pi_j>0$. But this term is the same term as the numerator so that $p_i=1$.

3. *A Generalized Linear Model for Probability Learning*

In Part II, Section 2.3, two distinct linear models for paired-associate learning were developed. We will apply the model exemplified in the matrix $\mathbf{L}^{(1)}$ of Equation (59) of the preceding section. The basic assumption behind matrix $\mathbf{L}^{(1)}$ is that the decrement in response probability of a response not reinforced on a trial was to be spread uniformly only among reinforced responses. As noted previously, with N response alternatives, A of which are reinforced on any trial, there are $\binom{N}{A}=J$ different reinforcement sets, k of which contain a_1, where $k=\binom{N-1}{A-1}$. Let us label the reinforcement sets in such a way that the first k contain a_1 then determine p_1, the asymptotic response probability for a_1. The probabilities of the J reinforcement sets are given by $\pi_1, \pi_2, \ldots, \pi_j$.

The recursion for $p_{1,n+1}$ is:

$$p_{1,n+1} = [(1-\alpha)\,p_{1,n}] \sum_{i=k+1}^{J} \pi_i + \left[p_{1,n} + \frac{\alpha(N-A)}{N-1}(1-p_{1,n})\right] \sum_{i=1}^{k} \pi_i . \tag{119}$$

The first term on the right-hand side represents $p_{1,n+1}$ given that a_1 was not reinforced on trial n times the probability that it was not reinforced; the second term is analogous except that it assumes a_1 *was* reinforced on n. The part in brackets in the second term of the right-hand side follows from (64).

We now define two terms:

$$r = \frac{N-A}{N-1} \quad \text{and} \quad \pi = \sum_{i=k+1}^{J} \pi_i , \tag{120}$$

from which it follows that $(1-\pi)=\sum_{i=1}^{k} \pi_i$. Here π is the probability that a_1 not be included in the reinforcement set and $1-\pi$ is the probability that it is included. We can now rewrite (119) as

$$\begin{aligned} p_{i,n+1} &= [(1-\alpha)\,p_{1,n}]\,\pi + [p_{1,n} + \alpha r(1-p_{1,n})]\,(1-\pi) \\ &= p_{1,n} - \alpha\pi p_{1,n} + [\alpha r(1-\pi)]\,(1-p_{1,n}) . \end{aligned} \tag{121}$$

In the limit, $p_1 = p_{1,n+1} = p_{1,n}$; hence,

$$(122) \qquad \alpha \pi p_{1,n} = \alpha r(1-\pi)(1-p_{1,n}).$$

From this it follows that:

$$(123) \qquad p_1 = \frac{r(1-\pi)}{\pi + r(1-\pi)}.$$

As a special case of the above, if $A=1$, then $r=1$ and $p_1 = 1-\pi$, giving probability matching.

This completes our discussion of probability learning with finite response sets. We have developed only a sample of the theories possible to obtain in analogy to the theories of paired-associate learning. It would seem profitable to obtain some data before continuing the theoretical development too far but, so far as we know, the only relevant data for $A \geqslant 2$ are from unpublished work of Michael Humphreys and David Rumelhart.

4. *Probability Learning With a Continuum of Responses*

The experiment discussed in Part II, Section 4 for a response continuum is an example of probability learning with a continuum of response and reinforcement possibilities. The next paradigm discussed also has a continuum of responses, but discrete reinforcement.

Probability learning with left-right reinforcement. Consider a task in which the subject is placed before a straight bar (perhaps 2 feet long) with a light bulb at either end. The subject is told that when he indicates a point on the bar at the beginning of each trial one of the lights will flash. His task is to minimize the average distance between the point he selects and the light flash on that trial. Clearly, this is a task with a continuum of response alternatives; it differs from the probability learning tasks to be described since there are only two reinforcing events. We shall call this task probability learning with left-right reinforcement. Reinforcement is determinate since, after one light flashes, the subject knows he should have selected that extreme end of the bar.

We first show that if the subject believes the probability of the left light flashing, $P(L)$, differs from .5, he should choose one extreme or the other.

Number the leftmost point the subject can select 0 and the rightmost point 1. Let r_n denote the subject's choice on trial n. Let k equal his loss. If the left light flashes $k = r_n$, and if the right light flashes $k = 1 - r_n$. His expected loss is given by:

$$(124) \qquad E(k) = P(L)\, r_n + [1 - P(L)]\,(1 - r_n).$$

Differentiating with respect to r_n we obtain,

$$(125) \qquad \frac{dE(k)}{dr_n} = 2P(L) - 1.$$

Assume that $P(L) > .5$; then the derivative of the subject's expected loss is strictly positive, that is, $E(k)$ is an increasing function of r_n so $E(k)$ is minimized by choosing $r_n = 0$. Exactly similar arguments hold if $P(L) < .5$.

The strategy just analyzed is an optimal strategy. Our belief, however, is that the subject's behavior will be analogous to the probability-matching behavior exhibited in finite probability learning situations. That is, we expect that r_n will approach $1 - \pi_L$ where π_L is the noncontingent probability that the left light will flash.

A simple linear model gives this result. Let r_{n+1} be given in terms of r_n:

$$(126) \qquad \begin{cases} r_{n+1} = (\theta r_n + 1 - \theta) & \text{if right light flashes} \\ \qquad\quad \theta r_n & \text{if left light flashes.} \end{cases}$$

It is then easy to show that

$$(127) \qquad \lim_{n \to \infty} r_n = 1 - \pi_L.$$

The linear model predicts, of course, considerable variation in r_n, even after its expected value reaches asymptote.

Let us now turn to a stimulus-sampling model that also gives probability matching, but that predicts decreasing motion around $1 - \pi_L$ as n increases. In the stimulus-sampling model, the subject is conditioned to one response on any given trial. He chooses his response, however, from some distribution 'smeared' about the response he is conditioned to. In most stimulus-sampling models this smearing distribution, $k(r|p)$ where p is a vector representing the parameters of the distribution, maintains a constant shape in the course of learning. In this model the shape of the distribution changes as does the response it is smeared around.

Specifically, the model assumes that k is a beta distribution with parameters α_n and β_n. The expected value of r_n is, then, $\int_0^1 rk(r|\alpha_n, \beta_n)\, dr$. Since k is a beta, this becomes

$$(128) \qquad E(r_n) = \frac{\alpha_n}{\alpha_n + \beta_n}.$$

The model further assumes that $\alpha_1 = \beta_1 = c_1$, where c_1 is a parameter to be estimated. The conditioning rule is:

$$(129) \qquad \alpha_{n+1} = \begin{cases} \alpha_n + c_2 & \text{if the right light flashes} \\ \alpha_n & \text{if the left light flashes} \end{cases}$$

and

$$(130) \qquad \beta_{n+1} = \begin{cases} \beta_n & \text{if the right light flashes} \\ \beta_n + c_2 & \text{if the left light flashes} \end{cases}$$

where c_2 is the second parameter of the theory.

For n large,

$$(131) \qquad \lim_{n\to\infty} E(r_n) \frac{c_1 + c_2 n(1 - \pi_L)}{2c_1 c_2 n(1 - \pi_L) + c_2 n \pi_2} = 1 - \pi_2,$$

which corresponds to probability matching. Assuming that the probability matching prediction is borne out, this model can be compared with the linear model on the basis of response variance for n large.

Modification of subjective probabilities. In estimating a probability a subject may be said to be responding from a continuum of alternatives. If he is then reinforced with new information relevant to the probability in question, the 'normative' prediction is that he will modify his probability estimate in accord with Bayes' theorem. It is our purpose in this subsection to look at one type of probability modification behavior from an explicitly learning-theoretic point of view.

Let the subject have some simple means of responding on the interval [0, 1]. Denote his response on trial n by p_n. The experimenter places before the subject a jar containing a large number of marbles, say 1000. He tells the subject that there are 1000 marbles in the jar and that the only colors the marbles may be are chartreuse (C) and heliotrope (H). The subject is told that there may be from 0 to 1000 of each color of marble. Under these circumstances Jamison and Kozielecki (1968) showed that subjects tend to have a uniform density for $P(\mathrm{C})$ where $P(\mathrm{C})$ represents the subject's estimate of the fraction of chartreuse marbles. Hence, it is natural to expect that p_1 will equal .5. In the experimental sequence, the subject responds with p_n; the experimenter fishes a marble from the jar, shows it to the subject, and replaces it; the subject responds with p_{n+1}. The similarity between this model and the left-right probability learning model mentioned previously is clear. Let us use the stimulus-sampling model developed for that situation (128)–(131). From Jamison and Kozielecki's observations it is natural to assume that the parameter c_1 of that model be equal to one. Results of data presented in Peterson and Phillips (1966) indicate that c_2 should be near one and observations by Phillips, Hays, and Edwards (1966) indicate that c_2 should be less than one. At any rate, after seeing n_C chartreuse and n_H heliotrope marbles, the density for p is:

$$(132) \qquad r(p_n) = \frac{1}{\beta(n_\mathrm{C}c_2 + 1,\, n_\mathrm{H}c_2 + 1)}\, p^{n_\mathrm{C}c_2}(1-p)^{n_\mathrm{H}c_2},$$

where $n = n_\mathrm{C} + n_\mathrm{H}$ and $\beta(\cdot, \cdot)$ denotes the beta function of those arguments. The expectation of this density is:

$$(133) \qquad E(p_n) = \frac{n_\mathrm{C}c_2 + 1}{n_\mathrm{C}c_2 + n_\mathrm{H}c_2 + 2}.$$

Asymptotically, this model implies that the subject will arrive at the correct probability. If $c_2 = 1$, the subject's behavior is normative throughout. Thus our learning model, if it gives an adequate account of this type of data, yields the same results as a Bayesian model. (In a sense, $c_2 \neq 1$ generalizes the normative model. See Suppes (1966) for an account of the one-element model viewed as a generalization of Bayesian updating.) What are the implications of this?

If we assume that the stimulus-sampling model is also adequate for the left-right probability learning situation, we have a single learning-theoretic model that accounts for behavior that in one case is normative and in the other case is not. Bayesian or degraded Bayesian models are adequate in some cases, *because* they approach the learning-theoretic models. The implication here is that our notion of optimality is very limited.

Multipoint reinforcement. We now consider a probability learning paradigm with a continuum of responses analogous to that with a finite response set, but A (the number of responses in a reinforcement) is greater than 1. There are A points on the circumference of the circle reinforced after the subject has set his pointer. With probability $1-\theta$ the mode, z, of his smearing distribution (defined prior to Equation (80)) is assumed to remain unchanged. With probability θ/A, z moves to any one of the points reinforced. Thus the recursion on the density for z is given by:

$$(134) \quad g_{n+1}(z) = (1-\theta)\, g_n(z) + \theta/A f_1(z) + \theta/A f_2(z) + \cdots + \theta/A f_A(z),$$

where $f_i(z)$ gives the density from which the ith reinforcement was drawn. For large n, $g_{n+1}(z) = g_n(z)$. Hence,

$$(135) \quad g_\infty(z) = g_\infty(z)(1-\theta) + \theta/A\,[f_1(z) + f_2(z) + \cdots + f_A(z)] = \frac{1}{A}\sum_{i=1}^{A} f_i(z).$$

In Suppes (1959), the asymptotic response density, $r_\infty(x)$, is derived from the above and shown to be:

$$(136) \quad r_\infty(x) = \frac{1}{A}\int_0^{2\pi} k(x \mid z) \sum_{i=1}^{A} f_i(z)\, dz\,.$$

The interesting prediction of this theory is that the same $r_\infty(x)$ is obtainable for multiple reinforcement as for single reinforcement, if the density for the single reinforcement is the average of the densities for multiple reinforcement.

Let us consider one other probability learning task. The subject is reinforced on each trial with a region of length 2α centered at y where y is a random variable with density $f(y)$. The simplest assumption is that if the subject becomes conditioned, he conditions to point y. If this is so, clearly, $r_\infty(x)$ must be given by:

$$(137) \quad r_\infty(x) = \int_0^{2\pi} k(x \mid z) f(z)\, dz\,.$$

This is somewhat counterintuitive since it is independent of α. Perhaps a more reasonable conditioning assumption would be that z is distributed uniformly over the reinforced region if conditioning occurs. Let us define $U(z|y, \alpha)$ to equal $1/2\alpha$ for $y-\alpha \leqslant z \leqslant y+\alpha$ and 0 elsewhere. The density for z on trial $n+1$, given that conditioning occurred, is denoted $U_\alpha(z)$; it is given by:

$$(138) \qquad U_\alpha(z) = \int_0^{2\pi} u(z \mid y, \alpha) f(y)\, dy .$$

The recursion for $g_n(z)$ is, then,

$$(139) \qquad g_{n+1}(z) = (1-\theta)\, g_n(z) + \theta U_\alpha(z),$$

and the asymptotic response density is:

$$(140) \qquad r_\infty(x) = \int_0^{2\pi} k(x \mid z)\, U_\alpha(z)\, dz .$$

We shall derive no further statistics for these models at this time.

IV. CONCLUDING COMMENTS: MORE GENERAL INFORMATION STRUCTURES

In the experimental paradigms discussed in this paper the set E of possible reinforcements can be divided into two subsets for each stimulus-response pair. One subset contains reinforcements that indicate the subject's response to the stimulus was 'correct'; the other contains reinforcements that indicate his response was 'incorrect'. By his design of the experiment, the experimenter chose a probability distribution for each *S-R* pair over the set of possible reinforcements; this distribution generates a distribution on the subsets 'correct' and 'incorrect'. If the subject can choose a response to each stimulus so that he is certain to receive a 'correct' reinforcement, we have the case defined previously in this paper as paired-associate learning. If the distribution on E depends only on the stimulus and not on trial number or the subject's response, the reinforcement is noncontingent. If the distribution on E is noncontingent, and there is no response that will insure the subject he is correct, we have probability learning.

Our purpose in this concluding section is to consider briefly the case where the set E has more than two subsets that are equivalence classes with respect to their value to the subject. To give a more concrete idea of what we have in mind, we will first discuss the experiment by Keller, Cole, Burke, and Estes (1965) that illustrates the notion of information via *differential reward.*

The subjects were faced with a paired-associate list of 25 items. There were two response alternatives and 5 possible reinforcements – the numbers 1, 2, 4, 6, and 8. One of these numbers was assigned to each S-R pair as its point value. The subject was told that his pay at the end of the session depended on the number of points he accumulated. So, for example, if the reward for pushing the left button, if *XAQ* were the stimulus, was 4, and the reward for pushing the right button was 1, the subject should learn to push only the left button. The experiment was run under two different conditions. In one the subject was told at the end of each trial the reward value for *both* of the possible responses; in the other he was told only the reward value for the response he had selected. In the latter case, since there were more than two possible reward values, knowing the value of one response gave only *partial information* concerning the optimal response. This is an example of information via differential reward.

Let us consider now information via differential reward in the context of alternative types of information a subject might receive. A learning experiment may include: (i) a set S of stimuli, (ii) a set R of response alternatives, (iii) a set E of reinforcements, (iv) a partition P of E into sets of reinforcements equivalent in value to the subject, and (v) an experimenter-determined function f from $S \times R$ into P_e, where P_e is the probability simplex in e dimensioned space and e is the cardinality of E. The probability that each reinforcement occurs is given by f as a function of the stimulus presented and the response selected. If e' is the number of members in P, f determines a function f' from $S \times R$ into $P_{e'}$, and f', then, gives the probability of each outcome value as a function of the stimulus and response chosen. *The subject's task in a learning experiment is to learn as much as is necessary about f' so that he may make the optimal response to each stimulus.*

The subject learns about f' from information provided him by the experimenter. We may classify this information into three broad types.

First, exogenous information is provided before the experiment begins. The subject learns what the responses are, what the stimuli are, whether reinforcement is contingent, possible reward values, number of trials, etc. Parts of this exogenous information might, of course, be deliberate *mis*information.

The second type consists of information concerning f' for a fixed stimulus. In a typical paired-associate experiment the subject receives complete information concerning f' on each trial for each stimulus. In the paradigms considered in Part II, subjects are given partial information by having E be the set of subsets of R (perhaps of fixed cardinality). The subject is told on each trial that the correct response was among those shown. Another type of information concerning the optimal response to a given stimulus is information via differential reward. Here the subject learns the rewards accruing to the members of the reinforcement set. *The forms of information of this type depend, then, on the structure of the reinforcement set.*

The third consists of information concerning f' for a fixed response. That is, does knowledge that response i is optimal for stimulus j give any information relevant to the optimal responses for other stimuli? This third type of information is obtained by 'concept formation', 'stimulus generalization', 'pattern recognition', 'recognition of universals', etc.; the term chosen depends on whether you are a psychologist, engineer, or philosopher.

Notice the symmetry between the second and third types of information a subject can be given. For a particular stimulus, the subject may receive knowledge relevant to the optimal response that concerns structure of the response set. For a particular response, the subject may receive information about the stimuli for which that response is optimal by placing structure on the stimulus set. The role of information via differential reward in this context is one way of placing structure on the reinforcement set; earlier sections of this paper considered other ways in detail.

For concept formation, there must be some sort of structure on the stimulus set. Roberts and Suppes (1967) and Jamison (1970) advanced quite different models for concept learning in which the basic structure on the stimulus set is of a particularly simple form, but they jointly assume that each stimulus is capable of being completely described by specifying for each of several attributes (e.g., color, size, ...) the value the stimulus takes on that attribute. We consider it an important theoretical task in learning theory to describe in detail other forms of structure that can be put on sets of stimuli.

As we indicated in the introduction, we view the results in this paper as simply a prolegomenon to detailed analysis of information structures in learning theory. Our results have been limited to rather special types of information structures placed on reinforcement sets. More general structures need to be considered and, more important, information structures on stimulus sets – concept learning – must be brought within the scope of the analysis.

APPENDIX: AN EXPERIMENT ON PAL WITH INCOMPLETE INFORMATION[3]

In Part II, Section 2, several models were discussed for the experimental paradigm of PAL with noncontingent, incomplete-information reinforcement. In the summer of 1967 such an experiment was performed at Stanford University with several different pairings of N and A, the response- and reinforcement-set cardinalities.

Method. Ten subjects from the undergraduate and recent graduate community at Stanford participated in the experiment for roughly an hour a day for 10 days within a period of two weeks. An on-line PDP-1 computer controlled all displays and data recording. The experimental equipment, a cathode-ray tube (CRT) with an electric typewriter keyboard placed directly below it, was housed in a sound-proof booth. The stimuli for any problem were the first N figures represented by the first N keys of the digit, or top letter, line on the keyboard; thus they were either 0, 1, ... or q, w, ... The $(N\!:\!A)$ pairings used on the first 4 days were (2:1), (6:1), (10:1), (10:3), (6:3), (10:5), (6:5), and (10:9). On days 5 to 10 the pairings were (10:3), (6:3), (10:5), (10:7), (6:5), and (10:9). Half the subjects received the conditions for each three cycles per day in the order given. The remaining half received a randomized order of conditions for each of the three cycles for the day. As shown later, the set order of conditions started with the easiest condition and ended with the hardest, as determined by the normative expected number of total errors. At any one time a subject worked on a solution to two problems, one with the displays on the top half of the CRT, the other with the displays on the bottom half. For half the subjects the top problem involved only the digits and the bottom problem only the letters. For the other subjects the top problem was letters and the bottom problem was digits. Trials on the two problems alternated. Both problems were on the same $(N:A)$ condition, and both had to be solved to a criterion of 4 successive correct responses to progress to the next condition. One pair of problems occurred on each $(N:A)$ condition in a single cycle. Unless stated otherwise successive trials on a problem from here on refer to successive trials on the same problem, not successive trials in the actual experiment.

On each trial the subject saw the display 'Respond From:' followed by the N possible responses. He pressed a key corresponding to the one he thought correct, and this response was displayed on the CRT below the response set. The feedback set of A responses, including the single correct response, then was displayed below the subject's response. The interval during which the feedback set was displayed is the study latency. Following onset of display of the feedback set the subject had up to 25 seconds to study the display, to press the carriage return key to conclude that trial, and to call for the next experimental trial. The 25-second limit between the cue to respond and the response, or the response latency, was never reached. The data for the first two days are not included in the results, and since the experimental design was completely explained to the subjects, they were well practiced on the procedure before the experiment began.

Results. The data from all experimental groups for days 3 through 10 are considered as a whole since none of the manipulations other than

TABLE III

Predicted and observed total errors

$N:A$	Observed mean total errors	Normative expected total errors
2:1	0.51	0.50
6:1	0.78	0.83
10:1	0.94	0.90
10:3	1.91	1.84
6:3	2.26	2.13
10:5	3.35	2.95
10:7	6.56	5.39
6:5	7.58	7.25
10:9	18.50	17.35

the $(N{:}A)$ pairings showed consistent differences. Table III shows the mean total errors for each condition. The normative expected total errors, determined by analytic methods for a derivation, or, in the more difficult cases, computer-run Monte Carlos, are also shown.

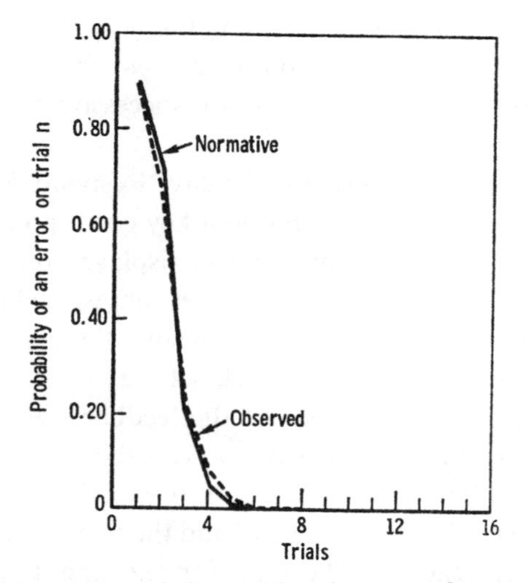

Fig. 6. Normative and observed learning curves for $N=10$, $A=3$.

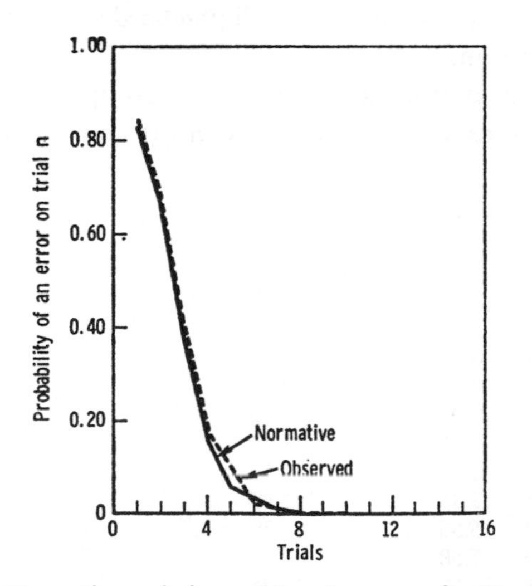

Fig. 7. Normative and observed learning curves for $N=6$, $A=3$.

The conditions with $A = 1$ were essentially cases of one-trial learning. Errors of chance happened on the first trial, and on succeeding trials the error frequency was less than .015. Thus the subjects performed essentially normatively on these 2-item list straightforward PAL tasks.

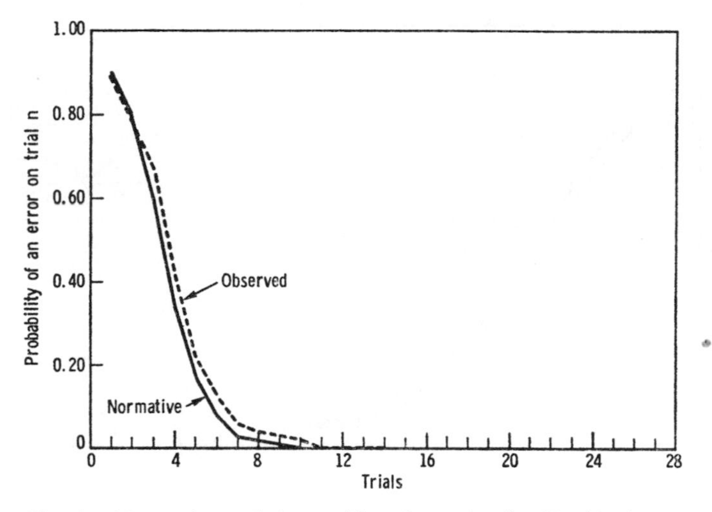

Fig. 8. Normative and observed learning curves for $N = 10$, $A = 5$.

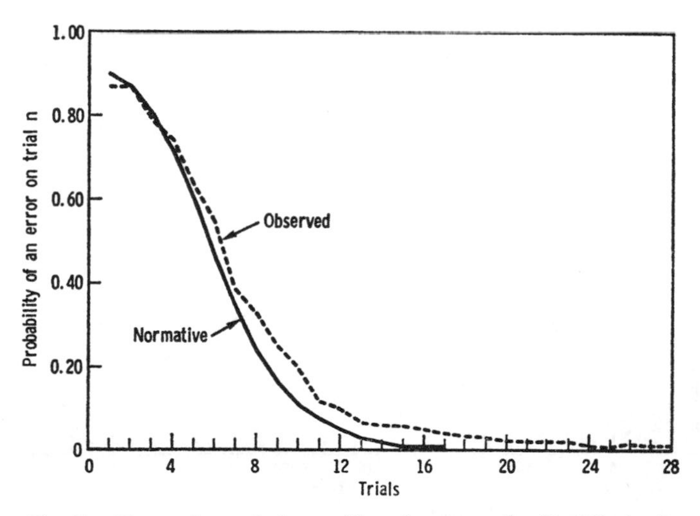

Fig. 9. Normative and observed learning curves for $N = 10$, $A = 7$.

The learning curves for the remaining six conditions with $A > 1$ are plotted in Figures 6 through 11, along with the normative learning curves. The normative error probabilities were not determined beyond the twentieth trial. The normative and observed learning curves are very

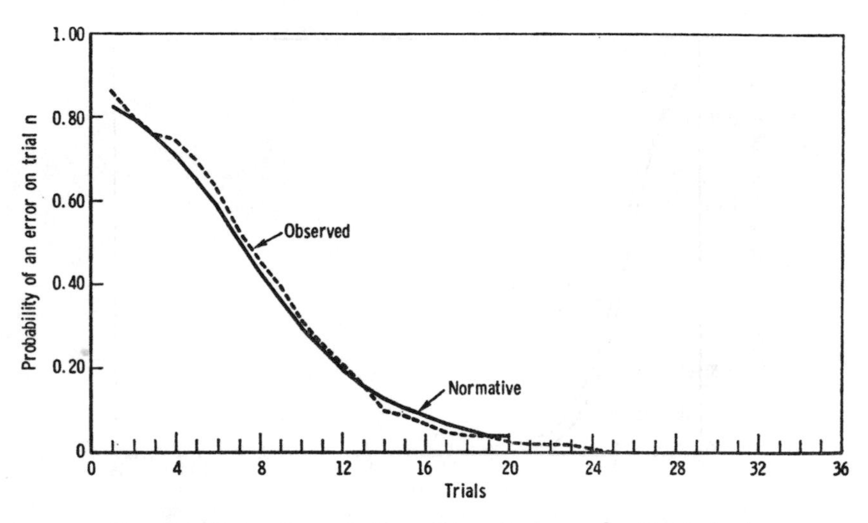

Fig. 10. Normative and observed learning curves for $N = 6$, $A = 5$.

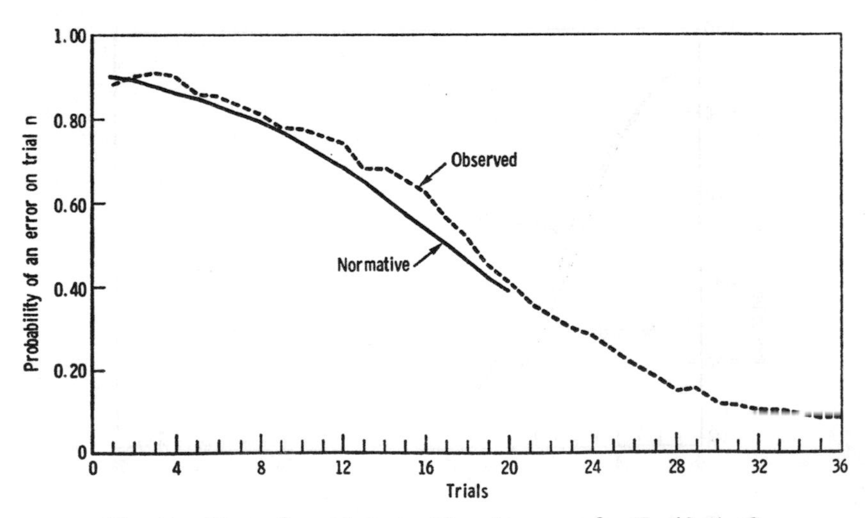

Fig. 11. Normative and observed learning curves for $N = 10$, $A = 9$.

close. Of some interest is the cross of normative and observed curves in the (6:5) condition (Figure 10). This better than normative performance is most likely due to the ability of subjects to use the 4-correct-response criterion to solve one problem once the other had been solved. This criterion use was not built into the normative model.

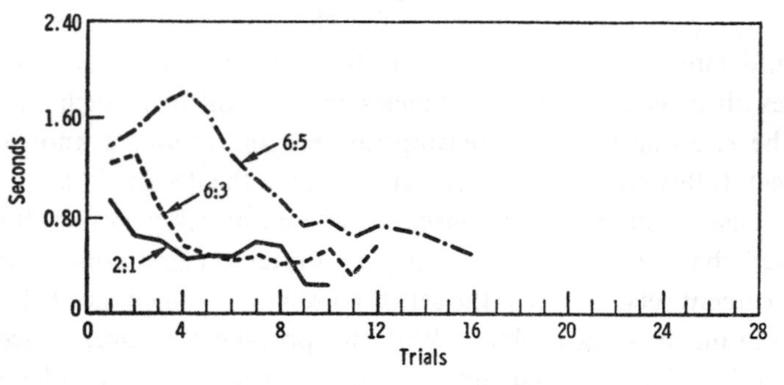

Fig. 12. Response latencies.

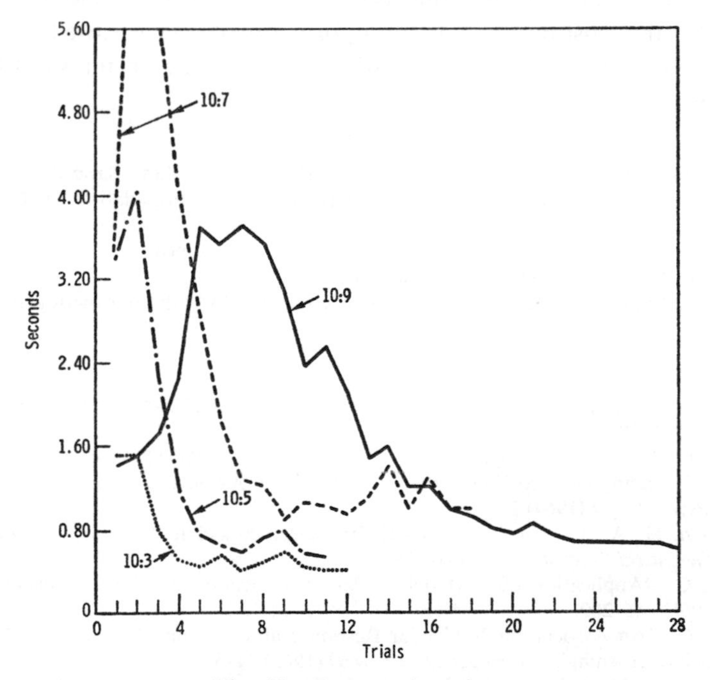

Fig. 13. Response latencies.

The study latencies are plotted in Figures 12–13. We do not know how to discuss these latencies meaningfully in a quantitative manner, but present them to call attention to a qualitative peculiarity. In the (10:9) and (6:5) conditions (Figure 13) a marked rise followed by a decline occurred in the study latency for several trials. In these conditions when the feedback set was close to the response set in size, subjects frequently said that they watched for the nonreinforced responses at first, and later switched to watching for the reinforced responses. The changes observed in the study latencies could result from such a practice, with the rise due to the increasing number of responses known to be incorrect, followed by the switchover, and then the decline in latency with the decreasing number of responses considered possibly correct. It should be noted that by using such a method to intersect first complements of reinforcement sets, and then the sets themselves, a subject needed at most 5 items in memory per problem. With two problems at once, he needed at most 10, but since the feedback sets on each problem were independent, the likelihood of both problems using maximum space at once was low. Thus for the most part all relevant information could be stored in fewer items than the 7 or 8 generally estimated as maximum for short-term memory.

REFERENCES

* This research was partially supported by Grant NSF GJ-443X, Basic Research in Computer-assisted Instruction, from the National Science Foundation to Stanford University.

[1] Howard, R. A., *Systems Analysis of Markov Processes*, to appear.

[2] The function $\delta(x)$ is the familiar Dirac delta function.

[3] Richard Freund of Stanford University assisted materially with this experiment.

BIBLIOGRAPHY

Atkinson, R. C., 'A Generalization of Stimulus Sampling Theory', *Psychometrika* **26** (1961) 281–290.

Atkinson, R. C. and Crothers, E. J., 'A Comparison of Paired-Associate Learning Models Having Different Acquisition and Retention Axioms', *Journal of Mathematical Psychology* **1** (1964) 285–315.

Bernbach, H. A., 'A Forgetting Model for Paired-Associate Learning', *Journal of Mathematical Psychology* **2** (1965) 128–144.

Bower, G., 'Application of a Model to Paired-Associate Learning', *Psychometrika* **26** (1961) 255–280.

Bower, G., 'An Association Model for Response and Training Variables in Paired-Associate Learning', *Psychological Review* **69** (1962) 34–53.

Bush, R. and Mosteller, F., *Stochastic Models for Learning*, Wiley, New York, 1955.

Calfee, R. C. and Atkinson, R. C., 'Paired-Associate Models and the Effects of List Length', *Journal of Mathematical Psychology* **2** (1965) 254–265.

Jamison, D., 'Bayesian Information Usage', in this volume (1970) pp. 28–57.

Keller, L., Cole, M., Burke, C. J., and Estes, W. K., 'Reward and Information Values of Trial Outcomes in Paired-Associate Learning'. *Psychological Monographs* **79** (1965) 1–21.

Millward, R., 'An All-or-None Model for Non-Correction Routines with Elimination of Correct Responses', *Journal of Mathematical Psychology* **1** (1964) 392–404.

Nahinsky, I. D., 'Statistics and Moments Parameter Estimates for a Duo-Process Paired-Associate Learning Model', *Journal of Mathematical Psychology* **4** (1967) 140–150.

Norman, M. F., 'Incremental Learning on Random Trials', *Journal of Mathematical Psychology* **1** (1964) 336–350. Referred to as 1964 (a).

Norman, M. F., 'A Two-Phase Model and an Application to Verbal Discrimination Learning', in *Studies in Mathematical Psychology* (ed. by R. C. Atkinson), Stanford University Press, Stanford, Calif., 1964. Referred to as 1964 (b).

Parzen, E., *Modern Probability Theory and Its Applications*, Wiley, New York, 1960.

Peterson, C. R. and Phillips, L. D., 'Revision of Continuous Subjective Probability Distributions', *IEEE Transactions on Human Factors in Electronics* **7** (1966) 19–22.

Phillips, L. D., Hays, W. L., and Edwards, W., 'Conservatism in Complex Probabilistic Inference', *IEEE Transactions on Human Factors in Electronics* **7** (1966) 7–18.

Roberts, F. and Suppes, P., 'Some Problems in the Geometry of Visual Perception', *Synthese* **17** (1967) 173–201.

Rumelhart, D. E., 'The Effects of Interpresentation Intervals on Performance in a Continuous Paired-Associate Task', Technical Report No. 116, August 11, 1967, Institute for Mathematical Studies in the Social Sciences, Stanford University, Stanford, Calif.

Suppes, P., 'A Linear Model for a Continuum of Responses', in *Studies in Mathematical Learning Theory* (ed. by R. R. Bush and W. K. Estes), Stanford University Press, Stanford, Calif., 1959, pp. 400–414.

Suppes, P., 'Stimulus Sampling Theory for a Continuum of Responses', in *Mathematical Methods in the Social Sciences* (ed. by K. J. Arrow, S. Karlin and P. Suppes), Stanford University Press, Stanford, Calif., 1960, pp. 348–365.

Suppes, P., 'Concept Formation and Bayesian Decisions', in *Aspects of Inductive Logic* (ed. by K. J. Hintikka and P. Suppes), North-Holland Publishing Company, Amsterdam, 1966, pp. 21–48.

Suppes, P. and Frankmann, R., 'Test of Stimulus Sampling Theory for a Continuum of Responses with Unimodal Noncontingent Determinate Reinforcement', *Journal of Experimental Psychology* **61** (1961) 122–132.

Suppes, P. and Ginsberg, R., 'A Fundamental Property of All-or-None Models, Binomial Distribution of Responses Prior to Conditioning, with Application to Concept Formation in Children', *Psychological Review* **70** (1963) 139–161.

Suppes, P., Rouanet, H., Levine, M., and Frankmann, R., 'Empirical Comparison of Models for a Continuum of Responses with Noncontingent Bimodal Reinforcement', in *Studies in Mathematical Psychology* (ed. by R. C. Atkinson), Stanford University Press, Stanford, Calif., 1964, pp. 358–379.

Wilks, S., *Mathematical Statistics*, Wiley, New York, 1962.

[illegible] R. G. and Atkinson, R. C., 'Finite-Markov Models and the Effects of [illegible]', *Journal of Mathematical Psychology* [illegible] 234–265.
[illegible] 1970, pp. 29–57.
[illegible]
[illegible]
[illegible] (1963) [illegible].
[illegible], 'The [illegible] Process' [illegible] *Journal of Mathematical Psychology* 4 (1967) [illegible]–150.
[illegible] Learning [illegible] *Journal of* [illegible]
Norman, M. F., [illegible] and [illegible], Discrimination Learning [illegible] (ed.), [illegible] Stanford University Press, Stanford, [illegible]
[illegible], New York, 19[illegible].
[illegible]
[illegible]
[illegible] (1969) [illegible].
[illegible] of Visual [illegible] (1967) [illegible].
[illegible] Information [illegible] Technical Report No. [illegible] Institute for Mathematical Studies in the Social Sciences, Stanford University, Stanford, [illegible].
[illegible] Linear Model for a Continuum of Responses', in [illegible] *Mathematical* [illegible] (ed. by R. R. Bush and W. K. Estes), Stanford University Press, Stanford, [illegible].
[illegible] Mathematical Research [illegible] in the [illegible] Stanford University Press, Stanford, [illegible].
[illegible] and Bayesian [illegible] *Aspects of* [illegible] (ed. by) K. J. Hintikka and [illegible], D. Reidel Publishing Company, Dordrecht, [illegible].
Suppes, P. and [illegible], R., [illegible] Continuum of Responses with [illegible] *Journal of* [illegible] *Psychology* [illegible].
[illegible]
[illegible]
[illegible]
Suppes, [illegible] and [illegible], [illegible]
[illegible]
[illegible]
[illegible]

PART IV

NEW APPLICATIONS OF INFORMATION CONCEPTS

JAAKKO HINTIKKA

SURFACE INFORMATION AND DEPTH INFORMATION

I. THE NATURE OF INFORMATION AND THE PROBLEMS OF MEASURING IT

In spite of its obvious importance, the notion of information has been strangely neglected by most contemporary epistemologists and philosophers of language. It seems to me more fruitful, however, to try to correct the situation than to lament it. In this paper, I shall point out some largely unexplored possibilities of using the concept of information in logic and especially in the philosophy of logic.

One reason for the hesitancy of recent philosophers to employ the term 'information' is perhaps the feeling that this term, at least in the phrase 'information theory', has been appropriated by communication theorists for their special purposes. In fact, the first systematic attempts to study the concept of information from a logical and philosophical point of view almost took the form of a conscious reaction to statistical information theory which was felt by the pioneers of this study to have neglected some of the most salient aspects of the idea of information.[1] These pioneers, especially Carnap and Bar-Hillel, called their study a theory of semantic information so as to distinguish it from what I shall here call statistical information theory.

I have become increasingly sceptical concerning the possibility of drawing a hard-and-fast boundary between statistical information theory and the theory of semantic information.[2] However, it also seems to me that the theory of semantic information is worth serious study on its own. Its starting-points are for obvious reasons quite different from those of statistical information theory, and they very soon lead to problems and results which a communications engineer is unlikely to be interested in but which are of the greatest importance to a logician, a student of language, and a theorist of human knowledge. It is especially interesting here to try to relate the theory of semantic information to the concept figuring in the very name of this theory, viz. to the concept of (linguistic) meaning.

J. Hintikka and P. Suppes (eds.), Information and Inference, 263–297.

The distinctive starting-point of a theory of semantic information was informally outlined already in the thirties by Sir Karl Popper.[3] This starting-point may perhaps be described as one way of spelling out the general idea that information equals elimination of uncertainty. In many interesting situations, perhaps in most of them, we can list all the different alternatives or possibilities that can be specified by certain resources of expression. Then each statement made by means of these resources will admit some of these basic alternatives, exclude the rest. Speaking roughly, the number of possibilities it admits of (suitably weighted) can be considered a measure of the uncertainty it leaves us with. The more alternatives a statement admits of, the more probable it also is in some purely logical sense of probability. Conversely, it is the more informative the more narrowly it restricts the range of those alternatives it still leaves open, i.e. the more alternatives it excludes. Thus probability and this kind of information (call it *semantic* information) are inversely related: the higher the information, the lower the probability, and *vice versa*.

If we want to speak more precisely, we have to specify several different things. *First*, we have to explain in the case of the languages we are interested in what the basic alternatives are. *Secondly*, we have to assign to them weights or *a priori* probabilities. *Thirdly*, we have to explain how our measures of information depend on the resulting measures of probability.

Of these problems, I am not in the present paper interested in the third one. For my purposes, either one of the most common definitions of information or content in terms of probability will usually do:

$$\text{(1)} \qquad \inf(s) = -\log p(s)$$

$$\text{(2)} \qquad \operatorname{cont}(s) = 1 - p(s)$$

I have dealt with this third task (by discussing the different kinds of relative information) in another paper, to which I shall here refer to for the requisite distinctions.[4] I am in principle greatly interested in the second task.[5] It seems to me that this problem very soon takes us away from purely logical and semantical considerations toward some sort of empiristic or subjectivistic concept of probability. However, in this paper I shall not try to say anything very much about this problem, either. Most of my discussion will be neutral with respect to all reasonable methods of assigning non-zero weights to whatever basic alternatives we are dealing with. By way of example, it often suffices to consider an even distribution of weights.

II. MONADIC FIRST-ORDER LANGUAGES

It is the first problem here, that of finding appropriate basic alternatives, that presents a challenge to a logician. By way of an example, let us first consider the case to which logicians have so far confined most of their attention, viz. the case in which our resources of expression consist of a number of one-place predicates (properties) $P_1(x), P_2(x), \ldots, P_k(x)$, an infinity of bindable variables $x, y, z, \ldots$, the two quantifiers $(\exists -)$ and $(-)$ where a bindable variable can be inserted in the argument-place, and the propositional connectives $\sim$, &, $\vee$ (plus whatever other connectives are definable in terms of them). These resources of expression define a monadic first-order language without individual constants.

It is often assumed that in order to be able to deal with this case in a satisfactory manner, we also have to possess a supply of individual constants (names), $a_1, a_2, \ldots, a_N$, one for each member of our domain of individuals. By means of them, we can form all the *atomic statements* $P_i(a_j)$. It might seem that the only natural method of specifying the desired basic alternatives here would be to form what Carnap has called state-descriptions.[6] They are just what the name promises: they describe the state of our domain of individuals completely in that they specify of each individual which predicates it has and which it does not have. A state-description is accordingly of the following form:

$$
\begin{aligned}
(3)\qquad & (\pm)P_1(a_1) \,\&\, (\pm)P_2(a_1) \,\&\cdots\&\, (\pm)P_k(a_1) \\
& \&\, (\pm)P_1(a_2) \,\&\, (\pm)P_2(a_2) \,\&\cdots\&\, (\pm)P_k(a_2) \\
& \&\, - - - - - \\
& \&\, (\pm)P_1(a_N) \,\&\, (\pm)P_2(a_N) \,\&\cdots\&\, (\pm)P_k(a_N),
\end{aligned}
$$

where each $(\pm)$ is to be replaced by $\sim$ or by nothing at all in all the different combinations. (We might say that each $(\pm)$ represents the answer 'yes' or 'no' to the question 'Does the individual no. j have the property no. i?')

However, if our aim is to have a satisfactory logical reconstruction of the procedures by means of which we come to have whatever information we have of reality, we cannot assume that we already know the whole universe of discourse well enough to have a name for each individual in it. We can hope to be able to approach this problem realistically only if we can restrict our attention to the less extensive resources of expression first indicated which do not include any names of individuals.[7]

How are we to find the basic alternatives that can be formed in terms of these resources of expression? They may be listed by means of a procedure which we shall later generalize.[8]

First, we list all the different *kinds of individuals* that can be specified in terms of our predicates and bindable variables. They are given by the following expressions (complex predicates):

$$(4) \qquad (\pm) P_1(x) \,\&\, (\pm) P_2(x) \,\&\cdots\&\, (\pm) P_k(x).$$

These are called by Carnap Q-predicates. Their number is $2^k = K$. I shall assume that they are, in some order,

$$(5) \qquad Ct_1(x), Ct_2(x), \ldots, Ct_K(x).$$

From these *kinds of individuals* we can get to the *kinds of possible worlds* which we can here specify by running through the list of all the Q-predicates (5) and for each of them indicating whether it is instantiated or not. The descriptions of these 'kinds of possible worlds' are therefore of the form

$$(6) \qquad (\pm)(\exists x)\, Ct_1(x) \,\&\, (\pm)(\exists x)\, Ct_2(x) \,\&\cdots\&\, (\pm)(\exists x)\, Ct_K(x).$$

These statements will be called *constituents*. They will be abbreviated $C_1, C_2, \ldots, C_{2^K}$. In the sequel, we shall assume that they have been rewritten in a more perspicious form. Instead of running through the list (5) of all the Q-predicates and specifying for each of them whether it is instantiated or not, it obviously suffices to list all those Q-predicates that in fact have instances and then to add that they are *all* the instantiated Q-predicates, i.e. that every individual of our domain exemplifies one of these instantiated Q-predicates. In other words, each constituent of the form (6) can be rewritten into the following form:

$$(7) \qquad \begin{aligned} &(\exists x)\, Ct_{i_1}(x) \,\&\, (\exists x)\, Ct_{i_2}(x) \,\&\cdots\&\, (\exists x)\, Ct_{i_w}(x) \\ &\&\, (x)\,[Ct_{i_1}(x) \vee Ct_{i_2}(x) \vee \cdots \vee Ct_{i_w}(x)], \end{aligned}$$

where $\{Ct_{i_1}(x), Ct_{i_2}(x), \ldots, Ct_{i_w}(x)\}$ is a subset of the set of all Q-predicates (5). By constituents (of a monadic language) we shall in what follows normally mean expressions of the form (7).

In forming our measure of 'purely logical' probability, we shall assume that all the constituents are given equal weights, or at least non-zero weights which add up to one. This second requirement will in fact be satisfied always when all closed statements of our language which are not logically false have a non-zero probability (weight). If we have names at our disposal, the weight of each constituent can be divided according to some suitable principles among all the different state-descriptions which it admits of. One reason why these principles are important is that by their means we can study how observations of individual cases affect the probabilities (relative probabilities, probabilities *a posteriori*) of generalizations. Hence they are vital in any inductive logic.[9] Here we are not especially interested in these principles, however.

That constituents (7) really describe the basic alternatives that can be formulated in our simple language is brought out by the fact that every consistent statement s of the language we are considering (by assumption s therefore contains no individual constants) can be represented in the form of a disjunction of some (maybe all) of the constituents:

$$(8) \qquad s = C_{i_1} \vee C_{i_2} \vee \cdots \vee C_{i_{w(s)}},$$

where $C_{i_1}, C_{i_2}, \ldots, C_{i_{w(s)}}$ is a subset of the set of all the constituents (7). The set $\{i_1, i_2, \ldots, i_{w(s)}\}$ is called the index set of s, and denoted by $I(s)$. The number $w(s)$ is called the width of s. The right-hand side of (8) is called the *(distributive) normal form* of s.

If the weight of C_i is $p(C_i)$, we have for the logical probability $p(s)$ of s the following expression:

$$(9) \qquad p(s) = \sum_{i \in I(s)} p(C_i)$$

From the probabilities we obtain measures of semantic information in the two ways given by (1) and (2).

III. GENERAL (POLYADIC) FIRST-ORDER LANGUAGES

This suffices to outline a theory of semantic information for a monadic first-order language, apart from whatever problems there are in assigning weights to different constituents and state-descriptions along the lines indicated. What I shall consider next are certain new features that appear when this approach is extended to a full first-order language, i.e. to a case in which our supply of primitive predicates may contain relations (polyadic predicates) in addition to (or instead of) properties (monadic predicates). I shall disregard individual constants for the time being. Statements containing no such constants will be said to be general.

Can we find a finite list of all basic alternatives in some natural sense in the case of full first-order logic? In absolute terms, the answer is clearly no. However, if we restrict our resources of expression further in one important respect, we can in fact find such a list.

This restriction can be described very simply both formally and intuitively. Formally, it amounts to imposing a finite upper bound on the number of layers of quantifiers or, which amounts to the same, on the lengths of nested sequences of quantifiers (i.e. sequences of quantifiers whose scopes are contained within the scopes of all the preceding ones). The maximal number of such layers of quantifiers in any part of a statement s will be called (tentatively) the *depth* of s, and designated $d(s)$.[10] Intuitively, we impose a finite limit on the number of individuals we are allowed to consider in their relation to one another at one and the same time.[11]

If this additional restriction is imposed on our statements, the situation in any polyadic language (without individual constants) is very much like the situation we encountered earlier in monadic first-order languages. The concept of a constituent can be generalized so as to apply to this more general case, and each general statement s with depth d (or less) can be effectively transformed into a disjunction of a number of constituents with depth d (and with the same predicates as s). The main difference is that the concept of a constituent is now relative to d. Each general statement s of depth d or less thus has a distributive normal form

$$(10) \qquad C_{i_1}^{(d)} \vee C_{i_2}^{(d)} \vee \cdots \vee C_{i_{w(s)}}^{(d)},$$

where $\{C_{i_1}^{(d)}, C_{i_2}^{(d)}, \ldots, C_{i_{w(s)}}^{(d)}\}$ is a subset of the set of all constituents with depth d (and with the appropriate predicates). For example, each constituent $C_i^{(d)}$ of depth d can be expressed as a disjunction of constituents of depth $d + e$. Such a disjunction is called the *expansion* of $C_i^{(d)}$ at depth $d + e$, and any constituent occurring in some expansion of $C_i^{(d)}$ is said to be *subordinate to* $C_i^{(d)}$.

IV. DEFINING CONSTITUENTS

Thus the main thing that remains to be explained here is what the appropriate constituents $C^{(d)}$ are. This can be done by generalizing the characterization I gave earlier of the constituents of a monadic first-order language.[12] In this generalization, we shall assume a fixed finite set of predicates. Instead of defining the list of all constituents, i.e. the list of all 'possible kinds of world' we can specify here, we shall first define recursively a list of all the kinds of individuals x that can be specified by means of d layers of quantifiers and by means of certain fixed reference-point individuals, say those named by $a_1, a_2, \ldots, a_m$. Let us suppose that these are given by the expressions $Ct_i^{(d)}(a_1, a_2, \ldots, a_m; x)$. We shall call these *attributive constituents.* The following figure illustrates the situation:

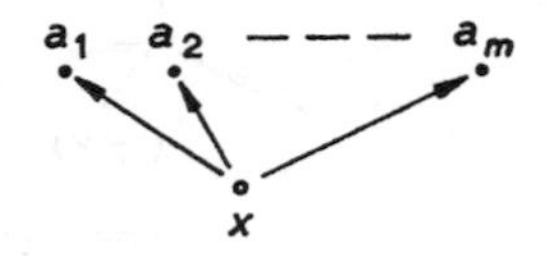

(characterized by means of d layers of quantifiers)

Fig. 1.

We can in fact proceed here essentially in the same way as in the monadic case. We can run through the list and specify which kinds of individuals are instantiated. We can also specify how a_m is related to $a_1, a_2, \ldots, a_{m-1}$. In other words, we can form all the different expressions of form

$$\begin{aligned}
(11) \qquad & (\pm)(\exists x)\, Ct_1^{(d)}(a_1, a_2, \ldots, a_m; x) \\
& \&\, (\pm)(\exists x)\, Ct_2^{(d)}(a_1, a_2, \ldots, a_m; x) \\
& ----- \\
& \&\, (\pm)\, A_1(a_m)\ \&\ (\pm)\, A_2(a_m)\ \&\cdots,
\end{aligned}$$

where $A_1(a_m), A_2(a_m), \ldots$ are all the atomic statements that can be formed from our predicates and from the constants $a_1, a_2, \ldots, a_m$ and which contain a_m. If we replace the constant a_m by a bindable variable y in (11), we obtain a list of all the different kinds of individuals that can be specified by means of $d + 1$ layers of quantifiers and by means of the smaller set of reference-point individuals named by $a_1, a_2, \ldots, a_{m-1}$. In other words, the expressions

$$\begin{aligned}(12)\qquad & (\pm)(\exists x)\,Ct_1^{(d)}(a_1, a_2, \ldots, y; x)\\ & \&\,(\pm)(\exists x)\,Ct_2^{(d)}(a_1, a_2, \ldots, y; x)\\ & \&\,-----\\ & \&\,(\pm)\,A_1(y)\,\&\,(\pm)\,A_2(y)\,\&\cdots\end{aligned}$$

are precisely what the expressions

$$(13)\qquad Ct_j^{(d+1)}(a_1, a_2, \ldots, a_{m-1}; y)$$

are supposed to stand for. The situation may be illustrated by means of Figure 2, which ought to be compared with Figure 1.

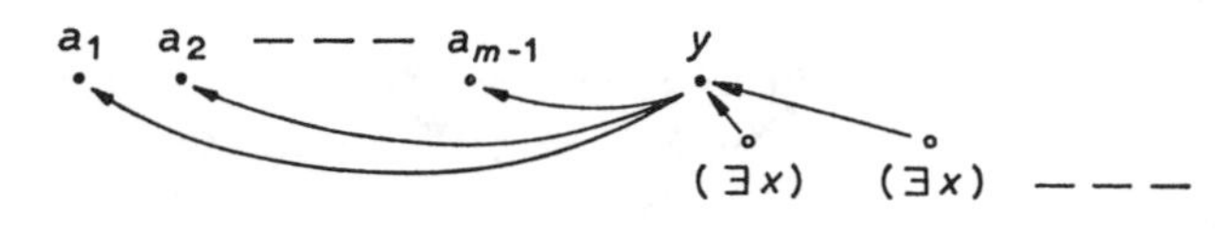

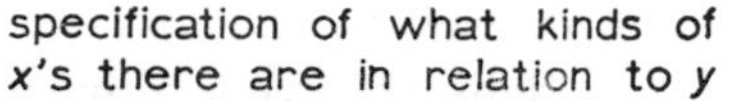

specification of what kinds of x's there are in relation to y

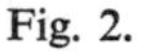

Fig. 2.

In the case $d = 0$ (12) contains the last line only, i.e. reduces to a constituent in the sense of propositional logic. This gives the first step in a recursive definition. From the expressions

$$(14)\qquad Ct_j^{(d-1)}(x)$$

we obtain constituents of depth d precisely in the same way as the constituents (6) of a monadic language were obtained from Q-predicates (5).

V. TRIVIALLY INCONSISTENT CONSTITUENTS

The resulting constituents as well as all the expressions (11)–(13) can be rewritten in the same way in which (6) was rewritten as (7). In the sequel, this reformulation is normally assumed to have been carried out. For instance, (12) will be assumed to be rewritten in the form

$$
\begin{array}{ll}
(12^*) & (\exists x)\, Ct_{i_1}^{(d)}(a_1, a_2, \ldots, a_{m-1}, y; x) \\
& \&\, (\exists x)\, Ct_{i_2}^{(d)}(a_1, a_2, \ldots, a_{m-1}, y; x) \\
& ----- \\
& \&\, (x)\, [Ct_{i_1}^{(d)}(a_1, a_2, \ldots, a_{m-1}, y; x) \\
& \vee \quad Ct_{i_2}^{(d)}(a_1, a_2, \ldots, a_{m-1}, y; x) \\
& \vee \qquad\qquad\qquad -----] \\
& \&\, (\pm)\, A_1(y)\, \&\, (\pm)\, A_2(y)\, \& \cdots
\end{array}
$$

where $\{Ct_{i_1}^{(d)}, Ct_{i_2}^{(d)}, \ldots\}$ is a subset of all the expressions of the form

$$Ct_{i_1}^{(d)}(a_1, a_2, \ldots, a_{m-1}, y; x),$$

likewise modified.

The resulting form is an especially simple one in that all negation-signs have a minimal scope which contains no connectives and no quantifiers. Since all the other logical operations are monotonic as far as the logical force of our expressions is concerned, any expression we obtain by omitting conjuncts in a constituent or in an expression of the form $Ct_i^{(d)}(a_1, a_2, \ldots, a_m; x)$ is implied by the original. Since a bindable variable z enters into any such expression only through minimal expressions ('atomic expressions') consisting of a predicate and a number of bindable variables or individual constants, and since these occur (negated or unnegated) in conjunctions only, we can in effect omit any one layer of quantifiers in a constituent or in (12) and obtain an expression which is implied by the original expression.

It is seen at once that the resulting expression is like a constituent or like (12*) except that here and there a constituent or an expression $Ct_i^{(d)}(y_1, y_2, \ldots, y_n; y_{n+1})$ has been replaced by a conjunction of similar expressions of depth $d-1$. Now different constituents and different expressions of form (12) or (12*) are clearly incompatible. Hence, two things have to be required for a constituent to be consistent:

(a) Omitting any layer of quantifiers in $C_i^{(d)}$ must yield (apart from repetitions) a constituent of depth $d-1$.

(b) The constituents obtainable by omitting different layers of quantifiers must all be identical. (We have assumed all along that the order of conjuncts and disjuncts in a constituent does not matter.)

A constituent may be inconsistent in other ways, too. Assume that

$Ct_i^{(d-1)}(z_1, z_2, \ldots, z_{m-1}; y)$ occurs in $C_j^{(e)}$, with $Ct_i^{(d-1)}(a_1, a_2, \ldots, a_{m-1}; y)$ given by (12*). Then it may happen that each disjunct of

$$\begin{aligned} (15) \qquad & Ct_{i_1}^{(d)}(a_1, a_2, \ldots, a_{m-1}, a_m; a_h) \\ & \vee\ Ct_{i_2}^{(d)}(a_1, a_2, \ldots, a_{m-1}, a_m; a_h) \\ & \vee\ ----- \end{aligned}$$

contains a conjunction with two directly contradictory conjuncts (one the negation of the other) where $1 \leqslant h \leqslant m$. Thus it has to be required that

(c) (15) must not be directly contradictory in the way just explained for any $C_i^{(d-1)}(z_1, z_2, \ldots, z_{m-1}; y)$ occurring in $C_j^{(e)}$.

If $C_j^{(e)}$ violates one of the requirements (a)–(c), it will be called *trivially* inconsistent. In a sense, different parts of such a $C_j^{(e)}$ contradict each other. The same remarks can be applied to expressions of form (12*).

VI. CONNECTIONS WITH DECISION PROBLEMS. COMPLETENESS

However, the fundamental difference between the polyadic and the monadic case is that in the polyadic case many constituents and expressions of form (12*) are inconsistent although they are not trivially inconsistent.[13] This can be seen by observing that there is a close connection between the inconsistency of constituents and the various decision problems that arise in first-order logic. To locate effectively all inconsistent constituents is to solve the decision problems for the first-order language with the same predicates, and the decision problem for a finitely axiomatizable theory with s as its sole non-logical axiom is tantamount to the problem of locating all the inconsistent constituents subordinate to one of the constituents of the normal form of s.

The former fact is obvious. In order to convince ourselves of the latter, let us first note that r is implied by s iff it is implied by all the constituents of the normal form of s (since these are pairwise exclusive). (It is of course assumed that the predicates and individual constants of r all occur in s.) Hence we can restrict our problem to testing whether r (of depth, e, say) is implied by a constituent $C_i^{(d)}$ of the same depth as s (say, of depth d). This can be done by expanding both $C_i^{(d)}$ and r into their respective normal forms at depth $\max(d, e)$. The implication holds iff each constituent which occurs in the former normal form but not in the latter is inconsistent. These constituents are *per definitionem* all subordinate to $C_i^{(d)}$. Being able to test these subordinate constituents for inconsistency therefore gives us the desired decision method.

More generally, it is seen what degree of unsolvability belongs to the problem of testing constituents subordinate to $C_i^{(d)}$ for inconsistency. This degree is the same as the Turing degree of the function $\varphi_{C_i^{(d)}}(e)$ which indicates as a function of e how many inconsistent constituents of depth $d + e$ there are subordinate to $C_i^{(d)}$. The fact that knowing this function would enable us to locate effectively all inconsistent constituents subordinate to $C_i^{(d)}$ is fairly obvious. In order to decide whether $C_j^{(d+e)}$ (subordinate to $C_i^{(d)}$) is inconsistent, we can simply list all theorems of first-order logic until they include either the negation of $C_j^{(d+e)}$ or the negations of precisely $\varphi_{C_i^{(d)}}(e)$ other constituents of depth $d + e$ subordinate to $C_i^{(d)}$. In the latter case, we know that these are *all* the inconsistent constituents of this kind, which implies that $C_j^{(d+e)}$ is consistent.

It follows that the degree of unsolvability of any finitely axiomatizable theory is the same as the degree of some finite sum of the functions $\varphi_{C_i^{(d)}}(e)$. Furthermore, it follows from an important result by Hanf that every recursively enumerable degree of unsolvability can be obtained in this way.[14]

Since many of the theories $T(C_i^{(d)})$ are unsolvable, we cannot effectively locate all inconsistent constituents. We have no mechanical way of identifying all of them. We have certain systematic ways of locating more and more inconsistent constituents, however. The simplest way (in principle) is perhaps to keep on expanding $C_i^{(d)}$ at greater and greater depths and to keep on testing the ensuing subordinate constituents for *trivial* inconsistency (for which I already gave a mechanical test). I have shown elsewhere[15] that this simplest method in a sense already gives us all we need: $C_i^{(d)}$ is inconsistent if and only if there is a depth (say $d + e$) at which all its subordinate constituents are *trivially* inconsistent. This result is valid although we have in general no effective ways of predicting what the number e might be. It represents a version of the completeness theorem for first-order logic – a version which is tailored to the special case of constituents.

It follows that all the usual types of logical argument – disproofs, proofs, proofs from premises, equivalence proofs, etc. – can all be carried out in the same way, all of them completely linearly. For instance, in order to prove that *r* and *s* are equivalent (for simplicity assume that they have the same predicates) one merely has to expand them to disjunctions of deeper and deeper constituents, omitting all the time every inconsistent constituent that appears. If they are equivalent, at some depth the two expansions will coincide.

As I have pointed out elsewhere[16], searches of interpolation formulas (in Craig's sense) can also take place in a similar way.

This presence of potential inconsistencies is a fact we have to live with. Any realistic account of our actual ways with logical expressions has to take it into account. It seems to me that we have to do this also in any satisfactory theory of semantic information.

VII. INDUCTIVE PROBABILITY VS. PRE-LOGICAL PROBABILITY

This has important implications for our idea of information. How are we to assign weights to the different constituents? Clearly, the sum of all of them at any depth cannot exceed 1 (for any given set of predicates). But can one of these weights equal zero? Does this happen in the case of inconsistent constituents? It is natural to give the weight zero to all *trivially* inconsistent constituents. I shall assume in the sequel that this has been done. (Alternatively, we may think of trivially inconsistent constituents as being systematically eliminated wherever they appear. This is possible in principle since there is an effective test of trivial inconsistency.) But what about non-trivially inconsistent ones? Are we to assign zero weights to them or not?

The answer depends, it seems to me, upon the kind of information we have in mind here. If what we have in mind is information concerning some kind of reality independent of our concepts and conceptual constructions, then we have to assign to them a zero weight. For an inconsistent constituent does not describe any real alternative concerning the world as it is independent of our concepts.[17] Hence admitting or excluding them should not make any difference to our concept of probability or information. I shall call the concept of information that results from this decision to assign the weight zero to all inconsistent constituents *depth information*. The corresponding sense of probability might be called *inductive probability*, for it is obtained by concentrating our attention, as it were, on what our statements can tell of objective reality. I am also tempted to dub it *post-logical probability*, for in a sense it presupposes that we have already done to our statements all that can be done to them in order to spell out their consequences as fully as possible. (This sense will be spelled out later in Section XII.)

It is important to realize, however, that even when we are dealing with inductive procedures, inductive probabilities and depth information are not what we can directly operate with. Consider, as an example, what happens if the weight (*a priori* probability) of each consistent constituent $C_i^{(d)}$ of depth d is divided evenly among all its subordinate constituents of depth $d + 1$. In order to know the weight of one of these we have to know how many consistent subordinate constituents of depth $d + 1$ there are. Generalizing this, it is seen that we can find out the weights of all constituents subordinate to $C_i^{(d)}$ if and only if we know the function $\varphi_{C_i^{(d)}}(e)$. But knowing this function was seen to be tantamount to knowing a decision method for the theory $T(C_i^{(d)})$. Hence very often we cannot calculate inductive probabilities and measures of depth information effectively.

If our inductive methods are formulated in terms of inductive probabilities or amounts of depth information, and if these are determined by some principle not radically different from even distribution, then our inductive policies are seldom computable. If the difficulty of inductions can be measured by the 'difficulty' of computing these policies, we might thus say that the difficulty of the general problem of induction (inductive generalization) in a first-order language is the same as the difficulty of the decision problem for this language, and that the difficulty of inductions in a world in which a first-order statement s is known to be true is the same as the difficulty of the decision problem for the theory $T(s)$ whose only non-logical axiom is s.[18]

It is perhaps to be expected that optimal inductive policies should not be computable.[19] After all, scientific discovery is often asserted to turn on insight and intuition rather than on explicit rules. What we have found perhaps gives a more precise formulation of a part of what is meant by these assertions.

It is also seen, however, that it would be false to say that scientific discovery, conceived of as a process of maximizing depth information or inductive probability or some balanced combination of these two, is not subject to rules. There may be highly important regularities to which they are subject. Only these regularities cannot in general be computable.

VIII. SURFACE INFORMATION

In order to be able to discuss in informational terms what we actually do in logic and in empirical science, and not just the ultimate *desiderata* of these pursuits, we must have at our disposal measures of information different from those of depth information – measures of information in which the impossibility of deciding which statements are consistent is taken into account. I shall call measures of this kind measures of *surface information*, and indicate them by a superscript as follows: $\inf^{surf}$ and cont^{surf}.

An interesting example of such a measure of information is obtained by assigning a non-zero weight to all constituents which are not *trivially* inconsistent. When we move from a given depth d to the next, the weight of a constituent $C_i^{(d)}$ of this depth is somehow divided between all those subordinate constituents $C_j^{(d+1)}$ which are not trivially inconsistent.

The main novel feature here is the fact that sometimes there are no such subordinate constituents although $C_i^{(d)}$ itself is not trivially inconsistent. Then the weight of $C_i^{(d)}$ either becomes 'idle' or is reassigned to constituents of depth $d + 1$ which are not subordinate to it. We shall first consider the former alternative (the weight of $C_i^{(d)}$ 'disappears').

The system of weights which goes together with surface information so defined creates a probability-like measure p' on first-order statements. It is not a probability in the normal sense, however, for it does not satisfy the usual axioms of probability calculus. For instance, $p'(s)$ may be > 0 even though s is inconsistent (as long as it is not trivially inconsistent). However, it is related to trivial inconsistency pretty much in the same way as inductive (post-logical) probability is related to ordinary inconsistency. We shall call logically valid statements depth tautologies, and we shall say that s is a surface tautology if and only if $\sim s$ is trivially inconsistent. Then the measure p' will be related to surface tautologies rather in the same way as normal (inductive) probability is related to depth tautologies.

It is tempting to call measures like p' *pre-logical* probability measures. This term, which admittedly has some connotations that are misleading here, is nevertheless motivated by the fact that $p'(s)$ might (tentatively) be thought of as the degree of belief which it is rational to associate with s before we have subjected it to the techniques that logic gives us for the purpose of bringing out its import more and more fully. (There are even better candidates for this role, however, as we shall see later. An obvious competitor, though not the strongest, is the relative weight $p'(s \mid t)$ where t is a surface tautology of the same depth as s.) Normally, we shall nevertheless call p' *surface probability*.

In terms of the probability-like measure p', we can define relative (conditional) surface probabilities in the usual way:

$$p'(s_1 \mid s_2) = \frac{p'(s_1 \,\&\, s_2)}{p'(s_2)}$$

assuming that $p'(s_2) \neq 0$.

Surface probability is easily seen to satisfy the following conditions *(inter alia)*:

(i') $p'(s) \geqslant 0$

(ii') $p'(s) = 0$ iff s is trivially inconsistent;

(iii') $p'(s_1 \vee s_2) = p'(s_1) + p'(s_2)$
provided that s_1 and s_2 are of the same depth and that $s_1 \,\&\, s_2$ is trivially inconsistent;

(iv') $p'(s \vee \sim s \mid t) = 1$
where t is an arbitrary surface tautology of the same depth as s;

(v') If $s_1 \equiv s_2$ is a surface tautology and if s_1 and s_2 are of the same depth, then s_1 and s_2 are intersubstitutable everywhere without affecting p'.

Instead of (iii'), we can have more generally

(iii'$_1$) $p'(s_1 \vee s_2) = p'(s_1 \,\&\, t) + p'(s_2 \,\&\, t) - p'(s_1 \,\&\, s_2)$ where t is an arbitrary surface tautology of the same depth as $(s_1 \,\&\, s_2)$.

From these observations it easily follows that for statements of depth d or less the relative weights $p(s \mid t^{(d)})$ satisfy all the usual axioms of probability calculus where $t^{(d)}$ is a surface tautology of depth d. In view of the well-known betting-theoretical justification of these axioms[20], it follows that the weights $p(s \mid t^{(d)})$ can be thought of as acceptable betting ratios for an agent who considers as possible all the 'events' expressed by such constituents of depth d as are not trivially inconsistent and who wants to avoid necessarily losing or non-winning betting systems ('Dutch

Books' in the wider sense). Conversely, for every such system of betting ratios, we can easily find a system of weights $p'(s)$ (down to statements s of depth d) which yields these betting ratios as the conditional probabilities $p'(s \mid t^{(d)})$. These simple observations perhaps illustrate the sense in which $p'(s \mid t^{(d)})$ can be thought of as the degree of belief which it is rational to associate with a statement s when the agent has carried out the elimination of trivially inconsistent constituents down to depth d.

A further argument for considering this kind of information might perhaps be conducted in terms of the concept of *uncertainty*. The amount of information a statement carries was tentatively identified above with the amount of uncertainty it eliminates. Now as long as we have not actually eliminated (found inconsistent) a given constituent (which is not trivially inconsistent but which could, for all that we know, be inconsistent), its presence in the normal form of a statement *increases* our *uncertainty* concerning which alternative admitted by the statement is true and hence *decreases* the *information* the statement gives us, even though the constituent might in the last analysis turn out to be inconsistent.

In view of observations of this kind, it is not surprising that some leading theorists of subjective probability have recently expressed interest in the kind of information my surface information exemplifies.[21]

The most important feature of surface information is its behavior when new layers of quantifiers are added. In order to discuss problems concerning this behavior, let us call the expansion (normal form) of a statement s at depth e, $E^{(e)}(s)$. Let us also call a statement s *trivially inconsistent at depth e* if and only if $E^{(e)}(s)$ is trivially inconsistent. (Trivial inconsistency of s *simpliciter* will then mean trivial inconsistency at its own depth.) If the depth of s is d, then the surface information of $E^{(d+e)}(s)$ grows when a new layer of quantifiers is added iff one of its constituents becomes trivially inconsistent for the first time at depth $d + e + 1$. In other words,

$$\inf^{surf}(E^{(d+e+1)}(s)) \geqslant \inf^{surf}(E^{(d+e)}(s)),$$

where the inequality is strict iff there is a non-trivially inconsistent constituent in $E^{(d+e)}(s)$ which is trivially inconsistent at depth $d + e + 1$.

In this sense, then, does the surface information of a statement grow (during the process of being expanded further) always when any non-trivially inconsistent constituent turns out to be inconsistent.

IX. SURFACE INFORMATION MADE MORE REALISTIC

This kind of surface information is not altogether very natural, however. Surely not all expansion of a given statement increases its information. Often, an addition in depth merely makes the normal form more complicated, without increasing our insight into the possibilities the statement rules out.[22] But when does the addition of a new layer of quantifiers to the normal form $E^{(d+e)}(s)$ of a statement (say s of depth d) add to its information? The most natural answer seems to be to say that information is increased by the step from depth $d + e$ to $d + e + 1$ if and only if at least one of the possibilities gets ruled out at this stage which s seemed to admit of in the first place while no comparable constituent is likewise ruled out. And the most natural interpretation of this statement is to take it to mean that one of the constituents of the normal form of $E^{(d)}(s)$ at the original depth d of s becomes trivially inconsistent at the depth $d + e + 1$ for the first time while no other constituent of the same depth d becomes trivially inconsistent at this depth. A measure – or, rather, a sequence of measures – which satisfies this requirement (with one natural exception) can be defined easily.

The principles which give rise to these measures can be seen from the special case of an even distribution of weights on subordinate constituents. Suppose we have eliminated all the trivially inconsistent constituents down to a certain depth d. This means that we have also eliminated many constituents of depth $< d$ as inconsistent although they are not themselves trivially inconsistent. They have been eliminated because all their subordinate constituents of depth d are trivially inconsistent.

An even distribution of weights on the remaining constituents will give us a probability-like measure which is relative to the given depth d. By an even distribution we of course mean a distribution which divides the weight of each constituent of depth e ($e < d$) which is not trivially inconsistent at depth d evenly among all its subordinate constituents of the next depth $e + 1$ (excluding those which are trivially inconsistent at depth d).

This is immediately generalized to the case in which we do not practice even distribution but rather specify the ratios of the weights of different constituents of depth $e + 1$, subordinate to the same constituent of depth e, provided they are not trivially inconsistent. We can stick to these prescribed ratios in the case of all those constituents of depth e which are not trivially inconsistent at depth d.

A probability-like measure which goes together in this sense with trivial inconsistency at depth d will be called $p^{(d)}$. For statements s, s_1, s_2 of depth d or less it will satisfy the following conditions:

(i$^{(d)}$) $p^{(d)}(s) \geqslant 0$

(ii$^{(d)}$) $p^{(d)}(s) = 0$
iff s is trivially inconsistent at depth d;

(iii$^{(d)}$) $p^{(d)}(s_1 \vee s_2) = p^{(d)}(s_1) + p^{(d)}(s_2)$
provided that $(s_1 \,\&\, s_2)$ is trivially inconsistent at depth d;

(iv$^{(d)}$) $p^{(d)}(t^{(d)}) = 1$

(v$^{(d)}$) If $E^{(d)}(s_1 \equiv s_2)$ is a surface tautology, s_1 and s_2 are intersubstitutable in expressions of depth d or less without affecting $p^{(d)}$.

At each depth $e < d$ the weights of all the constituents add up to one. We may either think of $p^{(d)}$ as being defined for statements of depth d at most, or else think of it (as we shall do) as being defined for constituents beyond the depth d in the same way as p'. (That is to say, beyond this depth the weight of each constituent of a given depth is somehow divided between all the constituents of next greater depth that are not trivially inconsistent, allowing the weight to get lost if there are none.)

The above results (i$^{(d)}$)–(v$^{(d)}$) show that intuitively the same thing can be said of $p^{(d)}(s)$ as was said above of $p'(s \mid t^{(d)})$. As far as statements of depth d or less are concerned, the measure $p^{(d)}(s)$ may be thought of as the degree of belief it is rational to associate with s at the moment when we have carried out our analysis of what first-order statements really say down to depth d, for instance, after we have expanded the negation $\sim s$ to the depth d.

It is obvious that we can identify measures $p^{(o)}$ and p'.

The above conditions (i$^{(d)}$)–(v$^{(d)}$) do not define $p^{(d)}$ fully, however. A definition may be given by recursion, by defining $p^{(d+1)}$ in terms of $p^{(d)}$.

If the ratios of the different constituents of depth d subordinate to the same constituent of depth $d - 1$ are thought of as being fixed but unspecified, then this kind of recursive characterization is obtained by adding the following requirements to the requirement that each measure satisfies (i$^{(d)}$)–(v$^{(d)}$):

(vi$^{(d)}$) If $C_1^{(d)}$ and $C_2^{(d)}$ are subordinate to the same $C_0^{(d-1)}$ and if neither of them is trivially inconsistent at depth $e > d$, then $p^{(e)}(C_1^{(d)})/p^{(e)}(C_2^{(d)}) = p^{(e-1)}(C_1^{(d)})/p^{(e-1)}(C_2^{(d)})$.

What this requirement amounts to is very natural. Its import is best described by stating what happens when one moves from $p^{(d)}$ to $p^{(d+1)}$. We restrict ourselves to statements of depth d or less. Since the weights always add up to one at these depths, no weight gets totally lost. Rather, when a constituent (of depth $\leqslant d$) becomes trivially inconsistent at depth $d + 1$ for the first time, its weight will be redistributed in a certain natural way. This way can be described in terms of the tree structure formed by all the different constituents which are not trivially inconsistent. One description is as follows:

Whenever all the subordinate constituents of a constituent $C_0^{(e)}$ $(e < d)$ become trivially inconsistent (for the first time) at depth $d + 1$, then we trace back the branch which led to $C_0^{(e)}$ to its last branching-point from which at least one branch emerges reaching down to depth $d + 1$, i.e. reaching to a constituent not trivially inconsistent at depth $d + 1$. The weight of $C_0^{(e)}$ is first distributed tentatively between all such branches emerging from that point in proportion to their earlier weights. The weights are carried downwards along these branches, until they reach their nodes at depth $d + 1$, on which these weights 'come to rest'.

X. RELATIVE PRE-LOGICAL PROBABILITY REDEFINED

In order to bring out the generality of the idea which underlies the measures $p^{(d)}$, it may be pointed out that an extremely natural measure of relative probability can be in terms of them. This measure is different from the earlier measure $p'(r \mid s)$, defined in the normal way $p'(r \mid s) = p'(r \,\&\, s) \,/\, p'(s)$. We shall call the new measure of relative weight (probability) $p'(r \parallel s)$, and define it as follows:

$$p'(r \parallel s) = \frac{p^{(d_o)}(r \,\&\, s)}{p^{(d_o)}(s)},$$

where d_o = the greater of the depths of r, s = the depth of $r \,\&\, s$. If you review this definition in the light of the explanations given above for $p^{(d)}(s)$, you will see that the idea underlying $p'(r \parallel s)$ is very natural, and that it could have been explained directly without using the measure $p^{(d_0)}$ as an intermediary. (Roughly speaking, $p'(r \parallel s)$ is the natural relative measure among those constituents which have not turned out to be trivially inconsistent by the time analysis is carried down to depth d_0 and which are also subordinate to at least one constituent of s.)

In terms of this new kind of relative weight, the sequence of measures $p^{(e)}$ $(e = 1, 2, \ldots)$ is readily defined. It is easily seen that we have, for an s of depth e,

$$p^{(d)}(s) = p'(s \parallel t^{(d)}),$$

where $t^{(d)}$ is (as usual) a surface tautology of depth d.

In a sense, we can therefore dispense with the sequence of different measures $p^{(d)}$, provided we introduce first the new kind of relative (prelogical) probability $p'(s \parallel r)$.

The intuitive difference between the two different senses of relative weight $p'(s \mid r)$ and $p'(s \parallel r)$ is readily seen from the case in which $r = t^{(d)}$ = surface tautology of depth d. Here we can think of the tree formed of all the constituents of depth d or less (which are not trivially inconsistent) as a sort of 'family tree'. Whenever the weight of a constituent (of depth $e < d$, say) gets 'lost' (because all its subordinate constituents of the next depth $e + 1$ are trivially inconsistent), it is in $p'(s \mid t^{(d)})$ divided among all the 'surviving' members of the family of depth d (i.e. among all constituents of depth d which are not trivially inconsistent) in proportion to their earlier weight. In $p'(s \parallel t^{(d)})$ its weight is only divided among the 'surviving' constituents of depth e which have the closest relation to the deceased family line. This perhaps also illustrates the (relative) 'naturalness' of $p'(s \parallel r)$ in general.

XI. INCREASING SURFACE INFORMATION

In terms of the sequence of measures $p^{(d)}$ we can define a corresponding sequence of measures of surface information. When d grows, this surface becomes increasingly less shallow. In terms of those measures of surface information we can do justice to the requirements which led us away from the original surface information based on p' in the first place. We shall designate these measures of information by $\text{inf}^{(d)}$ and $\text{cont}^{(d)}$. It is easily seen that the following result is valid:

(16) If s is of depth $e < d$, $\inf^{(d)}(s) < \inf^{(d+1)}(s)$ only if at least one of the constituents in the original normal form (of depth e) of s becomes trivially inconsistent *for the first time* at depth $d + 1$.

In brief, surface information of a statement grows when we move to a more deeply based measure of information (depth $d + 1$ instead of d) only if some of the possibilities the statement seemed to admit of in the first place gets eliminated (for the first time) at this move from depth d to depth $d + 1$.

This result can be partly inverted. If one of the constituents of the original normal form (of depth e) of s becomes trivially inconsistent for the first time at depth $d + 1$, then (16) holds unless (a) s is trivially implied at depth $d + 1$ by a constituent $C_0^{(f)}$ of depth f smaller than e (which is not trivially inconsistent at depth $d + 1$) or (b) at least one constituent $C_0^{(f)}$ of depth $f \leqslant e$ such that $s \supset \sim C_0^{(f)}$ is a surface tautology becomes inconsistent for the first time at depth $d + 1$.

In the case (a), one might say that all the various apparent possibilities admitted by s are 'really' expressed by the single shallower constituent $C_0^{(f)}$. At least this is the situation from the point of view of depth $d + 1$. Unless this single constituent also gets eliminated when we move from depth d to depth $d + 1$, no 'relevant' alternatives are eliminated at this move and no new information is gained. This is as it ought to be.

In the case (b), the elimination of one of the original alternatives admitted by s is counteracted by the elimination of a competing alternative. Whether our information grows or not depends on the weights of these respective alternatives. Again, this is as it ought to be.

XII. DEPTH INFORMATION AS A LIMIT OF SURFACE INFORMATION

In terms of our sequence of measures of surface information, we can state an important (but simple) result concerning the relation between depth information and surface information. Of course, not very much can be said with precision unless something further is assumed concerning the principles according to which the weights are determined on which surface information and depth information, respectively, are based. It is easy to see, however, that a relatively weak assumption concerning their interrelation suffices here. All we have to assume is that the ratios between the two are the same, in the following sense:

(17) Whenever $C_1^{(d)}$ and $C_2^{(d)}$ are consistent and subordinate to the same constituent $C_o^{(d)}$ of depth $d - 1$, then
$p^{(d)}(C_1^{(d)})/p^{(d)}(C_2^{(d)}) = p^{depth}(C_1^{(d)})/p^{depth}(C_2^{(d)})$
where p^{depth} is the probability-measure on which depth information is based.

This is a natural extension of our earlier requirement which pertained to the relation of the different measures of surface information.

Assuming (17), one can verify the following result:

(18) $$\lim_{e \to \infty} \inf^{(d+e)}(s) = \inf^{depth}(s)$$
where s is a statement of depth d.

Obviously, (18) is equivalent to the following:

(18*) $$\lim_{e \to \infty} p^{(d+e)}(s) = p^{depth}(s).$$

This result I dub the quantified completeness theorem. It may be called quantified because it pertains to numerical measures of information. Its relation to the completeness results of the first-order logic is indicated by the simple argument which shows its validity. This argument is as follows: By the variant of the completeness result mentioned earlier, there is a depth $d + e$ at which all the inconsistent constituents of depth d are trivially inconsistent. By our assumption (17), the values of p^{depth} and $p^{(d+e)}$ will then be identical for constituents of depth d, and hence also for all statements of this depth.

Equation (18) shows in what sense depth information can be thought of as a limit of surface information, thus fulfilling a promise given above. It also shows that in a sense depth information can be thought of as surface information at infinite depth, logical equivalence as trivial equivalence at infinite depth, logical implication as trivial implication at infinite depth, etc.

XIII. MEASURING THE INFORMATION YIELD OF LOGICAL ARGUMENTS

One of the most interesting uses of the concept of surface information is that it enables us to measure the information which a logical argument gives us. It was pointed out above that all the usual types of logical argument – disproofs, proofs, proofs from premises, equivalence proofs – can in principle be carried out in a linear form by first converting the statements in question into their normal forms and then expanding them until at great enough a depth the desired logical relation becomes apparent. In each case, there is a natural measure of the surface information gained in the course of the argument.

As an example, consider the disproof of a statement s of depth d. Its surface information at depth d is $\text{cont}^{(d)}(s) = 1 - p^{(d)}(s)$. If it is inconsistent, then at some depth $d + e$ all the constituents of $E^{(d+e)}(s)$ become trivially inconsistent for the first time. This means that $p^{(d+e)}(s) = 0$, $\text{cont}^{(d+e)}(s) = 1$. The change in information is

$$(19) \qquad \text{cont}^{(d+e)}(s) - \text{cont}^{(d)}(s) = p^{(d)}(s),$$

that is to say, precisely the surface probability of s at its own depth. This apparent probability (degree of rational belief *prima facie*) is eliminated in the course of the argument, and the information of s grows accordingly.

In order to prove s (of depth d_s) from r (of depth d_r), we have to show in some sense that all the possibilities admitted by r are among those admitted by s. In order to see precisely what this means, consider the cases (i) $d_r \geqslant d_s$ and (ii) $d_s > d_r$ separately.

(i) In this case, we have to show that all the constituents in $E^{(d_r)}(r)$ but not in $E^{(d_r)}(s)$ are inconsistent. If they are trivially inconsistent at depth d_r, then s is trivially implied by r at this depth. If this is not the case but if s is nevertheless logically implied by r, there is a depth e ($e > d_r$) such that all the constituents of $E^{(e)}(r)$ are among those of $E^{(e)}(s)$ (for the first time).

This means that the pre-logical probability of the critical constituents has become reassigned. The difference

$$(20) \qquad \text{cont}^{(e)}(r \,\&\sim s) - \text{cont}^{(d_r)}(r \,\&\sim s) = p^{(d_r)}(r \,\&\sim s)$$

measures the gain in our information. This gain is the total pre-logical probability at depth d_r of those constituents which at depth d_r seemed to violate the logical consequence from r to s.

(ii) When $d_s > d_r$, we have to show that those constituents of $E^{(d_s)}(r)$ which do not occur in $E^{(d_s)}(s)$ are all inconsistent. This involves the reassignment of pre-logical probability of the total

$$(21) \qquad p^{(d_s)}(r \,\&\, \sim s).$$

Moreover, in this case new information is also gained already by expanding r to the form $E^{(d_s)}(r)$. The information gain is

$$(22) \qquad p^{(d_r)}(r) - p^{(d_s)}(r).$$

The sum of (21) and (22) is

$$(23) \qquad p^{(d_r)}(r) - p^{(d_s)}(r) + p^{(d_s)}(r \,\&\, \sim s) = p^{(d_r)}(r) - p^{(d_s)}(r \,\&\, s)$$

which gives us the desired gain in surface information.

If we put $d_s = d_r$ in the expression, (23) becomes identical with (20). In a sense, both (i) and (ii) are thus covered by the expression

$$(24) \qquad p^{(d_r)}(r) - p^{(d_0)}(r \,\&\, s) = \mathrm{cont}^{(d_0)}(r \,\&\, s) - \mathrm{cont}^{(d_r)}(r),$$

where $d_0 = \max(d_r, d_s)$.

Only if this information is zero are we in a position to say that s is trivially implied by r *simpliciter*, i.e. without reference to the depth involved.

In analogy with the standard definition of relative content[23], (24) can be identified with

$$(24^*) \qquad \mathrm{cont}^{surf}_{add}(s \mid r)$$

i.e. with the informative content s adds to that of r. We could have postulated this simply as the desired expression, but then the justification of just this definition of the additional information would not have been obvious. What we have seen serves to vindicate (24*) as the appropriate definition of incremental relative surface content.

It may be observed that when $d_r = d_s$, the deduction of s from r is trivial (in the sense of yielding no new information of the kind we have discussed) if and only if the consequence-relation is seen already at their common depth, i.e. if and only if no deeper constituents are needed as intermediate stages of the proof. In terms of the intuitive idea of depth, this means that s can be deduced from r without considering in the intermediate stages of the proof any more individuals than we already considered in r or s.

Later (in Section XV) we shall find another intuitive interpretation of the difference between trivial and non-trivial proofs.

It is of some interest to notice that we can measure not only the gain in information which takes place when a logical argument (disproof, proof, proof from premises, etc.) is successfully carried out, but also the gain in information which results from an attempted but unsuccessful argument. For instance, if we are trying to disprove s (of depth d) and if we have carried out an expansion of s to depth $d + e$, then our gain in information so far is

$$\text{(25)} \qquad \text{cont}^{(d+e)}(s) - \text{cont}^{(d)}(s) = p^{(d)}(s) - p^{(d+e)}(s)$$

which equals $p^{(d)}(s)$ only if we have already succeeded in disproving s.

Suppose that in trying to derive s from r we have obtained the consequences $q_1, q_2, \ldots, q_i$ which we hope to be useful as intermediate steps in the disproof. Suppose further, for simplicity, that r and s are of the same depth d. Then the gain in information relevant to the desired implication is the sum of the pre-logical probabilities $p^{(d)}(C_i^{(d)})$ of all those constituents $C_i^{(d)}$ which occur in $E^{(d)}(r \,\&\, \sim s)$ but which are incompatible with one of the q's (say q_j) at its depth (say d_j), i.e. of all the constituents for which

$$\text{(26)} \qquad p^{(d_j)}(C^{(d)} \,\&\, q_j) = 0$$

with a suitably chosen q_j.

In these measures of partial success, the only thing that counts is the total elimination of some of the initial possibilities (for instance, constituents in $E^{(d)}(r \,\&\, \sim s)$ in the case of a proof from premises). However, we can also have other sorts of measures of partial success in which the elimination of some of the non-trivially inconsistent constituents subordinate to those of $E^{(d)}(r \,\&\, \sim s)$ also counts even though it does not bring about the (complete) elimination of any original constituent. In these new measures, a premium is thus put also on partial elimination. As a measure of partial success we can then use (in the case of a proof from premises)

$$\text{(27)} \qquad p^{(d)}(r \,\&\, \sim s) - p^{(d_0)}(r \,\&\, \sim s \,\&\, q_1 \,\&\, q_2 \,\&\ldots\&\, q_i)$$
$$\text{where } d_0 = \max(d, d_1, d_2, \ldots, d_i).$$

From a comparison with (23) it is seen that the information gain in this case is the same as the one involved in deducing q_1 & q_2 & ⋯ & q_i from r & ∼ s.

A generalization to the case in which the depth of $s (= d_s)$ may be greater than that of $r (= d_r)$ is obtained in the same way as in (24):

$$(28) \qquad p^{(d_r)}(r) - p^{(d_s)}(r \,\&\, s) - p^{(d_m)}(r \,\&\, \sim s \,\&\, q),$$

where $d_m = \max(d_r, d_s, d_1, d_2, \ldots, d_i)$.

Likewise, if in an attempted disproof you have arrived at a potential intermediate stage q (of depth $d + e$) between r (of depth d) and a trivial inconsistency, then your information gain in the present sense is

$$p^{(d)}(r) - p^{(d+e)}(r \,\&\, q).$$

Attempted proofs and equivalence proofs can be dealt with in a similar way.

This possibility of appraising attempted proofs for their informativeness suggests that our measures of surface information might have some uses in the systematic heuristics of logical proofs.

All these measures could be discussed at length and other ones could be defined. Here I have primarily wanted to give examples of what could be done in terms of information concepts in this area and not to exhaust the subject.

XIV. LOGICAL INFERENCE IS NOT TAUTOLOGICAL[24]

More important philosophically than any particular measure of the non-triviality of a logical proof is perhaps the possibility of having an objective measure of any kind in the first place. We are dealing with a question here that almost inevitably occurs to anyone who is familiar with logic and mathematics. A great deal of ingenuity often goes into a logical or mathematical argument. The result of such an argument often enlightens us, gives us new insights, informs us greatly. Yet many philosophers tell us that the truths of logic and perhaps also those of mathematics are only so many tautologies.[25] How are we to take these assertions? The philosophers in question have been impressed by a fact that is true and important, namely by the fact that in a sense a logical inference does not give us new information concerning the kind of reality the premises of the inference speak of. However, they leave open the question: what other kind of information do logical and mathematical arguments then give us? It seems to me that this is a question which logicians and philosophers have so far almost completely failed to answer. C. D. Broad has called the unsolved logical and philosophical problems concerning induction a scandal of philosophy. If I am allowed to exaggerate slightly – but only very slightly – I should like to say that there is, in addition to the scandal of induction, a closely related and equally disquieting scandal of deduction, viz. the failure of philosophers and logicians to answer the question: How does deductive reasoning add to our knowledge (information)?

Some philosophers have been bold enough to deny that there is a genuine problem here. In doing so, they have been forced to adopt a flagrantly subjectivistic and psychologistic view of the uses of deduction. They have said that the *only* sense in which a logical argument gives us new information is a purely psychological one. There is no *objective* sense in which the conclusion of a valid inference carries more information than the premises, they have repeatedly asserted, A logical argument only serves to spell out in a more accessible form what the premises already say. There cannot be any *objective* obstacles to seeing the conclusions right there in the premises, for if there were, their removal would constitute an objective gain in information. In fact, the bluntest and frankest of the philosophers taking this line, Ernst Mach, explicitly assented to this conclusion.[26]

The results we have reached belie this view completely. Our notion of surface information gives a measure of information that can be increased by logical and mathematical reasoning. Moreover, there is nothing subjective or psychological in this notion of surface information, nor therefore in the measures of the additional information which a logical argument gives us. One of the doctrines of the logical positivists has thus been given a definitive refutation. It is not just that we have found some farfetched sense in which there are truths that are logical and hence *a priori* and yet synthetic in the sense of containing information in some sense of the word. The sense we have given to the idea of information is one which was explicitly alleged to be impossible by several logical positivists.

XV. FREGE ON NON-TRIVIAL DEFINITION. OTHER HISTORICAL PERSPECTIVES

The distinction between inferences that correspond to surface tautologies and others receives an amusing sidelight when considered from the point of view of Venn diagrams. At each depth, the attributive constituents with just one free variable give us a partition (classification system) for all conceivable individuals. When we move to the next greater depth, the partition becomes finer (new boundaries between different classes of individuals are created). Moreover, some of the cells of the new partition turn out to be empty for reasons of trivial inconsistency.

In terms of these partitions, the difference between trivial implications (at a given depth) and others can be characterized rather strikingly. It follows from what was said above in Section XIII that, in order for the requisite relationship of class-inclusion to be established, all we need in the trivial case is to consider only the boundaries which already exist at depth d and simply to consider them from a suitable point of view. In contrast, in the second case (non-trivial implication), the class-inclusion cannot be seen without actually introducing new boundary lines.

This description is of some historical interest in that it matches almost completely Frege's informal explanation of the difference between trivial and non-trivial definitions in logic and mathematics.[27] Frege's distinction was perhaps intended to be merely informal and even metaphorical. We can see, however, that it turns out to be (when looked upon from a suitable point of view) an almost precise formal characterization of our distinction between surface tautologies (analytical steps of inference in one of the many senses of the word) and depth tautologies.

Many deeper historical and interpretative insights are obtained by recalling the connection mentioned above between the concept of depth and the intuitive idea of considering only a certain number of individuals in their relation to each other. I have explored some of these historical insights elsewhere.[28] They turn on the fact that an increase in depth means from our vantage point the introduction of a new individual to one's argument. The idea of such an introduction has played an extremely important role in the philosophy of mathematics. In geometrical proofs, such an introduction of a new geometrical entity goes by the name of a construction, and the role of such constructions exercised greatly the interest of philosophers from Plato and Aristotle to Kant. A generalization of this notion is at the bottom of Kant's philosophy of mathematics, I have argued. Here we shall not try to trace these fascinating historical relationships, however.

The idea of having to increase the number of individuals considered together in an argument indicates what the obstacles are that make it impossible to see the conclusion of a logical argument as being already present in the premises. If some possibilities apparently admitted by the premises can only be seen to be inconsistent by considering more individuals in their relation to each other than are considered in the premises, no amount of merely psychological conditioning and removal of mental blocks will enable us to see the conclusion as being already present in the premises. We have to 'construct' (bring in) the new individuals in order to reach the conclusion. The necessity of doing so is part and parcel of the logical situation and not due merely to our psychological inability to look at the premises from the right point of view. It is one reason why the new information which a logical argument can yield is completely objective.

XVI. SURFACE INFORMATION AS INFORMATION CONCERNING OUR CONCEPTUAL SYSTEM. THE INEXTRICABILITY OF SURFACE INFORMATION FROM DEPTH INFORMATION

What has been said so far leaves open an important question. All talk about information is apt to provoke the question: Information about what? Elsewhere, I have discussed this question at some length in the case of surface information.[29] Here, I shall only indicate some of the main points. It is *prima facie* tempting to suggest something along the following lines: Depth information is information about the reality which our statements speak of, whereas surface information is somehow information about the conceptual system we use to deal with this reality (or aspect of reality).

Although this suggestion is perhaps not altogether wrong, it is seriously oversimplified. The fundamental feature of the situation one faces, say, when receiving an item of news expressed by a first-order statement *s* is that one cannot in general tell the 'real' alternatives admitted by this statement from those 'apparent' alternatives which have nothing to do with reality but are, rather, contributed by the conceptual system we use to express our knowledge. The better we master the ramifications of our own conceptual system, i.e. the more surface information we have, the more of these merely apparent alternatives we can eliminate. In a very real sense, this *ipso facto* gives us a better grasp of what *s* tells about the 'external' reality and hence in this sense also yields information (relief from uncertainty!) concerning this mind-independent reality. In another sense, this new information is of course information concerning our conceptual system. For instance, it grows in proportion to our grasp of the consequences of the basic assumptions of this conceptual system.

Thus in the messages (couched in first-order terms) which we send, receive, or store, information concerning the world and information concerning our own concepts are inextricably intertwined. This inextricability is a consequence of the undecidability of first-order logic in that this undecidability shows that depth information is not in general effectively computable. This undecidability thus turns out to have interesting (and philosophically highly relevant) consequences for our concepts of information.

XVII. SURFACE INFORMATION AND THE CONCEPT OF MEANING

The same failure of the effective computability of depth information has interesting consequences for the concept of meaning.[30] Whatever the meaning of a sentence is or may be, it seems to me that the (literal) meaning of a (grammatically correct) sentence has to be something that anyone who knows the language in question can effectively find out.[31] If so, the meaning of a general statement cannot go together with depth information. For instance, it cannot be identified with the set of all consistent constituents the statement admits of. Relations of logical implication cannot be relations of meaning inclusion; and so on.

A much better way of trying to explicate the problematic notion of linguistic meaning seems to be to relate it to surface information and to sets of constituents not trivially inconsistent. Trivial implication seems to me a much better explication of the idea of meaning inclusion than logical implication. In this way a partial merger of a theory of semantic information with a theory of semantics could perhaps be brought about.

This possibility is closely related to another promising possibility which opens here. One of the main problems in trying to analyze such philosophically important concepts as knowledge, belief, memory, perception, and the other so-called propositional attitudes is their behavior *vis-à-vis* logical implication. If a man believes that p and if p logically implies q, does he believe that q? Surely not. Yet it is not quite easy to build a logic in which this implication does not hold (at least *prima facie*).[32]

The above considerations suggest a way out of this difficulty. Surely, one is inclined to say, if q not only is implied by p but part of what p means, then clearly believing that p will imply believing that q. If the concept of meaning is connected with surface information in the way indicated, then we end up saying that

'a believes that p'

implies logically

'a believes that q'

if and only if q is trivially implied by p in the sense indicated earlier.

This suggestion seems to eliminate all the problems that are connected with the invariance (or its failure) of belief (and other propositional attitudes) with respect to logical equivalence. You might express my point by saying that logical equivalence has to be replaced here by a genuine equivalence of meaning, i.e. by a trivial equivalence.[33]

It may of course happen that someone who says that he believes that r will nevertheless deny that he believes that s even when s is trivially implied by r. What I just said presupposes that we can somehow rule out the possibility that such a person really believes and fails to believe what he says he does. How can we possibly rule this out? By showing that our man cannot fully understand the meaning of the statements in question, and by showing that this is shown by his very pair of assertions. (For simplicity, we may assume that r and s have the same predicates and individual constants and that they are of the same depth d.)

Quite independently of our theory of surface information, how can we go about finding out what someone means by a statement he utters? One very natural general suggestion is to take all the different possibilities concerning the world and see which of these possibilities his statement admits of and which of them it excludes. Now in order to avoid going beyond what our man says we must restrict ourselves to those resources of expression which he already employs in his own utterance. In the case of first-order statements one of the most important restrictions on one's expressive powers is of course the number of individuals one is simultaneously considering in their relation to each other, i.e. the depth of one's statements. Hence the basic alternatives we must consider here are specified by constituents of the same depth *d* as one's initial statement. But if we remain at this depth, we cannot hope to rule out those inconsistent constituents which are not trivially inconsistent at depth *d*, and we must therefore recognize all the constituents that are not trivially inconsistent at this depth as specifying *bona fide* possibilities for our current purposes.

This suggests the following procedure for spelling out the meaning of *r*: We run trough the list of all the different alternatives concerning the world specified by constituents which are of the same depth *d* as *r*, which contain the same predicates and individual constants as *r*, and which are not trivially inconsistent at this depth; and we specify for each of them whether it is compatible with *r* or not. In short, we transform *r* into its distributive normal form at depth *d*. The same goes for *s*. Grasping the meaning of *r* and *s*, I just suggested, is closely related to being able to do this (in principle). But if our man can do this, and if *s* is indeed trivially implied by *r*, then he cannot (in principle) fail to see that the alternatives admitted by *r* are all among those admitted by *s*, too, i.e. that *s* is implied by *r*. The only reason why he does not see this is his failure to appreciate fully the meaning of *r* and *s*, i.e. his failure to understand fully what he says he believes and does not believe.

The elimination of trivially inconsistent constituents does not affect this line of thought. It goes by the board completely, however, if in order to obtain *s* from *r* through our normal forms – or in any other way, for that matter – one has to consider as an intermediate stage constituents or other expressions of a depth greater than *d*. Familiarity with such expressions is not (according to our lights) required to grasp the meaning of *r* and *s*, and in them one is considering more individuals in their relations to each other than in *r* or *s*. Hence in such circumstances one can understand fully *r* and *s* and yet fail to perceive the implication from the former to the latter, i.e. one may believe the former but not the latter while understanding completely the meaning of the sentences in question.

Several interesting possibilities for further development thus seem to be opened by our observations.

University of Helsinki
and Stanford University

REFERENCES

[1] The basic references are R. Carnap and Y. Bar-Hillel, *An Outline of a Theory of Semantic Information*, Technical Report No. 247, M.I.T., Research Laboratory of Electronics, 1950; reprinted in Y. Bar-Hillel, *Language and Information*, Addison-Wesley, Reading, Mass., 1964, pp. 221–274; and Y. Bar-Hillel and R. Carnap, 'Semantic Information', *British Journal for the Philosophy of Science* **4** (1953) 147–157. (Slightly different version also in *Communication Theory: Papers Read at a Symposium on Applications of Communication Theory* [ed. by W. Jackson], London 1953, pp. 503–511.)
[2] Cf. my paper 'On Semantic Information' in the present volume (forthcoming in *Physics, Logic, and History* [ed. by W. Yourgrau], The Plenum Press, New York 1970), last few pages.
[3] Karl R. Popper, *Logik der Forschung*, Springer-Verlag, Vienna, 1934. (English translation, with additions, as *Logic of Scientific Discovery*, Hutchinson's, London, 1959.)
[4] See my paper, 'The Varieties of Information and Scientific Explanation', in *Logic, Methodology and Philosophy of Science III, Proceedings of the 1967 International Congress* (ed. by B. van Rootselaar and J. F. Staal), North-Holland Publishing Company, Amsterdam, 1968, pp. 151–171.
[5] Some aspects of it are dealt with in my paper 'On Semantic Information' (present volume) and in the earlier papers of mine on which it is based.
[6] See R. Carnap, *Logical Foundations of Probability*, University of Chicago Press, Chicago 1950 (2nd ed. 1962).
[7] Cf. in the connection my paper 'Are Logical Truths Tautological?', in *Deskription, Analytizität und Existenz* (ed. by P. Weingartner), A. Pustet, Salzburg and Munich, 1966, pp. 215–233, especially pp. 221–223.
[8] With the following discussion, cf. my paper 'Distributive Normal Forms in First-Order Logic' in *Formal Systems and Recursive Functions, Proceedings of the Eighth Logic Colloquium, Oxford, July 1963* (ed. by J. N. Crossley and M. A. E. Dummett), North-Holland Publishing Company, Amsterdam 1965, pp. 47–90.
[9] Cf. my papers 'Toward a Theory of Inductive Generalization', in *Logic, Methodology, and Philosophy of Science, Proceedings of the 1964 International Congress* (ed. by Y. Bar-Hillel), North-Holland Publishing Company, Amsterdam 1965, pp. 274–288; and 'A Two-Dimensional Continuum of Inductive Methods', in *Aspects of Inductive Logic* (ed. by Jaakko Hintikka and Patrick Suppes), North-Holland Publishing Company, Amsterdam 1966, pp. 113–132.
[10] This definition may be sharpened somewhat. Let us consider two quantifiers which occur in the same statement s, let us assume that they contain the bound variables x and y, respectively, and let us assume that the latter quantifier occurs within the scope of the former. Then we shall say that the two quantifiers are *connected* iff there are within the scope of the former quantifiers (say containing the variables $z_1, z_2, \ldots, z_k$, respectively, such that x and z_1, z_i and z_{i+1} $(i = 1, 2, \ldots, k - 1)$, and z_k and y occur

in the same atomic expression in s. The depth of s can now be defined as the length of the longest chain of nested *and connected* quantifiers in s. This sharpened definition does not imply any changes in the subsequent discussion.

[11] For this intuitive meaning of depth, see also my papers 'Are Logical Truths Analytic?', *Philosophical Review* **74** (1965) 178–203; and 'An Analysis of Analyticity', in *Deskription, Analytizität und Existenz* (ed. by P. Weingartner), A. Pustet, Salzburg and Munich 1966, pp. 193–214.

[12] See once again my paper 'Distributive Normal Forms in First-Order Logic' (reference 8 above).

[13] No constituent was inconsistent in the monadic case. If expressions of form (12*) are considered in the monadic case, however, some of them turn out to be trivially inconsistent. But even in this case the only clause that is needed to locate all inconsistencies is the relatively uninteresting third clause (c), which can in fact be dispensed with by treating somewhat differently the relation of quantifiers to identity.

[14] William Hanf, 'Degrees of Finitely Axiomatizable Theories', *Notices of the American Mathematical Society* **9** (1962) 127–128 (abstract).

[15] In 'Distributive Normal Forms' (reference 8 above).

[16] In 'Distributive Normal Forms and Deductive Interpolation', *Zeitschrift für mathematische Logik und Grundlagen der Mathematik* **10** (1964) 185–191. Furthermore, all relations of definability among the predicates of one's finitely axiomatizable theory can be found in this way.

[17] The nature of constituents in general and of inconsistent constituents in particular is briefly discussed in my paper 'Are Logical Truths Analytic?' (reference 11 above).

[18] All this is of course conditional on the assumption that even distribution or some similar weighing principle is used. It seems to me, however, that most natural principles of weighing will yield similar results.

[19] A technical argument to the effect that they cannot be has been given by Hilary Putnam in his contribution to the volume on Carnap in the Library of Living Philosophers, entitled 'Degree of Confirmation and Inductive Logic', in *The Philosophy of Rudolf Carnap* (ed. by P. A. Schilpp), Open Court, La Salle, Ill., 1963. Cf. also Hilary Putnam, 'Probability and Confirmation', in *Philosophy of Science Today* (ed. by S. Morgenbesser), Basic Books, New York 1967.

[20] The basic ideas of this justification go back to Frank Ramsey and Bruno de Finetti, whose fundamental papers are conveniently reprinted in H. E. Kyburg and H. E. Smokler (eds.), *Studies in Subjective Probability*, John Wiley, New York 1964. More recent treatments are contained in the following papers: A. Shimony, 'Coherence and the Axioms of Confirmation', *Journal of Symbolic Logic* **20** (1955) 1–28; R. S. Lehman, 'On Confirmation and Rational Betting', *ibid.*, 251–262; J. G. Kemeny, 'Fair Bets and Inductive Probabilities', *ibid.*, 263–273.

[21] Leonard J. Savage, 'Difficulties in the Theory of Personal Probability', *Philosophy of Science* **34** (1967) 305–310; cf. Ian Hacking, 'Slightly More Realistic Personal Probability', *ibid.*, 311–325.

[22] Further arguments to the same effect can be given. Consider, for instance, the disjunction of all consistent constituents of a given depth. At this depth, it appears to convey some information, for it excludes *prima facie* acceptable alternatives (viz. those specified by inconsistent but not trivially inconsistent constituents of this depth). When the depth of our constituents is increased, it gradually becomes clearer and clearer that our disjunction does not exclude any consistent alternatives and contains therefore no (depth) information.

[23] See my paper 'The Varieties of Information' (reference 4 above).
[24] With this section, cf. my papers in *Deskription, Analytizität und Existenz* (ed. by P. Weingartner), A. Pustet, Salzburg and Munich, 1966, especially 'Are Logical Truths Tautologies?' (pp. 213–233) and 'Kant Vindicated' (pp. 234–253).
[25] Cf. e.g. Carl G. Hempel, 'Geometry and Empirical Science', *American Mathematical Monthly* **52** (1946), reprinted in *Readings in Philosophical Analysis*, New York 1949, pp. 238–249 (especially p. 241); A. J. Ayer, *Language, Truth and Logic*, London 1936, p. 80.
[26] Ernst Mach, *Erkenntnis und Irrtum: Skizzen zur Psychologie der Forschung*, Leipzig 1905, pp. 300 and 302.
[27] See Gottlob Frege, *Die Grundlagen der Arithmetik*, Wilhelm Koebner, Breslau 1884 (§ 88, pp. 100–101).
[28] In addition to the papers mentioned earlier, see 'Kant and the Tradition of Analysis' in *Deskription, Analytizität und Existenz* (references 11 and 24 above), pp. 254–272.
[29] In my contribution to the symposium 'Are Mathematical Truths Synthetic A Priori?', *Journal of Philosophy* **65** (1968) 640–651.
[30] Here the familiar distinction between information and the amount of information seems to be needed. It does not make any essential difference for my purposes, however.
[31] If this is not the case, the very concept of meaning will lose much of its usefulness. Its connection with what someone means or can mean is broken, and so is all connection with the concept of intention, for intention and intended meaning are something which one actually has or at least can always in principle recall to one's mind. An entity that cannot be effectively found cannot play this role.
[32] I have discussed these difficulties in *Knowledge and Belief*, Cornell University Press, Ithaca, New York, 1962, chapter ii. The solution I proposed there does not involve the unrealistic idealizing assumption that people always believe (hope, know, remember, etc.) all the logical consequences of what they believe (hope, know, remember, etc.). However, it makes the limits of applicability of one's epistemic logic to what people actually believe unclear, and hence leaves a great deal to be desired.
[33] It might even be suggested that the failure of an operator to be invariant with respect to logical equivalence is an indication of the intentional (or, as some philosophers prefer to say, 'psychological') character expressed by this operator. This suggestion would certainly deserve further attention.

JAAKKO HINTIKKA AND RAIMO TUOMELA

TOWARDS A GENERAL THEORY OF AUXILIARY CONCEPTS AND DEFINABILITY IN FIRST-ORDER THEORIES

I. A THEORY OF AUXILIARY CONCEPTS IS NEEDED FOR METHODOLOGICAL PURPOSES

There is a fairly extensive philosophical and methodological literature dealing with the role of auxiliary ('theoretical') terms in scientific theories. In this literature, logical and foundational ideas play a surprisingly small role, despite the wealth of results concerning definability and related concepts which logicians have established. Virtually the only non-trivial result cited is Craig's (general) elimination theorem, and the purpose in bringing it up is all too often to deny its relevance. The possible importance of Craig's less general, but in certain respects more informative, interpolation theorem has not caught the fancy of philosophers of science, whose store of systematic logical results and techniques seems to be often rather restricted.

The aim of the present paper is to enrich methodologists' diet of concepts and viewpoints concerning the role and uses of auxiliary concepts. We shall try to accomplish this not so much by bringing well-known logical results and techniques to bear on the methodologists' problems. Many of these techniques seem too sophisticated for the relatively simple-minded methodological problems in question. Instead, we shall try to define a number of simpler concepts, some of them apparently new, that are applicable to first-order theories and to establish a few results that hold for them. Most of these results are quite elementary. That they are established only for first-order theories may be considered a restriction on their applicability to interesting methodological situations. However, it seems to us that some of these new concepts help put into a new and clearer perspective several aspects of the role which auxiliary concepts play or may play in scientific theories.

Not unexpectedly, some of the new concepts which seem useful in this area are closely related to the logical theory of definition. The later parts

J. Hintikka and P. Suppes (eds.), Information and Inference, 298–330.

of the present paper may therefore be viewed as an attempt to contribute to the theory of definition in first-order logic.

We shall speak of *auxiliary* rather than *theoretical* terms because most of what we say will be independent of that interpretation of these terms as non-observational terms which is likely to be uppermost in a methodologist's mind.

We shall assume some normal formulation of first-order logic, and consider on this basis some given applied first-order language. Our notation and other prerequisites will be announced as we proceed. We shall formulate our results for the case in which no individual constants and no identities are present. However, all our results can be proved also in their presence, unless otherwise stated. Functors will be assumed to have been replaced by predicates; hence we shall assume that they are not explicitly present at any stage of our discussion. We shall systematically employ the theory of distributive normal forms developed by Hintikka. For an exposition of the fundamentals of this theory, see Hintikka (1964 and 1965a).

Let us indicate briefly what these distributive normal forms look like when considered from the point of view of graph-theoretical semantics, and what role they play in our first-order language. For simplicity, we restrict our attention to the set of closed sentences of a first-order language with a fixed finite set of extralogical predicates (not all of which are monadic). The structure of this set can be described in terms of an (inverted) tree whose nodes are certain sentences called *constituents*. Each constituent belongs to certain level in the tree, and this level is called its *depth*. If one constituent can be reached from another by going down some branch or other, the former is said to be *subordinate* to the latter.

Each closed sentence s of our language is characterized by the number of layers of quantifiers it contains at its deepest; this parameter will be called the depth of s in our discussion. The constituents which are located at depth d in our tree structure all turn out to have the very same depth d according to this new definition of depth. Our two concepts of depth are thus essentially equivalent. Furthermore, each sentence s of depth d can be effectively transformed into a (finite) disjunction of constituents of the depth d. This is what gives the tree structure its significance for the whole language. The representation of a sentence s of depth d as a disjunction of constituents of this depth d is called the complete disjunctive *distribu-*

tive normal form of s. We denote it by $s^{(d)}$. Its representation as a disjunction of constituents of some greater depth $d+e$ is called its *expansion* at this depth. A constituent is equivalent to the disjunction of all its subordinate constituents of any fixed greater depth. Any two constituents of the same depth are mutually incompatible. Hence all the constituents compatible with a given one belong to branches passing through it.

This suffices to indicate the role of constituents and distributive normal forms in first-order languages. The structure of a constituent $C^{(d)}$ of depth d is given by a finite set of finite trees of length d. Each node of each of these trees describes what attributes an individual has and how it is related to all the individuals lower in the same branch. Each branch (from bottom to top) thus specifies a sequence of individuals having certain properties and related to each other in certain ways. These sequences are precisely those that exist in a world in which $C^{(d)}$ is true. The tree structure indicates the interrelations of the different kinds of sequences of individuals one may find in such a world.

II. A 'DIRECT' ELIMINATION OF AUXILIARY CONCEPTS

Suppose we are given a first-order theory, axiomatized by $t(\lambda+\mu)$, whose non-logical concepts can be dichotomized into proper concepts and auxiliary concepts. Let the class of the former be λ and the class of the latter μ. (Thus $\lambda \cap \mu = \emptyset$.)

A natural question to ask is what our theory $t(\lambda+\mu)$ says that is exclusively about the members of λ. The sense of 'saying' is not immediately clear here, however. Often, it is assumed that the only sensible answer is the set of deductive consequences of $t(\lambda+\mu)$ which contain no members of μ. But this is not the only relevant question or the only relevant answer. We must also ask, how efficiently and perspicuously the consequences exclusively about members of λ would be formulated by rival theories. We must ask how much more clearly λ and μ together (connected by $t(\lambda+\mu)$) enable us to talk about λ than λ alone does. Speaking in methodological terms, a wide variety of questions falling under the notion of deductive systematization belong here. For instance, problems concerning efficiency and rapidity in testing and confirming scientific theories are cases in point.

To discuss these questions, we must try to spell out what it is that

$t(\lambda+\mu)$ says explicitly about the members of λ. If this can be accomplished, questions of the sort just mentioned can be discussed by comparing the *unequal total* consequences concerning λ of different theories $t(\lambda+\mu)$ which have the *same explicit* consequences concerning λ.

A natural candidate for the explicit content of $t(\lambda+\mu)$ about λ is obtained by taking $t(\lambda+\mu)$ (here assumed to be the only non-logical axiom of the theory in question) and modifying it syntactically so as to put aside all the information that is not obviously about the members of λ. The result may be called the *direct reduct* of $t(\lambda+\mu)$ into the vocabulary of λ.

How the direct reduct $\mathrm{r}(s)$ of an arbitrary sentence in the terminology $\lambda+\mu$ can be defined recursively is perhaps seen most clearly if we drive all negation-signs as deeply into the sentences as they go – that is, so as to precede immediately an atomic sentence (or an identity when the symbol of identity is present). In this case, the recursive definition will be the following, assuming that u, u_1, u_2, ... are arbitrary sentences, that s is an arbitrary sentence which contains no occurrences of any member of μ, and that t is a negated or unnegated atomic sentence or identity which contains at least one occurrence of some member of μ or other:

$$\begin{aligned} \mathrm{r}(u_1 \,\&\, u_2) &= \mathrm{r}(u_1) \,\&\, \mathrm{r}(u_2) \\ \mathrm{r}(u_1 \vee u_2) &= \mathrm{r}(u_1) \vee \mathrm{r}(u_2) \\ \mathrm{r}((Ex)\,u) &= (Ex)\,\mathrm{r}(u) \\ \mathrm{r}((Ux)\,u) &= (Ux)\,\mathrm{r}(u) \\ \mathrm{r}(s) &= s \\ \mathrm{r}(t) &= \mathrm{T} \end{aligned}$$

These clauses are assumed to be applied starting from the outside. It is assumed that &, $\vee$, and $\sim$ are the only sentential connectives present. T is the symbol for truth (a propositional constant, one of our logical constants.) It is assumed that identically $(Ex)\mathrm{T}=\mathrm{T}=(Ux)\mathrm{T}$, $u \,\&\, \mathrm{T}=u$, $u \vee \mathrm{T}=\mathrm{T}$. It is easily seen that we can obtain the same result (after the preliminary dislocation of negation signs) simply by replacing each negated or unnegated atomic sentence (or an identity) which contains a member of μ by T. If we do not want to undertake the preliminary simplification, we can extend our recursive definition and put

$$\mathrm{r}(\sim u) = \sim \mathrm{r}(u).$$

provided that u is not atomic, and (instead of the last clause above)

$$\mathrm{r}(t) = \mathrm{T} \text{ or } \sim \mathrm{T}$$

depending on whether t occurs within the scope of an even or odd number of negation signs.

A more straightforward reduction is hard to think of. All clauses of the recursive definition are completely trivial except the last, and this only instructs us to disregard all information that depends on the members of μ. It might thus seem that $\mathrm{r}(u)$ is a workable explication of what u says explicitly about the members of λ; for, note that $\mathrm{r}(u)$ is always logically implied by u.

In the case of distributive normal forms, the direct reduction assumes an especially dramatic form. We simply omit from the normal form all atomic sentences containing members of μ (together with the connectives which thereby become idle).[1] (For further details, see Hintikka (1965a), pp. 57–59.)

However, it turns out that the direct reduction just defined is not after all a very natural and explicit explication of the idea we are trying to get at. Not only does our direct reduction fail to be invariant with respect to logical equivalence. It turns out that the direct reduct $\mathrm{r}(u)$ of a sentence u can even be altered by entirely trivial equivalence transformations – so trivial, indeed, that they can scarcely be said to make a difference to what s directly tells us concerning the members of λ. For instance, although

$$\mathrm{r}(u \mathbin{\&} (t \vee \sim t)) = \mathrm{r}(u) \mathbin{\&} ((\mathrm{r}(t) \vee \mathrm{r}(\sim t))) = \mathrm{r}(u).$$

as might be hoped, we also have

$$\mathrm{r}(u \vee (t \mathbin{\&} \sim t)) = \mathrm{r}(u) \vee (\mathrm{r}(t) \mathbin{\&} \mathrm{r}(\sim t)) = \mathrm{T}$$

which does not normally coincide with $\mathrm{r}(u)$. Likewise we have

$$\mathrm{r}((Ex)(u(x) \vee t(x)) \mathbin{\&} (Ux)(\sim u(x) \mathbin{\&} \sim t(x))) \\ = (Ux)\, \mathrm{r}(\sim u(x)),$$

which usually is not contradictory.

It seems to us that a natural way of improving our notion is suggested by what has already been said. Suppose, for instance, that all the members of $\lambda + \mu$ are one-place (monadic) predicates. Then all that $t(\lambda + \mu)$ can in any case say of the world is that the Q-predicates instantiated in it belong to a certain set of combinations of Q-predicates (defined by reference to $\lambda+\mu$). What these admissible combinations are is spelled out by the several constituents in the distributive normal form of $t(\lambda+\mu)$. One of these constituents might, for example, say that the following Q-predicates (and only they) are instantiated: $Q_1(x, \lambda, \mu)$, $Q_2(x, \lambda, \mu)$,..., $Q_j(x, \lambda, \mu)$. What this constituent says of the members of λ is completely obvious: it says that those and only those poorer Q-predicates are instantiated that are obtained as direct reducts of the Q_1,..., Q_j to the vocabulary of λ. Thus in a clear-cut and intuitive sense the direct reduct of the distributive normal form of $t(\lambda+\mu)$ spells out precisely what $t(\lambda+\mu)$ says of the members of λ. In other words, $\mathrm{r}(t^{(1)}(\lambda+\mu))$ (where $t^{(1)}(\lambda+\mu)$ is the distributive normal form of $t(\lambda+\mu)$) is in the monadic case an excellent explication of the idea the direct reduction r was supposed to define.

Somewhat surprisingly, for logical equivalents t' of $t(\lambda+\mu)$, $\mathrm{r}(t')$ need not be equivalent to $\mathrm{r}(t^{(1)}(\lambda+\mu))$. Thus it seems necessary to impose a certain amount of normalization on our idea of reduction. We define what will be called the *reduct proper* $\mathrm{tr}(s)$ of an arbitrary formula $s(\lambda+\mu)$ of depth d as the direct reduct $\mathrm{r}(s^{(d)}(\lambda+\mu))$ of the normal form $s^{(d)}(\lambda+\mu)$ of s at its own depth.

The naturalness, not to say inevitability, of this definition can be argued for in the general (relational) case in the same way it was argued for above in the monadic case. A formula s of depth d says that combinations of certain ramified sequences of individuals can be found (i.e. in a world in which s is true). (For an elaboration of this point, see Hintikka (1970).) As was pointed out above, each such combination of sequences is specified by a constituent $C^{(d)}(\lambda+\mu)$ of depth d (in the vocabulary $\lambda+\mu$). What $C^{(d)}(\lambda+\mu)$ says of the members of λ is of course that such interrelated sequences of individuals can be found as are obtained from those specified by $C^{(d)}(\lambda+\mu)$ by omitting all reference to the members of μ. That is to say, $\mathrm{r}(s^{(d)}(\lambda+\mu))=\mathrm{tr}(s)$ spells out what s says of the members of λ, just as we suggested.

Here a certain intended ambiguity is present. We must assume that all trivially inconsistent constituents are eliminated from the normal form $s^{(d)}(\lambda+\mu)$ of s. However, sometimes it is useful to assume that *all* inconsistent (and not just all trivially inconsistent) constituents have been omitted. In fact, we shall in the sequel normally assume the latter. However, the main difference here is only the question of what can be done recursively (effectively), as will be indicated in Section V below (and elsewhere).

It is clear that $\mathrm{tr}(s)$ is less sensitive to equivalence-preserving transformations than $\mathrm{r}(s)$ is, but it is not invariant under all such. For instance, $\mathrm{tr}(s)$ is in general sensitive to tautological introduction of quantified subsentences, and thus to increase in depth. Hence normally we have, where d is the depth of s,

$$\mathrm{tr}(s) = \mathrm{tr}(s^{(d)}) \neq \mathrm{tr}(s^{(d+e)}).$$

Whenever we want to indicate explicitly the non-logical vocabulary which is preserved in the direct reduction, we shall use a more perspicuous notation, writing $u[\lambda]$ for $\mathrm{r}(u)$. When we want to indicate the non-logical vocabulary used in a normal form, say in the normal form of u, we shall signal it by writing this normal form (of depth d, say) as $u^{(d)}(\lambda+\mu)$ (where the vocabulary in question consists of the members of $\lambda+\mu$).

The notation $u[\lambda]$ will be used also for the reduct proper, hopefully without causing confusion.

III. WHAT ARE THE ADVANTAGES OBTAINED FROM INTRODUCING AUXILIARY CONCEPTS INTO A THEORY?

The primary (though admittedly partial) answer which we propose to offer to this question is the following: When auxiliary terms (new non-logical constants, say the members of μ) are introduced into a statement s (over and above its old non-logical constants, the set of which is λ), we can formulate statements t whose *direct import* for the members of λ is the same as that of s (that is, $\mathrm{tr}(t)$ is equivalent to s) but whose total deductive consequences, even as far as λ only is concerned, are greater than those of s (that is, there are statements whose non-logical constants all belong to λ which are logically implied by t but not by s). The methodological interest of this advantage depends on the methodological status of our reduction proper. Some comments on this subject will be made below (in Sections VIII and IX).

From the theory of distributive normal forms it is known that the greater logical force of t as compared to s can be brought out by expanding t into deeper and deeper normal forms $t^{(d+e)}(\lambda+\mu)$. (See Hintikka (1965a and 1964).) The corresponding reducts (that is, the sentences

$\mathrm{tr}(t^{(d+e)}(\lambda+\mu))$, which may be written simply $t^{(d+e)}(\lambda+\mu)[\lambda]$), will then constitute a basis for the theory which consists of those consequences of t whose non-logical vocabulary belongs to λ. The situation can therefore be illustrated by means of Figure 1.

$$\begin{array}{ccccccc} t^{(d)}(\lambda+\mu) & \leftrightarrow & t^{(d+1)}(\lambda+\mu) & \leftrightarrow \cdots \leftrightarrow & t^{(d+e)}(\lambda+\mu) & \leftarrow \cdots \\ \downarrow & & \downarrow & & \downarrow & \\ (t^{(d)}(\lambda+\mu))[\lambda] & \leftarrow & (t^{(d+1)}(\lambda+\mu))[\lambda] & \leftarrow \cdots \leftarrow & (t^{(d+e)}(\lambda+\mu))[\lambda] & \leftarrow \cdots \end{array}$$

Fig. 1.

Here double arrows indicate logical equivalences and ordinary arrows (vertical and horizontal) indicate logical implications. Single arrows cannot in general be replaced by double ones. It is known that each logical consequence of t (i.e. of $t^{(d)}(\lambda+\mu)$) whose non-logical vocabulary belongs to λ is implied by some member of the lower row in Figure 1. (Cf. Section IV below.)

In Figure 1 $(t^{(d)}(\lambda+\mu))[\lambda]$ is of course assumed to be identical with $s^{(d)}(\lambda)$ and therefore equivalent with s, that statement we tried to 'improve' by introducing the members of μ as auxiliary concepts. Notice also that t is not uniquely determined by s and μ; there usually are a great many statements (whose non-logical vocabulary is $\lambda+\mu$) which can play the role of t with respect to a given s in Figure 1. This indicates that the construction of a richer theory is not a mechanical procedure but depends on the skill and insight (and the luck) of the investigator. It also means that the theory axiomatized by $t^{(d)}(\lambda+\mu)$ is usually not a conservative (i.e. non-creative) extension of the theory axiomatized by $s^{(d)}(\lambda)=(t^{(d)}(\lambda+\mu))[\lambda]$, although it is a conservative extension of the theory based upon the whole of the lower row.

IV. THE LOGIC OF REDUCTION

Some further comments on our Figure 1 above are in order. The equivalences in the upper row are trivial. The reverse implications in the lower row of the figure are also obvious. The implications between the two rows are due to the omission lemma of Hintikka (1965a). The fact that the members of the lower row axiomatize the set of all those consequences of $t^{(d)}(\lambda+\mu)$ which contain members of λ only follows from the following separation lemma for first-order predicate logic with or without identity (Hintikka (1964), Craig (1960)):

Two first-order sentences $t = t^{(d_t)}(\lambda + \mu)$ and $s = s^{(d_s)}(\lambda + \eta)$ (with depths and extralogical constants as indicated and assuming that $\mu \cap \lambda = \mu \cap \eta = \lambda \cap \eta = \emptyset$) are incompatible if and only if for some d_0 their expansions at depth d_0 are separated with respect to λ.

We say that two sentences t and s are separated with respect to λ at depth d_0 if for each arbitrarily chosen pair $C_t(\lambda + \mu)$, $C_s(\lambda + \eta)$ of their constituents the reducts $(C_t(\lambda + \mu))[\lambda]$ and $(C_s(\lambda + \eta))[\lambda]$ are (not merely notationally) different at depth d_0.

Note for futher reference that the deduction of a formula in the vocabulary of λ depends in only one respect of the upper row – namely, the separation depth d_0. The expansion process in the upper row does not depend on the formula in the vocabulary of λ to be deduced. This makes it more systematic than any other known method of building the subbasis.

V. WHAT CAN BE DONE EFFECTIVELY

What we have said so far is ambiguous in an important but innocuous way. For, when speaking of the distributive normal form, we usually assume that this form can always be reached effectively (recursively).

We will assume in the sequel that all trivially inconsistent constituents have been located and eliminated. This can of course be accomplished recursively. However, often it is convenient to assume, by way of idealization, that *all* the inconsistent constituents, not just trivially inconsistent ones, have been omitted from the normal form. How is Figure 1 to be interpreted on these two different assumptions?

Since the members of the upper row are simply the increasingly deeper normal forms of one and the same sentence $t^{(d)}(\lambda + \mu)$ and since it does not affect the logical strength of one's expressions to omit inconsistent constituents (whether inconsistent trivially or non-trivially) from these normal forms, the members of the upper row are all logically equivalent and have precisely the same logical power no matter whether all inconsistent constituents or merely all trivially inconsistent constituents are assumed to have been eliminated from them.

However, the lower row will be different in the two cases, for the reduct proper of a non-trivially inconsistent constituent need not be inconsistent.

Hence the disjunctions of the lower row will in general be longer (i.e. logically weaker) if only trivially inconsistent constituents are assumed to have been eliminated from the normal forms of the upper row.

This makes no difference in the long run, for it follows from the completeness result established in Hintikka (1965a) that all the subordinate constituents of each inconsistent constituent of the upper row will be trivially inconsistent at a sufficiently great depth, that is if we move far enough right. At this point the corresponding reducts therefore disappear also from the lower row.

What this means is that although the members of the lower row are generally weaker if only the trivially inconsistent constituents are eliminated, in the long run this makes no difference, and amounts only to a slower convergence in the lower row. Each consequence of the sentences of the upper row which contains only members of λ is still implied by some members of the lower row, although this member is usually much farther right if only trivially inconsistent constituents are eliminated from our normal forms. The sentences of the lower row thus constitute in either case a basis for that subtheory which consists of those consequences of $t^{(d)}(\lambda+\mu)$ which only contain members of λ. When only trivially inconsistent constituents are assumed to be eliminated, we obtain a recursive criterion of membership in the basis and hence a set of axioms for the subtheory.

It may of course happen that the subtheory is finitely axiomatizable. Then (and only then) the implications of the lower row in Figure 1 will all turn into equivalences from some point on, however the figure is interpreted. If only trivially inconsistent constituents are assumed eliminated, the critical point of course normally comes much later. However, even on this interpretation such a point will always come in the case of finite axiomatizability of the subtheory.

That the subtheory cannot be finitely axiomatized if the implications of the lower row do not eventually turn into equivalences follows from a familiar result (see for example Robinson (1963) p. 36) according to which the set of members of the lower row cannot then be equivalent to any single sentence. The completeness result mentioned above told us that they form an axiom system for the subtheory; so this subtheory cannot in general be finitely axiomatized.

VI. HOW ARE WE TO MEASURE THE GAIN OBTAINED BY USING AUXILIARY CONCEPTS?

Natural ways of measuring the gain which one obtains by introducing auxiliary concepts, that is, by moving from $(t^{(d)}(\lambda+\mu))[\lambda]$ to $t^{(d)}(\lambda+\mu)$ (see Figure 1) are obtained if we have a strictly positive probability measure defined for all the consistent closed sentences (sentences without free variables or constants) of our (fixed) first-order language. Let us assume that such a measure p is defined; in terms of it, various measures of information can be defined, for example $\text{cont}(s)=1-p(s)$.

Using such measures of information the gain that accrues from the use of auxiliary concepts can be expressed in a natural way. Using the terms used in Figure 1, we can, for instance, identify this gain with

$$(1)\qquad \lim_{e\to\infty} \text{cont}\left(\left(t^{(d+e)}(\lambda+\mu)\right)[\lambda]\right) - \text{cont}\left(\left(t^{(d)}(\lambda+\mu)\right)[\lambda]\right).$$

Here the second term equals the informative content (substantial information) of the original sentence supposedly 'improved' by introducing the auxiliary concepts in μ. From (1) we can see how the gain which the introduction of auxiliary concepts yields can be understood as an increase in information.

What (1) gives us is the additional power of the subtheory obtained from $t^{(d)}(\lambda+\mu)$ over and above the logical power of $(t^{(d)}(\lambda+\mu))[\lambda]$. Since this logical power comes from an involvement of the new concepts with the old ones in the richer new theory, whose normal form is just $t^{(d)}(\lambda+\mu)$, (1) can also be thought of as a measure of this involvement. If the new concepts are introduced, but not mixed with the old ones in any way, that is, if the involvement is zero, then no gain of the kind expressed by (1) is obtained.

VII. ANOTHER KIND OF GAIN

This does not yet exhaust the possible gains offered by auxiliary concepts. In what follows we shall call the gain measured by (1) *gain of the first kind*. We may obtain another kind of advantage by going from the poorer theory (axiomatized by the sentence s, whose distributive normal form is $(t^{(d)}(\lambda+\mu))[\lambda]$) to the richer one (axiomatized by $t^{(d)}(\lambda+\mu)$). Even when there is no sentence equivalent to s of a smaller depth than the depth d

of s, as we shall assume in what follows that there is not, there may very well be shallower sentences equivalent to $t^{(d)}(\lambda+\mu)$. Let the shallowest sentence of this kind be of depth $c \leqslant d$. Then the situation can be represented (after all sentences are rendered into their distributive normal forms) by Figure 2, which amplifies Figure 1.

$$\begin{array}{ccccc}
t^{(c)}(\lambda+\mu) & \leftrightarrow & t^{(c+1)}(\lambda+\mu) & \leftrightarrow\cdots\leftrightarrow & t^{(d)}(\lambda+\mu) \\
\downarrow & & \downarrow & & \downarrow \\
(t^{(c)}(\lambda+\mu))[\lambda] & \leftarrow & (t^{(c+1)}(\lambda+\mu))[\lambda] & \leftarrow\cdots\leftarrow & (t^{(d)}(\lambda+\mu))[\lambda]
\end{array}$$

$$\begin{array}{ccccc}
\leftrightarrow & t^{(d+1)}(\lambda+\mu) & \leftrightarrow\cdots\leftrightarrow & t^{(d+e)}(\lambda+\mu) & \leftrightarrow\cdots \\
 & \downarrow & & \downarrow & \\
\leftarrow & (t^{(d+1)}(\lambda+\mu))[\lambda] & \leftarrow\cdots\leftarrow & (t^{(d+e)}(\lambda+\mu))[\lambda] & \leftarrow\cdots
\end{array}$$

Fig. 2.

The notation is here the same as in Figure 1.

In analogy with the gain of the first kind, the gain due to pushing down the depth of sentences (by means of auxiliary concepts) may naturally be measured by

$$(2) \qquad \text{cont}\left((t^{(d)}(\lambda+\mu))[\lambda]\right) - \text{cont}\left((t^{(c)}(\lambda+\mu))[\lambda]\right).$$

This will be called the *gain of the second kind* from using auxiliary concepts in $t^{(d)}(\lambda+\mu)$. The *total gain* is the sum of the gains of the first and the second kinds. It depends of course crucially on the richer theory we are dealing with, that is, on the particular sentence whose distributive normal form is $t^{(d)}(\lambda+\mu)$. As pointed out earlier, this sentence is not uniquely determined by s or $s^{(d)} = (t^{(d)}(\lambda+\mu))[\lambda]$.

As a rougher measure of a gain of the second kind we could also use the difference $d-c$ (but then, of course, the total gain cannot be defined as before).

VIII. METHODOLOGICAL PERSPECTIVES ON THE USE OF AUXILIARY CONCEPTS

At this stage, a skeptical reader may very well still doubt the reality of the gains measured in the last few sections, and supposed to be provided by auxiliary concepts. The definitions of the various kinds of so-called gains may be formally all right, he may aver, but are they worth anything?

In both kinds of gains, and especially in a gain of the second kind, the crucial advantage which auxiliary concepts offer us is essentially due to the possibility of getting along with shallower axioms (and other assumptions). This is patent in the case of a gain of the second kind. But in the case of a gain of the first kind, too, we can in a sense say that the essential advantage is that we can get along with the sentence t (or with its normal form $t^{(d)}(\lambda+\mu)$) of the fixed depth d so that we can obtain from it various consequences concerning the members of λ which not only did not follow from s (that is $(t^{(d)}(\lambda+\mu))[\lambda]$) alone but also could not be formulated, in many cases, without resort to much deeper sentences.

What is the intuitive meaning of this 'reduction' of the depth of one's axioms? Elsewhere (Hintikka (1965b and 1966)) Hintikka has suggested (with a few minor qualifications which we can here disregard) that the intuitive meaning of the concept of depth of a sentence s is the number of individuals considered together (in their relation to each other) in the deepest part of s, of course over and above the indidivuals referred to by the individual constants.

Now this idea of 'considering so many individuals together' has a clear methodological significance. It is not much of an exaggeration to say that what an experimental scientist often – perhaps typically – does is to study the interrelations of a relatively small number of entities involved in an experimental situation which he can control. By discovering their interdependencies he hopes to obtain the means of theoretically mastering other, often much more complicated, situations. (See, for example, Newton's classic description of his 'method of analysis' in the General Corollary to his *Opticks*.)

To the extent that this enterprise can be described (even approximately) in first-order terms, it seems to amount to an attempt to discover relatively shallow (in the technical sense) laws whose deeper (technical sense again) logical consequences nevertheless match the behavior of more complicated situations (that is, situations involving more individuals). The (minimal) depth of the non-logical axioms of an empirical theory is thus roughly proportional to the complexity of the empirical situations in which this theory can come into the play and in which it therefore can (in principle) be tested and perhaps corroborated. It is obvious that *ceteris paribus* it is highly desirable to have this number as small as possible. In fact, it is clearly one of the main factors that seem to influence our ideas of the simplicity and easy testability of an empirical theory.

Hence if this complexity of a theory can be reduced by introducing new concepts, such reduction represents (*ceteris paribus* again) a definite and often very real methodological gain. It may not be too far-fetched to speak of a kind of *ontological economy* of a theory whose depth is small, though this sense has nothing to do with the size of the domain needed in order for the theory to be true. We might say, and not entirely as a pun, that if to be is to be a value of a bound variable, then theories with few interrelated (nested) bound variables exhibit greater ontological economy than those with many layers of quantifiers.

This may be highlighted by consideration of a hypothetical special case in which at any one time we can (perhaps because of restricted possibilities of experimentation) observe at most d individuals in their relations to one another. Then we have no possibilities of telling whether such hypotheses as can be distinguished from each other only by means of sentences of a depth $>d$ are true or not, unless we introduce new (auxiliary) concepts. What we have seen is that by so doing we can in fact formulate hypotheses in which only d individuals are considered in their relation to each other and from which consequences nevertheless ensue concerning the old concepts which cannot be formulated otherwise except by going to depths greater than d. Thus the auxiliary concepts may bring essentially deeper hypotheses within the purview of testing and corroboration even when this testing is of a restricted complexity.

From these remarks it perhaps also appears what our discussion in the first place serves to elucidate. Our remarks are not geared to the special case in which the new (auxiliary) terms express theoretical (unobservable) concepts. (The only respect in which the new concepts are treated as inferior is that we are only interested in those consequences of the richer theory which are in terms of the *old* concepts.) What we are trying to understand are the advantages of introducing new concepts of any kind, observable or unobservable.

Many philosophers and scientists (among them Nagel, Campbell, Mach; see, for example, Nagel (1957)) seem to think that 'pure' theories (sentences containing auxiliary concepts only) play a special role in systems containing such terms. As we saw at the end of Section IV, purely theoretical statements in a fixed and 'finished' richer theory have a no more important function within deductive systematization than do 'correspondence rules' (statements where members of both λ and μ occur). (A similar remark is to be found in Craig (1960).) What is crucial here is as much the way old and new concepts are interwoven (involved with each other) in the richer theory as the assumptions concerning the auxiliary concepts alone.

IX. COMMENTS AND ILLUSTRATIONS

As was mentioned above, we do not attempt here anything like an exhaustive answer to the question: Why are theoretical (auxiliary) concepts needed in science? First, we have restricted ourselves mainly to the deductive systematization of non-auxiliary statements; thus we do not here investigate the role of auxiliary concepts within inductive systematization, the context in which Hempel (among others) thinks that their indispensability can best be seen. But even within deductive systematization there are many other important properties of theories which we will not directly deal with here. For example, it is claimed that theoretical concepts may in some pragmatic sense give us deeper understanding and better explanation of the phenomenon under investigation than do statements in the vocabulary of λ only. We certainly agree with this. Moreover, it may (or may not) be the case that theories using auxiliary concepts are heuristically more fruitful, more manageable and suggestive, and also simpler (in some sense) than purely observational statements. Again, there may be strong ontological reasons for using auxiliary concepts; we may be interested in theoretical concepts like fields or unconscious wishes, for example, because we believe that there are such entitities and that they can be studied for their own sake.

Despite the restricted scope of our discussion, we believe that we have already brought out something interesting. There are two main aims which auxiliary terms are typically supposed to serve. They are (observational) *richness* and *economy*. However, in the literature of philosophy of science we find next to no insights as to *how* they serve these purposes, and not even much indication that they in fact succeed in doing so.

Now it seems to us obvious that these two aims can be partly identified with gains of the first and of the second kind, respectively. Our analysis shows that these aims are achieved (in first-order theories), and to some extent also how they are achieved. (That a gain of the first kind means increased observational richness is obvious, since it means a greater wealth of observational consequences. The connection of a gain of the second kind with the notion of economy was already commented on above.) In particular, we have seen that by using auxiliary (theoretical) concepts we obtain (in the first-order case) something that Barker (1957) for one claims to be unattainable:

> By introducing theoretical expressions one cannot obtain a system which is richer than would otherwise be possible with regard to the observational statements derivable from it (richer in its 'cash-value') or more economical than would otherwise be possible with regard to its array of primitive terms. Whatever degree of richness and economy may be desired, these always can be secured at least as well through the adoption of a system containing no theoretical expressions whatever (p. 147).

We have arrived at a diametrically opposite conclusion by using essentially the same basic logical result due to Craig as Barker is here commenting on, in the case of richness as well as in that of economy.

Let us try to illustrate the usefulness of auxiliary concepts. First, we consider an artificial example best described by reference to elementary arithmetic. Suppose that we are in a position to make observations concerning a single monadic property P of individuals in an ordering. The ordering is given by a two-place relation S to be interpreted as denoting immediate succession. The first-order axioms of S are given in a first-order language with identity as follows:

A1. $(Ux)(Ey)\,S(x, y);$

A2. $(Ux)(Uy)(S(x, y) \supset \sim S(y, x));$

A3. $(Ux)(Uy)(Uz)(S(x, y)\ \&\ S(x, z) \supset y = z);$

A4. $(Ux)(Uy)(Uz)(S(y, x)\ \&\ S(z, x) \supset z = y);$

A5. $(Ex)(Uy) \sim S(y, x);$

A6. $(Ux)(Uz)(\sim (Ey)\,S(y, x)\ \&\ \sim (Ey)\,S(y, z) \supset x = z);$

These axioms describe an immediate succession-ordering with a unique first element but no last element. The immediate and distant (finite) successors of the first element can be correlated with the natural numbers 2, 3, 4, ⋯ .

Let us assume that a theorist can 'observe' which individuals have the property P. Assume that, starting from the beginning, individuals no. 2, 3, 5, 7, 11, and 13 are found to have this property out of the 13 first individuals. How could one try to 'systematize' or to 'explain' these observations? How can one obtain predictions concerning new individuals? It is intuitively obvious what the tempting guess here is, *viz.* that the predicate P belongs to those and only those distant and immediate successors of the first individual which are correlated with prime numbers. It is also clear that this conjecture cannot be expressed in finite terms solely by means of P and the successor relation. Hence we have here a clear-cut (though artificial) example of the need to resort to 'auxiliary' concepts.

One thing our imaginary theoretician could do here is to expand our vocabulary so as to be able to formulate a fragment of elementary number theory. To our observational theory (formulated in the vocabulary $\lambda = \{P, S\}$) he adds a set μ consisting of two new three-place predicates A and M representing the addition and multiplication operations, and a few new axioms to obtain a richer theory $t(\{P, S\} + \{A, M\})$. Four of these axioms will state that A and M are functions defined everywhere, and the others can be modified versions of the usual recursion equations for addition and multiplication (let us call them A7–A14) plus the following:

A15. $(Ux)(P(x) \equiv (Uy)(Uz)((y \cdot z = x \supset y = 1 \vee z = 1) \,\&\, x \neq 1))$

Here we have of course used the dot as a shorthand which can be eliminated in favor of the relation M. By means of axiom A11 we predict of for any given future individual in our 'observational' sequence whether it has the property P or not.

Our richer theory $t(\{P, S\} + \{A, M\})$ with its fifteen axioms (the deepest has depth 5) has thus been specified clearly enough. Here we will not undertake the laborious task of writing it out in its distributive normal form, nor is this needed for our purposes. Neither can the axioms of the 'observational' theory axiomatized by the successive reducts

$$(t^{(d)}(\{P, S\} + \{A, M\}))\,[\{P, S\}]\ (d = 5, 6, 7, 8, \ldots)$$

be easily transformed into their distributive normal forms. However, we can see what their essential features would be like. In order to distinguish individual no. d from the others (so as to be able to say whether it has predicate P or not) by means of the successor relation alone we need $d+1$ layers of quantifiers. Thus an essential part of what $(t^{(d)}(\{P, S\} + \{A, M\}))$

$[\{P, S\}]$ says is to list which individuals among those numbered 1, 2,..., $d-1$ are P and which ones are not. In brief, the poorer theory $(t^{(d)}(\{P, S\} + \{A, M\}))\,[\{P, S\}]$ $(d=5, 6, 7,\ldots)$ specifies which individuals have P, by enumerating them one by one, while the richer theory presents us with definite law governing the distribution of P.

It is thus striking how the introduction of auxiliary concepts can yield some real gain. The richer theory has deductive consequences concerning P and S beyond those of the reducts of any of its successive expansions. If one starts from any of these as one's poorer theory, a gain of the first kind is obtained. (For instance, in the 'experimental' situation described in our fictional example, a natural starting-point would have been $(t^{(14)}(\{P, S\}+\{A, M\}))\,[\{P, S\}]$. Furthermore, starting from any reduct with depth >5, a gain in economy (gain of the second kind) is obtained also, in that the richer theory can be axiomatized with fewer layers of quantifiers. It seems to us that the feeling of much greater 'insight into' and 'appreciation of' the situation that the richer theory seems to give us here is partly a reflection of these clearly definable features of the underlying logical situation.

Although the confirmation theory of relational (polyadic) generalizations is notoriously underdeveloped, it is clear that the poorer theory (lower row) in which we present our general hypothesis so to speak by enumerating the different cases cannot in any case be highly confirmed by finite evidence, for such evidence cannot distinguish between this theory and a great many competing theories. In contrast, any adequate confirmation theory presumably ought to show how the richer theory is confirmed by its instances. Thus the gain obtained by enriching the conceptual basis may in the last analysis turn out to be as closely connected with the inductive properties of theories as with their purely deductive ones.

It is possible to axiomatize some standard version of set theory in first-order logic. As a great part of mathematics can be developed within set theory we can in principle formalize almost all actual scientific theories, even if the task may seem quite hopeless to carry out in practice. Here it must suffice to refer briefly to only one simple illustration taken from actual science. It is offered to us by certain sociometric investigations (and attempts to explain the resulting sociograms) conducted by Sherif and Sherif. In these investigations contacts between individuals (a two-place relation) are recorded to form a sociogram, together with certain further properties of these individuals. Then new concepts incorporated in certain new axioms are introduced to explain these observations rather in the manner suggested above (see Sherif and Sherif (1953)). However, in this connection we cannot undertake a detailed methodological investigation of these sociometric studies.

X. WHAT IS GAINED BY INTRODUCING EXPLICITLY DEFINABLE CONCEPTS?

A special but important case of the introduction of new concepts is by explicit definition. Here the richer theory is axiomatized simply by $s(\lambda)$ together with the explicit definitions of the members of μ in terms of those of λ. If none of the explicit definitions has a depth greater than the depth d of s, our richer theory $t^{(d)}(\lambda+\mu)$ may be simply the distributive normal form of the conjunction of $s(\lambda)$ and all these definitions. Let us consider this possibility first. How does it fit into our discussion?

In virtue of the non-creative character of explicit definitions, we have for each sentence $v(\lambda)$ whose non-logical constants are all in λ:

$$\text{(3)} \qquad \vdash t^{(d)}(\lambda+\mu) \supset v(\lambda) \quad \text{iff} \quad \vdash s^{(d)}(\lambda) \supset v(\lambda)$$

Here $s^{(d)}(\lambda)$ is of course by assumption $(t^{(d)}(\lambda+\mu))[\lambda]$. This means that all the implications of the lower row in Figure 1 are equivalences, and the gain of the first kind is zero.

Does this mean that the introduction of new concepts by explicit definition is pointless? The answer is no, for it is perfectly possible that we obtain a gain of the second kind (and hence a non-zero total gain). In fact, there are interesting connections between this gain and the nature of the explicit definitions that mark the transition from s to the richer theory. These definitions are of the form

$$\text{(4)} \qquad (Ux_1)(Ux_2)\dots(Ux_k)(P(x_1, x_2, \dots, x_k) \equiv v(\lambda, x_1, x_2, \dots, x_k)$$

where the definiendum $P \in \mu$ is a k-place predicate ($k \geqslant 1$); and where v(= here, the definiens) is a sentence whose extralogical constants are all in λ, and whose free variables are $x_1, x_2, \dots, x_k$. Likewise, assuming that identity is present, the definition of an individual constant $a \in \mu$ will be of the form

(5) $(Ux)((x = a) \equiv v(\lambda, x))$

where it is assumed that

$$\vdash (s \supset (Ux)(Uy)(v(\lambda, x) \,\&\, v(\lambda, y) \supset x = y)).$$

In the sequel we shall assume that both in (4) and in (5) the definiens v has been brought to a form in which its depth is minimal (i.e. that there is no sentence equivalent to v with the same constants and free variables).

The simplest of the connections mentioned above is probably the inequality

(6) $d - c \leqslant m$

where m is the maximum of the depths of the definienses of the definitions of the members of μ in terms of those of λ.

To show this, let us assume that in $t^{(c)}(\lambda+\mu)$ (see Figure 2) each member of μ is replaced by its definiens. The depth of the resulting sentence is at most $c+m$. This resulting sentence is clearly equivalent with the original s. Since it was assumed that s has no equivalents of smaller depth, we must have $c+m \geqslant d$.

As a corollary it follows that if each definiens of a member of μ in terms of those of λ is of depth zero ($m=0$), as may very well be the case, $c=d$ and no gain of the second kind (and no total gain) is obtained. Such definitions may naturally be dubbed (to give a traditional term a modern though not completely new use) *nominal* definitions. No gain whatsoever accrues from such definitions. A gain can be expected from the introduction of explicit definitions only if at least one of these definitions presents a genuine quantificational analysis of the definiendum, in the sense that its definiens has one or more (irreducible) layers of quantifiers ($m \geqslant 1$). Such definitions seem to correspond to the old idea of 'real' definitions. (This connection between modern and traditional ideas might merit some further attention.) It may be pointed out that our distinction between nominal and real definitions coincides with Frege's distinction between trivial and non-trivial definitions in *Grundlagen der Arithmetik* pp. 100–101.

Furthermore, from (6) it is seen that the deeper the analyses are (deeper both in the literal and in the metaphorical sense) which our explicit definitions give of their definienda (in the sense that the definienses have greater depths), the greater information gain is (by and large) likely to result from these definitions.

These observations are at once generalized to serial definitions (each new concept of a sequence is defined explicitly in terms of some of the earlier ones).

If the depth of the conjunction t of the explicit definitions of the members of μ in terms of those of λ is higher than that of s (of depth d), no gain is obtained unless the conjunction s & t (of depth $e>d$) has an equivalent of depth less than d. In that case, the gain can be measured as above.

XI. HOW EXPLICIT DEFINABILITY IS MANIFESTED

The relation of distributive normal forms to definability requires a few additional comments. For simplicity, consider the case of just one new concept, which is assumed to be a monadic predicate P. Consider a constituent $C_i^{(d+e)}$ occurring somewhere in the upper row of Figure 2, that is in one of the normal forms $t^{(d+e)}(\lambda+\mu)=t^{(d+e)}(\lambda+\{P\})$. If all occurrences of P at depths greater than one are omitted, the result (which is of course implied by $C_i^{(d+e)}$) will be of the following form:

(7)
$$\begin{aligned}&(Ex)(Px \,\&\, Ct_{i1}(\lambda, x)) \,\&\, (Ex)(Px \,\&\, Ct_{i2}(\lambda, x))\\ &\&\ldots \&\, (Ex)(\sim Px \,\&\, Ct'_{i1}(\lambda, x)) \,\&\, (Ex)(\sim Px \,\&\, Ct'_{i2}(\lambda, x))\\ &\&\ldots \&\, ((Ux)[(Px \,\&\, Ct_{i1}(\lambda, x)) \vee (Px \,\&\, Ct_{i2}(\lambda, x))\\ &\vee\cdots\vee(\sim Px \,\&\, Ct'_{i1}(\lambda, x)) \vee (\sim Px \,\&\, Ct'_{i2}(\lambda, x)) \vee\ldots]\end{aligned}$$

Here the depth of all the attributive constituents Ct, Ct' is of course $d+e-1$. They contain only members of λ plus the free variable x. For simplicity, let us put $\{Ct_{i1}, Ct_{i2},\ldots\}=\alpha_i$, $\{Ct'_{i1}, Ct'_{i2},\ldots\}=\beta_i$ and assume α_i and β_i are non-empty.

The crucial uncertainty factor is then the intersection

(8)
$$\begin{aligned}\alpha_i\cap\beta_i &= \{Ct_{i1}, Ct_{i2},\ldots\}\cap\{Ct'_{i1}, Ct'_{i2},\ldots\}\\ &= \{Ct^*_{i1}, Ct^*_{i2},\ldots\} = \gamma_i\end{aligned}$$

If γ_i is empty, $C_i^{(d+e)}$ implies an explicit definition of P in terms of the members of λ. For instance, we can put:

(9)
$$(Ux)(Px \equiv (Ct_{i1}(\lambda, x) \vee Ct_{i2}(\lambda, x) \vee\ldots))$$

In fact, we can add to the disjunction in (9) any attributive constituent $Ct(\lambda, x)$ which does not occur in $\{Ct'_{i1}, Ct'_{i2}, \ldots\}$.

From (9) it is seen that the resulting definition has a definiens of depth $d+e-1$.

Now $C_i^{(d+e)}$ is normally much stronger than our richer theory $t^{(d)}(\lambda+\mu)=t^{(d)}(\lambda+\{P\})$ and hence the definability of P on the basis of the former does not imply definability on the basis of the latter. But assuming that $t^{(d)}(\lambda+\{P\})$ does imply an explicit definition of P in terms of the members of λ, we can see that a situation of the kind we just considered will occur in *all* the constituents of some sufficiently deep $t^{(d+e)}(\lambda+\{P\})$.

In order to see this, suppose that there is an explicit definition v of P of depth $d+e$. (If a definition of a depth lower than d exists, then so does one of the depth d, and we can hence assume that $e \geqslant 0$). Then we can form the conjunction v & $t^{(d+e)}(\lambda+\{P\})$ which is equivalent to a disjunction of conjunctions of the form

$$(10) \qquad C_i^{(d+e)} \,\&\, (Ux)\,[Px \equiv (Ct_1(\lambda, x) \vee Ct_2(\lambda, x) \vee \ldots)]$$

where $C_i^{(d+e)}$ is as before, and where the Ct_i are of depth $d+e-1$. Notice that $v=(Ux)\,[Px\equiv(Ct_i(\lambda, x) \vee Ct_2(\lambda, x) \vee \ldots)]$ is of course independent of i. By way of abbreviation, let us put

$$\{Ct_1, Ct_2, \ldots\} = \delta$$

Since $C_i^{(d+e)}$ implies (7), it is easily seen that (10) is inconsistent unless, for this i,

$$(11) \qquad \text{(a) } \delta \supseteq \alpha_i \qquad \text{(b) } \delta \cap \beta_i = \emptyset$$

But this implies that γ_i is empty for each i, as we required. For were this not the case with a disjunct, it would be eliminated when (10) is converted into its distributive normal form, and hence this normal form will consist solely constituents which do make γ_i empty. But since v(= the second conjunct of (10)) is implied by $t^{(d)}(\lambda+\{P\})$ at the depth $d+e$, the distributive normal form of their conjunction must be identical with $t^{(d+e)}(\lambda+\{P\})$, assuming that all inconsistent constituents are eliminated.

Thus for explicit definability we must have not only $\alpha_i \cap \beta_i = \emptyset$ for each i, but a *uniform* separation of α_i and β_i for all i in the sense that (11) holds for any i and a *constant* set δ.

Thus the definability of P will eventually be betrayed by the fact that all the consistent constituents of $t^{(d+e)}(\lambda+\{P\})$ satisfy (11), and thereby explicitly show the definability of P. If only trivially inconsistent constituents are assumed to be eliminated in the disjunctions $t^{(d+e)}(\lambda+\{P\})$, the same statement will hold, although the definability will usually manifest itself in this way much later.

Thus the standard procedure of converting a sentence into its normal form, increasing its depth and eliminating trivially inconsistent constituents is seen to have one more important systematic use (in principle). All relations of explicit definability are eventually brought into the open by this procedure. (We have proved this only for monadic predicates so far, but the generalization to other kinds of concepts is straightforward.)

It may happen that γ_i is not empty in *any* constituent $C_i^{(d+e)}$ occurring in the upper row of Figure 1; then we may call P *essentially undefinable* in terms of λ in $t^{(d)}(\lambda+\{P\})$. Such a P is undefinable in terms of λ in any theory compatible with $t^{(d)}(\lambda+\{P\})$ whose non-logical vocabulary is $\lambda+\{P\}$.

XII. CONDITIONAL DEFINABILITY

It may happen that each $C_i^{(d+e)}$ of $t^{(d+e)}(\lambda+\mu)=t^{(d+e)}(\lambda+\{P\})$ satisfies only a part of that crucial requirement (11) which guarantees explicit definability, in that we have $\alpha_i \cap \beta_i=\emptyset$ for each i, but no uniform separation which would satisfy (11) for some constant set δ of relevant attributive constituents. If this happens for some e, we shall say that P is *conditionally definable* in terms of λ at depth $d+e$ on the basis of $t^{(d)}(\lambda+\{P\})$. (Explicit definability will thus be considered a special case of conditional definability.) If P is conditionally definable at some depth (on the basis of $t^{(d)}(\lambda+\{P\})$ we shall simply say that it is conditionally definable (on this basis). Then each constituent $C_i^{(d+e)}$ of $t^{(d+e)}(\lambda+\{P\})$ implies an explicit definition of form (9) of P in terms of λ. The definiens is here of course of depth $d+e-1$. As was seen, unless (11) is satisfied for a constant δ, P is not explicitly definable in terms of λ.

Since $t^{(d+e)}(\lambda+\{P\})$ is equivalent to the disjunction of the $C_i^{(d+e)}$, it implies in these circumstances a disjunction of different explicit definitions. However, unless we have (11), there is no way of 'reconciling' or 'putting together' these definitions.

By an obvious extension of the argument we used above concerning explicit definability, it can conversely be shown that whenever $t^{(d)}(\lambda+\{P\})$ implies a finite disjunction of different definitions of form (9), P is conditionally definable in terms of the members of λ on the basis of $t^{(d)}(\lambda+\{P\})$. Here, then, we have a criterion of conditional definability which is independent of distributive normal forms.

Other criteria are not hard to come by. If P is conditionally definable in terms of λ on the basis of $t^{(d)}(\lambda+\{P\})$ at depth $d+e$, each model $\mathcal{M}$ of $t^{(d)}(\lambda+\{P\})$ satisfies exactly one of the $C_i^{(d+e)}$. Hence, the corresponding explicit definition (9) is true in $\mathcal{M}$, which means that P is in the usual sense of the word *definable in* $\mathcal{M}$ in terms of the members of λ.

Conversely, assume that P is definable in terms of λ in each model $\mathcal{M}$ of $t^{(d)}(\lambda+\{P\})$. Let the complete theory compatible with $t^{(d)}(\lambda+\{P\})$ which holds in $\mathcal{M}$ be defined by the sequence of constituents

$$C_i^{(d+f)}(\lambda + \{P\})\,(f = 0, 1, 2, \ldots).$$

If the explicit definition which holds in $\mathcal{M}$ is of depth $d+e$, $C_i^{(d+e)}(\lambda+\{P\})$ must satisfy $\alpha_i \cap \beta_i = \emptyset$. By letting the model $\mathcal{M}$ vary (to go through all the models of $t^{(d)}(\lambda+\{P\})$), and considering at one time each such complete theory determined by $\mathcal{M}$, we find that each such sequence of constituents (of the above kind) compatible with each other and with $t^{(d)}(\lambda+\{P\})$ gives rise to this situation. By the tree theorem (König's lemma), *all* of them will do this at some depth $d+e$. At this depth, P is thus conditionally definable in terms of λ in the theory $t^{(d)}(\lambda+\{P\})$.[2]

Thus we have obtained a clear-cut syntactical criterion for a non-logical predicate to be definable in each model of a first-order theory t in terms of (say) λ. A necessary and sufficient condition for this is that it be conditionally definable in terms of λ given t.

The concept of conditional definability seems to have considerable general theoretical interest. As examples of the results that can be established for it, we mention here the following simple theorems:

In a monadic first-order theory $t(\lambda+\mu)$ without identity, the members of μ are Ramsey-eliminable iff they are conditionally definable in terms of λ.

If the members of μ are conditionally definable in terms of the members of λ in a first-order theory $t^{(d)}(\lambda+\mu)$ (with or without identity), then they are Ramsey-eliminable.

If the members of μ are conditionally definable in terms of λ in a first-order theory $t^{(d)}(\lambda+\mu)$ (with or without identity), then the subtheory becomes finitely axiomatizable with $t^{(d+e)}(\lambda+\mu)[\lambda]$ as its axiom for some $e \geqslant 0$.

In these theorems we call a concept of μ Ramsey-eliminable from the given richer theory iff an axiomatizable poorer theory can be found such that all the models of the poorer theory have an expansion (for this notion see, for example, Shoenfield (1967), ch. IV)[3] which is a model of the richer theory.

It is important to note that what is here called conditional definability is a more general notion than what is sometimes referred to by this term. When the members of μ are conditionally definable in terms of λ on the basis of $t(\lambda+\mu)$, the following is logically true for some suitable conditions $C_1, C_2, \ldots, C_n$ which are mutually exclusive and collectively equivalent to $t(\lambda+\mu)$ (that is, their disjunction is logically equivalent to $t(\lambda+\mu)$):

$$\begin{aligned} &[t(\lambda + \mu) \supset (C_1 \supset Df_1)] \\ \& &[t(\lambda + \mu) \supset (C_2 \supset Df_2)] \\ \& &--- \\ \& &[t(\lambda + \mu) \supset (C_n \supset Df_n)] \end{aligned}$$

where each Df_i is a conjunction of explicit definitions of the members of μ in terms of the members of λ. If the special case in which the C_i contain no members of μ, we can turn this bunch of 'conditional definitions' into an explicit one. Our definition of conditional definability allows for the possibility, however, that the C_i contain members of μ in addition to those of λ. When this is irreducibly the case – as it often is – conditional definability does not reduce to explicit definability.

Thus it might even be advisable to make a terminological distinction between our notion of conditional definability and the ideas which are traditionally discussed under this heading, and to call the former for instance piecewise definability rather than conditional definability. On the other hand, conditional definability in our sense satisfies several of the traditional requirements on satisfactory definition, including the requirements of *non-creativity* and *eliminability*.

We shall later discuss the methodological significance of the notion of conditional definability. Here we only mention a mathematical example of conditional definability. In this example, our richer theory $t(\lambda+\mu)$ is a system of equations which has a finite number of solutions (>1) for the members of μ in terms of those of λ.

XIII. PARTIAL DEFINABILITY

It may of course be the case that $\gamma_i (=(8))$ is not empty in the case of some constituents $C_i^{(d+e)}$. Then P is not even conditionally definable in terms of λ at depth $d+e$. Even when this happens, we can see how close γ_i is to being empty, and use the answer to define measures of partial (conditional) definability.

To do this systematically, let us assume that a strictly positive probability measure is defined on all the consistent sentences of our language (with λ as its non-logical vocabulary) which contain just one free variable, namely x. Let us denote this measure, too, by p. (No danger of confusion lurks here, for the arguments of the old p and the new p will be different.) This measure will be defined, among other things, on attributive constituents with one free variable (x). Then we can call

$$(12) \qquad \mathrm{p}(Ct_{i1}^*(x) \vee Ct_{i2}^*(x) \vee \ldots)$$

a measure of the indeterminacy (7) leaves us with respect to P. The (absolute) *degree of definability* of P in $t^{(d)}(\lambda+\{P\})$ at depth $d+e$ will be defined so as to be proportional to the sum of all terms of the form

$$(13) \qquad \mathrm{p}(C_i^{(d+e)})\,[1-\mathrm{p}(Ct_{i1}^*(x) \vee Ct_{i2}^*(x) \vee \ldots)],$$

where the first and the second p of course have arguments of different kinds. The different terms are obtained from the different constituents $C_i^{(d+e)}(\lambda+\mu)$ that occur in $t^{(d+e)}(\lambda+\{P\})$. The proportionality factor that has to be used in (13) is of course $1/\sigma$ where σ is the sum of all the terms $p(C_i^{(d+e)})$ with $C_i^{(d+e)}$ in $t^{(d+e)}(\lambda+\{P\})$.

From what has been said above it follows that P is conditionally definable on the basis of $t^{(d)}(\lambda+\{P\})$ at depth $d+e$ iff the degree of (partial) definability of P in $t^{(d+e)}(\lambda+\{P\})$ is one. It can easily be seen that in general the degree of definability of a concept at depth $d+e+1$ is at least equal to its degree of definability at depth $d+e$. Hence

$$(14) \qquad \lim_{e\to\infty} (\text{degree of definability of } P \text{ in } t^{(d)}(\lambda+\{P\}) \text{ at depth } d+e)$$

is always well defined, and can be used as an (unqualified) measure of the (absolute) degree of definability of P on the basis of $t^{(d)}(\lambda+\{P\})$ (or of any sentence of depth d whose distributive normal form this is).

If P is conditionally definable in terms of the members of λ, this degree is one. The converse does not always hold. Whether it holds or not depends essentially on the particular probability measure p used in (12).

This dependence might seem to be a reason for doubting the general theoretical interest of our notion of partial definability. However, this dependence may perhaps be viewed as a sign of flexibility rather than vagueness. By choosing the probability measure in different ways we can express different assumptions concerning the interest and importance of different kinds of individuals, and take these differences into account in evaluating partial definability. For instance, in some cases it is possible to choose the measure so that the degree of definability of a new term equals one if and only if this term is definable in the standard model or models. Thus a suitable choice of the probability measure may reflect our special interest in standard models.

For many purposes, closely connected relative measures of partial definability seem more important. (The measures defined above cannot equal zero unless all attributive constituents of the given depth that are in the vocabulary of λ occur in (8) and hence in (9).) Instead of (12) we now have

$$(12^*)\quad \frac{\mathrm{p}[Ct^*_{i1}(x) \vee Ct^*_{i2}(x) \vee \ldots]}{\mathrm{p}[Ct_{i1}(x) \vee Ct_{i2}(x) \vee \cdots \vee Ct'_{i1}(x) \vee Ct'_{i2}(x) \vee \ldots]}$$
$$= \mathrm{p}[(Ct^*_{i1}(x) \vee Ct^*_{i2}(x) \vee \ldots) \mid (Ct_{i1}(x) \vee Ct_{i2}(x) \vee \ldots \vee Ct'_{i1}(x) \vee Ct'_{i2}(x) \vee \ldots)]$$

where relative probability is defined in the usual way.

Instead of (13) we then have a sum of terms of the form

$$(13^*)\quad \frac{1}{\sigma}\,\mathrm{p}(C_i^{(d+e)}) \left[1 - \frac{\mathrm{p}[Ct^*_{i1}(x) \vee Ct^*_{i2}(x) \vee \ldots]}{\mathrm{p}(Ct_{i1}(x) \vee Ct_{i2}(x) \vee \cdots \vee Ct'_{i1}(x) \vee Ct'_{i2}(x) \vee \ldots)}\right]$$

Here σ has the same meaning as before, and the sum is taken over all constituents $C_i^{(d+e)}$ occurring in $t^{(d+e)}(\lambda+\{P\})$. We shall call this sum the *relative* degree of partial definability of P in $t^{(d)}(\lambda+\{P\})$ at depth $d+e$. From what has been said it follows that for some sufficiently large

e this equals one whenever P is conditionally definable in terms of the members of λ on the basis of $t^{(d)}(\lambda+\{P\})$.

We conjecture that the degree of relative definability at depth $d+e$ always converges when $e \to \infty$, but no proof is attempted here. If the conjecture holds, the limit serves as an unqualified relative degree of the partial definability of P in terms of λ in $t^{(d)}(\lambda+\{P\})$ (or in any sentence having this as its distributive normal form). If the conjecture does not hold, this concept will not be defined in all cases.

It may happen that the relative partial definability of P in terms of λ will be zero from some depth $d+e$ on. (If so, the unqualified degree of relative partial definability is also zero.) Then there is no kind (type) of individuals definable in terms of λ which is compatible with $t^{(d)}(\lambda+\{P\})$ and whose members must all satisfy P or must all satisfy $\sim P$ if $t^{(d)}(\lambda+\{P\})$ is true.

An additional notion of partial definability arises from model-theoretic considerations, for we can now explicate what is meant by the partial definability of P in terms of λ given $t^{(d)}(\lambda+\{P\})$ and a model $\mathscr{M}$ of its reduct at depth $d+e$.

Let us thus assume that we have been given a model $\mathscr{M}$ which satisfies all the sentences $(t^{(d+e)}(\lambda+\{P\}))[\lambda]$ of the lower row of our Figure 1. The constituents (in terms of λ) which it makes true form a sequence of never weaker (and normally increasingly stronger) constituents $C_0^{(d+e)}(\lambda)(e=0, 1, 2, \ldots)$. Let us consider those constituents

$$C_i^{(d+e)}(\lambda + \{P\})$$

which have the property that

$$(C_i^{(d+e)}(\lambda + \{P\}))\,[\lambda] = C_0^{(d+e)}(\lambda)$$

and which occur in the disjunction

$$t^{(d+e)}(\lambda + \{P\})$$

of the upper row of Figure 1. Then we form the sum of all terms of the form (13*) resulting from such constituents $C_i^{(d+e)}$. This serves as a measure of the partial definability of P in terms of λ on the basis of $\mathscr{M}$ and $t^{(d)}(\lambda+\{P\})$ at depth $d+e$. The reference to a particular depth can be omitted if we take the limit of this measure when $e \to \infty$ (assuming that the limit exists).

XIV. DIFFERENT KINDS OF DEFINABILITY IN FIRST-ORDER THEORIES

We have been led to distinguish different cases of definability as follows: (1) explicit definability, (2) conditional definability, and (3) partial definability (underdefinability).

When discussing definitions, philosophers and methodologists have so far almost exclusively restricted their attention to the notion of explicit definability and to certain kinds of underdefinability (for example, underdefinability due to the introduction of new concepts by means of reduction sentences). However, it seems to us that conditional definability is quite as important as the others. Let us try to indicate some of our reasons for this claim by discussing a few of the methodologically interesting properties of the notion of conditional definability.

In the literature of the philosophy of science one often finds statements of the following kind: Even if we cannot define a concept explicitly, it may still be defined by means of the context in which it occurs, perhaps so that different contexts define (in some sense) the concept in different ways. (See, for example, Pap (1962), Simon (1959), p. 446.) However, it seems to us that what can be meant by such contextual definitions has never been spelled out in a satisfactory way, even though some attempts to make it precise have been made. (See Simon (1959), pp. 446, 449, Simon (1967), p. 13.) The basic idea behind these attempts to weaken the notion of (explicit) definability seems to have been to introduce a notion of definability which implies explicit definability provided that certain special cases or singularities are excluded. However, it may very well happen that these special cases are specifiable only by means of the auxiliary concepts to be defined. Thus we are led to conditional definability rather than explicit definability. For one interesting feature of our notion of conditional definition – one already commented on – is that the conditions themselves are stated in terms of auxiliary expressions where the concepts to be defined occur. For instance, if we had conditional definitions for, say, component forces in a first-order formalization of particle mechanics, the conditions might be force laws. (See Simon (1947), p. 903.)

Some philosophers of science have emphasized the importance of employing 'open' theoretical concepts because – amongst other things –

this enables one to use the advantages of inductive systematization (Hempel (1965), pp. 214–215), because it makes possible the growth of science (Braithwaite (1953), pp. 76–77), or because this openness is essential, owing to the meanings of the concepts involved. (See Pap (1962), p. 70 on psychological concepts.) In the last two cases what is primarily meant is openness in the sense of underdefinability, but in the first case conditional definability might perhaps be relevant, too. However, a satisfactory answer to these openness problems requires systematic investigation which nobody appears to have undertaken so far. Suffice it here to comment on one special case.

Consider an inductive systematization where we are trying to find the value of a theoretical concept P for each individual (Hempel (1965), pp. 214–215). Given the value of P (that is, given that an individual has this property, or that it does not) new observational predictions can be made concerning this individual on the basis of the given theory $t(\lambda+\{P\})$. Hempel considers it very important that this task be carried out by using an (underdefined) theoretical concept rather than by employing, for instance, direct inductive inference. The value of P for each individual is thought of as being somehow obtained on the basis of the theory $t(\lambda+\{P\})$ and of certain observable properties (*Ct*-predicates) which the observed individuals are known to have. The problem is whether we can find a (unique) systematic way of doing this. Obviously this question can be approached from the point of view of the definability of P.

Let us first consider the case where P is underdefined. Assume that we are given an individual known to satisfy certain *Ct*-predicate (or a disjunction of them) which uses the vocabulary of λ only and which occurs in the reduct of a constituent C_i of $t(\lambda+\{P\})$ to this vocabulary. Then it will generally be the case that there is no systematic method for solving our problem. For the exemplified *Ct*-predicate may belong to a non-empty uncertainty set γ_i. Then we know that either P or $\sim P$ (but not both at the same time) may be associated with our exemplified *Ct*-predicates. A non-arbitrary estimation method may be obtained only if some (possibly inductive) reasons independent of our theory $t(\lambda+\{P\})$ can be found to determine the choice of a unique estimation method for P in the constituent C_i.

But even when we have such a extratheoretical estimation method, there is another source of ambiguity remaining. This can be illustrated as follows. Let us denote the set of *Ct*-predicates exemplified by our evidence by ρ. Now it may happen that $\alpha_i \supseteq \rho$ in a constituent C_i and $\beta_j \supseteq \rho$ in some other constituent C_j (using the terminology of Section XI). Hence we obtain P as our 'estimate' on the basis of C_i but $\sim P$ on the basis of C_j.

We thus see that, while both of the above kinds of ambiguity are present when P is partially defined, only the second kind of uncertainty can occur when P is conditionally definable. (If P is explicitly definable, neither trouble exists.) Hence, if some kind of openness is wanted (for reasons not to be discussed here), conditional definability is the likely candidate in our case of inductive systematization (in view of the advantages it was just found to possess in estimation problems). It must be admitted, however, that this kind of predictive problem constitutes a very demanding task and need not be representative of the other jobs one is likely to expect a scientific theory to do.

The situation we have considered is closely analogous to that prediction task of a structural linear econometric theory which is called structural estimation. (See for example Valavanis (1959), pp. 90–91.) In this situation one is given certain inputs (values of exogenous variables) at the present moment of time, and one wants to predict future outputs (values of endogenous variables) in terms of a mechanism or structure (determined by the structural parameters and the disturbances) which is known to change with time. Here underdefinability (or underidentifiability), explicit definability (unique identifiability), and conditional definability (overidentifiability) of the structural parameters give rise to estimation difficulties similar to those discussed above. (See Valavanis (1959), p. 103.) In particular, we suggest that econometricians do not seem to have realized that one kind of case of overidentifiability can in principle be discussed in very general terms without reference to any statistical sampling problems or to any particular estimation methods connected with them, and that a closer conceptual study is needed to clarify the estimation problems connected with overidentifiability.

From the point of view of a scientist it may be natural to suggest that the following kind of openness of concepts is of importance: in certain central and often realized empirical circumstances there should be an explicit definition for a theoretical concept, whereas in some other, so far empirically rare or technically unrealizable circumstances no definition is implied by the theory. In other words, the theoretical concept is definable in some or most (but not necessarily all) models of a theory. This represents an instance of partial definability. The more developed a science is, the higher should the degree of definability of a theoretical concept be (as measured by (13*), for instance). This is one of the reasons why it is an interesting task to measure degrees of definability. It is important to realize, however, that what we can in the first place naturally measure is how close we are to *conditional*, not how close we are to *explicit* definability (contrary to what Simon seems to think; see for example Simon (1959), p. 459). This is what such measures as (13*) indicate.

Within our framework we can also perhaps understand certain other aspects of the 'growth of a theory'. By the possibility of such growth some philosophers (e.g. Braithwaite) have meant only the logical possibility of obtaining new observational and possibly also other kinds of new statements by deducing them from a new richer theory. Braithwaite tries to convince us by lengthy examples formulated (in effect) in monadic predicate logic that such a growth is possible only by means of theoretical concepts which are not explicitly defined in terms of observables. In terms of our Figure 1 Braithwaite's main point amounts to saying that in the case of explicit definability no gain of the first kind can be obtained, just as we pointed out in Section X. One possible line of further growth of a theory is to add new statements in the vocabulary $\lambda+\mu$ (and perhaps additional concepts) to the richer theory. In view of the distributive properties of our reduction operation r (see Section II) we can see that the logical situation will not be essentially changed and the gains of both kinds can be measured as before.[4]

BIBLIOGRAPHY

Barker, S. F., *Induction and Hypothesis*, Ithaca, N.Y., 1957.

Braithwaite, R., *Scientific Explanation*, Cambridge 1953.

Craig, W., 'Bases for First-Order Theories and Subtheories', *Journal of Symbolic Logic* **25** (1960) 97–142.

Hempel, C. G., *Aspects of Scientific Explanation*, New York 1965.

Hintikka, K. J., 'Distributive Normal Forms and Deductive Interpolation', *Zeitschrift für mathematische Logik und Grundlagen der Mathematik* **10** (1964) 185–191.

Hintikka, K. J., 'Distributive Normal Forms in First-Order Logic' in *Formal Systems and Recursive Functions* (ed. by J. Crossley and M. Dummett), Amsterdam 1965, pp. 47–90. (Referred to as (1965a).)

Hintikka, K. J., 'Are Logical Truths Analytic?', *Philosophical Review* **74** (1965) 178–203. (Referred to as (1965b).)
Hintikka, K. J., 'An Analysis of Analyticity', in *Deskription, Analytizität und Existenz* (ed. by P. Weingartner), 3–4 Forschungsgespräch des Internationalen Forschungszentrums für Grundfragen der Wissenschaften Salzburg, Munich and Salzburg 1966, pp. 193–214.
Hintikka, K. J., 'Information, Deduction, and the A Priori', forthcoming in *Noûs* (1970).
Nagel, E., 'A Budget of Problems in the Philosophy of Science', *Philosophical Review* **66** (1957) 205–225.
Pap, A., *An Introduction to the Philosophy of Science*, New York 1962.
Robinson, A., *Introduction to Model Theory*, Amsterdam 1963.
Sherif, M., and Sherif, C., *Harmony and Tension in Groups*, New York 1953.
Shoenfield, J., *Mathematical Logic*, Reading, Mass., 1967.
Simon, H. A., 'The Axioms of Newtonian Mechanics', *Philosophical Magazine*, ser. 7, **33** (1947) 888–905.
Simon, H. A., 'Definable Terms and Primitives in Axiom Systems' in *The Axiomatic Method* (ed. by L. Henkin, P. Suppes, and A. Tarski), Amsterdam 1959.
Simon, H. A., 'The Axiomatization of Physical Theories' *Philosophy of Science* **37** (1970) 16–26.
Svenonius, L., 'A Theorem on Permutations in Models', *Theoria* **25** (1959) 173–178.
Valavanis, S., *Econometrics*, New York 1959.

REFERENCES

[1] When identities are present, an exclusive interpretation of quantifiers must be assumed in the normal forms. Then predicates can be omitted as before in the direct reduction. However, it is obvious that a modification is needed when an individual constant $a \in \mu$ is omitted as a part of the reduction. For when no longer explicitly mentioned, it becomes a possible substitution value of the quantifiers in a way it was not one before.

It is clear enough what we must do. Let us assume that we want to omit a from a constituent $C_0^{(d)}$. Then, over and above omitting all atomic sentences containing a, we must change each part of $C_0^{(d)}$ of the form

$$(Ex)\, Ct_1^{(c)}(x) \,\&\, (Ex)\, Ct_2^{(c)}(x) \,\&\, \cdots \,\&\, (Ux)\, [Ct_1^{(c)}(x) \vee Ct_2^{(c)}(x) \vee \cdots],$$

by adding to the conjunction a new $(Ex)\, Ct_0^{(c)}(x)$ and to the disjunction a new member $Ct_0^{(c)}(x)$. Intuitively speaking, $Ct_0^{(c)}(x)$ says here of x (in relation to the individuals mentioned in it) the same thing as $C_0^{(d)}$ said of a. It is not hard to see how $Ct_0^{(c)}(x)$ can be syntactically constructed from $C_0^{(d)}$, although the details of the recipe are somewhat complicated.

[2] The latter part of this result amounts to the theorem on permutations in models which is proved in Svenonius (1959).

[3] This meaning of 'expansion' is of course completely different from that of an expansion of a normal form into a deeper one.

[4] We are indebted to Mr, David Miller for valuable suggestions concerning expositional and stylistic matters.

INDEX OF NAMES

INDEX OF SUBJECTS

SYNTHESE LIBRARY

Monographs on Epistemology, Logic, Methodology,
Philosophy of Science, Sociology of Science and of Knowledge, and on the
Mathematical Methods of Social and Behavioral Sciences

‡KAREL LAMBERT, *Philosophical Problems in Logic. Some Recent Developments*. 1970, VII+176 pp. Dfl. 38.—

P. V. TAVANEC (ed.), *Problems of the Logic of Scientific Knowledge*. 1969, XII+429 pp. Dfl. 95.—

‡ROBERT S. COHEN and RAYMOND J. SEEGER (eds.), *Boston Studies in the Philosophy of Science*. Volume VI: *Ernst Mach: Physicist and Philosopher*. 1970, VIII+295 pp. Dfl. 38.—

‡MARSHALL SWAIN (ed.), *Induction, Acceptance, and Rational Belief*. 1970, VII+232 pp. Dfl. 40.—

‡NICHOLAS RESCHER *et al. (eds.)*, *Essays in Honor of Carl G. Hempel. A Tribute on the Occasion of his Sixty-Fifth Birthday*. 1969, VII + 272 pp. Dfl. 46.—

‡PATRICK SUPPES, *Studies in the Methodology and Foundations of Science. Selected Papers from 1951 to 1969*. 1969, XII + 473 pp. Dfl. 72.—

‡JAAKKO HINTIKKA, *Models for Modalities. Selected Essays*. 1969, IX + 220 pp. Dfl. 34.—

‡D. DAVIDSON and J. HINTIKKA: (eds.), *Words and Objections: Essays on the Work of W. V. Quine*. 1969, VIII + 366 pp. Dfl. 48.—

‡J. W. DAVIS, D. J. HOCKNEY, and W. K. WILSON (eds.), *Philosophical Logic*. 1969, VIII + 277 pp. Dfl. 45.—

‡ROBERT S. COHEN and MARX W. WARTOFSKY (eds.), *Boston Studies in the Philosophy of Science*. Volume V: *Proceedings of the Boston Colloquium for the Philosophy of Science 1966/1968*. 1969, VIII + 482 pp. Dfl. 58.—

‡ROBERT S. COHEN and MARX W. WARTOFSKY (eds.), *Boston Studies in the Philosophy of Science*. Volume IV: *Proceedings of the Boston Colloquium for the Philosophy of Science 1966/1968*. 1969, VIII + 537 pp. Dfl. 69.—

‡NICHOLAS RESCHER, *Topics in Philosophical Logic*. 1968, XIV + 347 pp. Dfl. 62.—

p.t.o.

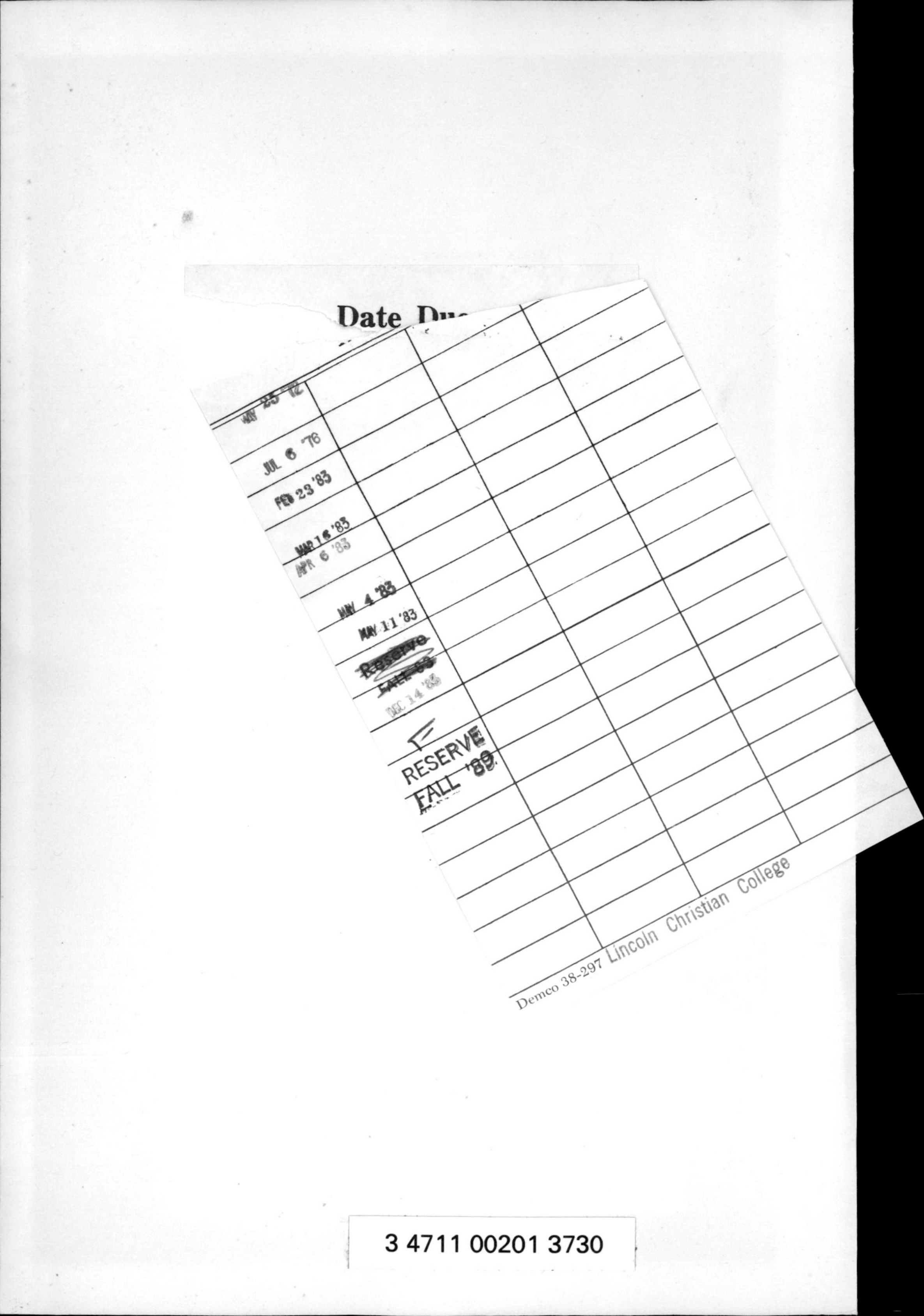
Date Due
MAY 25 '72
JUL 6 '76
FEB 23 '83
MAR 16 '83
APR 6 '83
MAY 4 '83
MAY 11 '83
Reserve
FALL 83
DEC 14 '83
RESERVE
FALL '89
Demco 38-297 Lincoln Christian College